MW01502538

DESIGN
AND ANALYSIS
OF BIOAVAILABILITY
AND BIOEQUIVALENCE
STUDIES

STATISTICS: Textbooks and Monographs

A Series Edited by

D. B. Owen, Founding Editor, 1972–1991

W. R. Schucany, Coordinating Editor

Department of Statistics
Southern Methodist University
Dallas, Texas

Additional Volumes in Preparation

DESIGN AND ANALYSIS OF BIOAVAILABILITY AND BIOEQUIVALENCE STUDIES

SHEIN-CHUNG CHOW

Biostatistics Department
Bristol-Myers Squibb Company
Plainsboro, New Jersey

JEN-PEI LIU

Biostatistics Department
Berlex Laboratories, Inc.
Wayne, New Jersey

Marcel Dekker, Inc.   New York • Basel • Hong Kong

Library of Congress Cataloging-in-Publication Data

Chow, Shein-Chung,
  Design and analysis of bioavailability and bioequivalence studies
  / Shein-Chung Chow, Jen-pei Liu.
      p.   cm. -- (Statistics, textbooks and monographs ; 133)
  Includes bibliographical references and index.
  ISBN 0-8247-8682-3 (acid-free paper)
  1. Bioavailability--Research--Methodology. 2. Drugs--therapeutic
equivalency--Research--Methodology. I. Liu, Jen-pei,
II. Title. III. Series: Statistics, textbooks and monographs ; v.
133.
RM301.6.C46 1992
615'.7--dc20   92-12142
                                                             CIP

This book is printed on acid-free paper.

MARCEL DEKKER, INC.
270 Madison Avenue, New York, New York 10016

Current printing (last digit):
10 9 8 7 6 5 4 3 2 1

PRINTED IN THE UNITED STATES OF AMERICA

Preface

In recent years, bioavailability studies for assessment of bioequivalence between formulations of drug products have become very popular in drug development. Since the early 1970s, the vast and continuously growing literature on statistical methodology for bioavailability and bioequivalence studies has appeared in statistical, medical, pharmacokinetic, and other related journals. In response to this rapid development, this book provides a comprehensive and unified summarization of statistical design and analysis of bioavailability and bioequivalence studies for both clinical scientists and statisticians. This book is written from a practical viewpoint at a basic mathematical and statistical level, but at the same time it introduces important statistical concepts for assessing bioequivalence through real examples. An introductory course in statistics that covers the concepts of probability, sampling distribution, estimation, and hypothesis testing will be helpful for the reader. No knowledge of mathematics beyond ordinary algebra is required, although, for completeness, some mathematical derivations and proofs of the results are included. Readers are encouraged to pay attention to the applications of different procedures in the examples provided.

The scope of this book is restricted to statistical assessment of bioequivalence between formulations of a drug product based on data collected from human subjects. Chapter 1 includes the definition, history, decision rules, and general statistical considerations for design and data analysis of bioavailability and bioequivalence studies. In Chapter 2, basic design considerations at the planning stage of a bioavailability or bioequivalence study are discussed. The standard 2

$\times$ 2 crossover design is the most commonly used design, and is also acceptable to the FDA for assessment of bioequivalence between two formulations of a drug product. Chapters 3 through 8 are devoted entirely to the analysis of data from this design. We examine statistical inferences for various effects from a standard 2 $\times$ 2 crossover design in Chapter 3. Chapter 4 provides a comprehensive review of current statistical methods, including parametric, nonparametric, and Bayesian methods for assessment of bioequivalence in average bioavailability. Power and sample size determinations for interval hypotheses based on the two one-sided tests procedure are discussed in Chapter 5.

In Chapter 6, we explore the relationship between an additive model and a multiplicative model. The minimum variance unbiased estimator of the direct formulation effect on the original scale under a multiplicative model is included in this chapter. Chapter 7 focuses on the assessment of inter- and intra-subject variabilities which include point, interval estimation and two test procedures for assessing equivalence of variability of bioavailability. Statistical tests for verification of model assumptions and for detection of outlying data are described in Chapter 8. Chapter 9 provides statistical methods for assessing bioequivalence for two formulations under a higher-order crossover design. Assessment of bioequivalence for more than two formulations is outlined in Chapter 10. Chapter 11 gives an introduction to assessment of bioequivalence based on clinical endpoints such as therapeutic responses and time to the onset of a therapeutic response when the drug products produce negligible plasma concentrations. Some related topics in bioavailability and bioequivalence studies such as dose proportionality studies, drug interaction studies, steady state analyses, and population pharmacokinetics are briefly described in the last chapter.

All computations in this book were performed using version 6.06 of SAS. Other statistical packages such as MINITAB and BMDP may also be applied. It should be noted that most methods for assessment of bioequivalence discussed in this book can be carried out easily with a pocket calculator and the statistical tables provided in Appendix A. It is our hope that this book will not only be a useful reference book for scientists, clinicians, and statisticians in the pharmaceutical industry, regulatory agencies, and academia but also serve as a textbook for graduate courses related to the topic of bioavailability and bioequivalence in the area of pharmacokinetics, clinical pharmacology, and biostatistics.

We would like to thank Ms. M. Allegra of Marcel Dekker, Inc., for providing us with the opportunity to work on this book. We are deeply indebted to the Bristol-Myers Squibb Company and Berlex Laboratories, Inc., for their support, in particular to Drs. S. A. Henry, M. P. Rogan, S. Weiss, and W. Holyoak. We are grateful to Drs. C. S. Weng, K. Phillips, C. M. Metzler, and Mr. J. S. Lin for providing SAS programs for procedures in Chapter 4 and the figures in Chapter 5. We are also grateful to Drs. J. T. Whitmer, H. M. Yang, M. Emery, C. Yeh, and Mr. A. Berin for many helpful discussions and for reviewing portions of the manuscript. We also wish to thank Drs. E. Nordbrock and W. S. Mullican for their support and encouragement. The first author also wishes to express his appreciation to his wife, Yueh-Ji, and two daughters, Emily and Lilly, for their patience and understanding during the preparation of this book.

Finally, we are fully responsible for any errors remaining in the book. The views expressed in this book are those of the authors and are not necessarily those of Bristol-Myers Squibb Company and Berlex Laboratories, Inc.

Shein-Chung Chow
Jen-pei Liu

Contents

1
Introduction

1.1. HISTORY OF BIOAVAILABILITY STUDIES

The term *bioavailability* is a contraction for "biological availability" (Metzler and Huang, 1983). The definition of bioavailability has evolved over time with different meanings by different individuals and organizations. For example, differences are evident in the definitions by the Academy of Pharmaceutical Sciences (1972), the Office of Technology Assessment (OTA) of the Congress of the United States (1974), and Wagner (1975). Throughout this book, however, the definitions adopted by the United States Food and Drug Administration (FDA) for bioavailability and some related terms will be used (21 CFR, Part 320.1, 1983).

The bioavailability of a drug is defined as the rate and extent to which the active drug ingredient or therapeutic moiety is absorbed and becomes available at the site of drug action. A *comparative bioavailability* study refers to the comparison of bioavailabilities of different formulations of the same drug or different drug products. When two formulations of the same drug or two drug products are claimed *bioequivalent,* it is assumed that they will provide the same therapeutic effect or that they are therapeutically equivalent. Two drug products are considered *pharmaceutical equivalents* if they contain identical amounts of the same active ingredient. Two drugs are identified as *pharmaceutical alternatives* to each other if both contain an identical therapeutic moiety, but not necessarily in the same amount or dosage form or as the same salt or ester. Two

1

drug products are said to be bioequivalent if they are pharmaceutical equivalents (i.e., similar dosage forms made, perhaps, by different manufacturers) or pharmaceutical alternatives (i.e., different dosage forms) and if their rates and extents of absorption do not show a significant difference when administered at the same molar dose of the therapeutic moiety under similar experimental conditions. For more discussion regarding the definition of bioavailability, see Balant (1991).

The study of absorption of an exogenously administered compound (sodium iodide) can be traced back to 1912 (Wagner, 1971). The concept of bioavailability, however, was not introduced until some thirty years later. Oser et al. (1945) studied the relative absorption of vitamins from pharmaceutical products and referred to such relative absorption as "physiological bioavailability."

In recent years, generic drug products (i.e., products manufactured by other than the innovator) have become very popular. Bioavailability/bioequivalence studies are of particular interest to the innovator and the generic drug companies in the following two ways. First, for the approval of a generic drug product, the FDA usually does not require a regular new drug application (NDA) submission, which demonstrates the efficacy, safety, and benefit/risk of the drug product, if the generic drug companies can provide the evidence of bioequivalence between the generic drug product and the innovator through bioavailability studies in a so-called "abbreviated new drug application" (ANDA). Second, when a new formulation of a drug product is developed, the FDA requires that a bioavailability study be conducted to assess its bioequivalence to the standard (or reference) marketed formulation of the drug product. Thus, bioavailability studies are important since an NDA submission includes the results from phases I–III clinical trials which are very time consuming and/or costly to obtain.

The concept of bioavailability/bioequivalence became a public issue in the late 1960's because of the concern that a generic drug product might not be as bioavailable as that manufactured by the innovator. These concerns rose from clinical observations in humans together with the ability to quantitate minute quantities of drugs in biologic fluids. In 1970, the FDA began to ask for evidence of "biological availability" in applications submitted for approval of certain new drugs. In 1974, a Drug Bioequivalence Study Panel was formed by the OTA to examine the relationship between the chemical and therapeutic equivalence of drug products. Based on the recommendations in the OTA report, the FDA published a set of regulations for the submission of bioavailability data in certain new drug applications. These regulations became effective on July 1, 1977 and are currently codified in 21 CFR Part 320.

In 1971, by the time the FDA began to require evidence of bioavailability for NDA of some drug products, the Biopharmaceutical Subsection of the American Statistical Association simultaneously formed a Bioavailability Committee to investigate the statistical components for the assessment of bioequivalence. Metzler (1974) summarized the efforts by the Committee and addressed several concerns regarding some statistical issues in bioavailability studies. During the decade proceeding the early 1980's, the search for statistical methods for the assessment of bioequivalence received tremendous attention. Several methods which met the FDA requirements for statistical evidence of bioequivalence were proposed. These methods included an *a posteriori* power approach, reformulation of bioequivalence hypotheses (Anderson and Hauck, 1983; Schuirmann, 1981), a confidence interval approach (Metzler, 1974; Westlake, 1972, 1976, 1979), and a Bayesian approach (Rodda and Davis, 1980; Mandallaz and Mau, 1981). A detailed discussion of these statistical developments during this period can be found in Metzler and Huang (1983).

In 1984, the FDA was authorized to approve generic drug products under the Drug Price Competition and Patent Term Restoration Act. However, as more generic products become available, the following concerns were raised:

(i) Whether generic drug products are comparable in quality to the innovator drug product and,
(ii) Whether the generic copies of innovator drug products have comparable therapeutic effect.

To address these concerns, a 1986 hearing on Bioequivalence of Solid Oral Dosage Forms was conducted by the FDA during September 29–October 1 in Washington, D.C. As a consequence of the hearing, a Bioequivalence Task Force was formed to examine the current procedures adapted by the FDA for the assessment of bioequivalence between immediate solid oral dosage forms. Some efforts were also directed at investigating the statistical issues that often occur in various stages of design and data analysis in bioavailability/bioequivalence studies. A report from the Bioequivalence Task Force was released in January 1988. Several statistical issues related to the assessment of bioequivalence are summarized below:

(1) Lot-to-lot uniformity;
(2) Alternative statistical designs for intra-subject variability;
(3) Statistical methodology in decisional criteria for bioequivalence;
(4) Product-to-product variability;
(5) Detection and treatment of outlying data.

Recently, several statistical methods have been developed which provide some answers to the above statistical questions. For example, for the evaluation of lot-to-lot (or batch-to-batch) uniformity, Chow and Shao (1989) proposed several statistical tests for batch-to-batch variability. To account for the heterogeneity of intra-subject variability, an estimation procedure for the assessment of intra-subject variability assuming that the coefficient of variation (CV) is the same from subject to subject was proposed by Chow and Tse (1990b) under a conditional random effects model. Chow (1989) and Chow and Shao (1991) compared the decision rules under lognormality assumption. Chow and Shao (1990) also proposed an alternative approach for assessing bioequivalence using the idea of a confidence region. The proposed procedure was shown to rigorously meet the FDA's requirements. For outlier detection, Chow and Tse (1990a) proposed two tests using the idea of likelihood distance and estimates distance for detection of a possible outlying subject. The same problem was also examined by Liu and Weng (1991).

1.2. FORMULATIONS AND ROUTES OF ADMINISTRATION

When a drug is administered to a human subject, the drug generally passes through an absorption phase, distribution phase, metabolism phase, and finally an elimination phase within the body. As mentioned in the previous section, the bioavailability of a drug is defined as the rate and extent to which the active ingredient of the drug is absorbed and becomes available to the body. Since clinical effects may be associated with blood or plasma levels of the drug, the information of bioavailability is useful for the assessment of efficacy and safety of the drug. Bioavailability is usually determined by some pharmacokinetic measurements which can be estimated based on the blood or plasma concentration-time curve obtained following drug administration. The blood or plasma concentration-time curve, however, is dependent in part on the dosage form and/or the route of administration.

In the pharmaceutical industry, when a new drug is discovered, it is important to design an appropriate dosage form for the drug so that it can be delivered to the body efficiently for optimal therapeutic effect. The dosage form, however, should also account for the acceptability to the patients. Dosage forms such as tablet, capsule, solution, powder, and liquid suspension are usually considered. For a given drug product, several dosage forms may be designed for different purposes. For example, solution and liquid suspension dosage forms may be more appropriate than solid dosage forms for children and elderly patients. However, in practice, most drugs are taken orally in solid dosage forms (e.g., tablet and capsule). Generally, solid dosage forms have to dissolve to be absorbed. The dissolution of the drug depends on the particle size. The reduction of particle size may increase the bioavailability of the drug. Examples of drugs whose bioavailability have been increased as a result of particle size reduction are aspirin and estradiol. See, e.g., Dare (1964).

The route of administration can certainly affect the bioavailability of a drug. Different routes of administration may result in a significant differences in bioavailability. For example, a study of kanamycin (Kunin, 1966) demonstrated that the oral administration has extremely low bioavailability (about 0.7%). In contrast, the bioavailability of intramuscularly administered kanamycin is much greater (about 40%–80%).

Basically, there are several routes at which drugs are commonly administered. These routes may be classified as either intravascular or extravascular. Intravascular administration refers to giving the drug directly into the blood either intravenously or intra-arterially. Extravascular administration includes the oral, intramuscular, subcutaneous, sublingual, buccal, pulmonary, rectal, vaginal, and transdermal routes. Drugs administered extravascularly must be absorbed to enter the blood.

Since different dosage forms may affect the bioavailability of the drug, they may exhibit marked variability in their absorption. Thus, before a drug can be released for medical use, the FDA requires that the drug be tested in vitro in compliance with United States Pharmacopeia (USP) specifications to ensure that the drug contains the labeled active ingredient within an acceptable variation. The USP standards for the evaluation of the drug include potency testing, content uniformity testing, dissolution testing, disintegration testing, and weight variation testing. In addition, a bioavailability study is also required by the FDA. The assay method used for the active ingredient(s) to quantitate the drug must be validated in terms of the closeness of the test results obtained from the assay method to the true values (accuracy) and the degree of closeness of the test results to the true values (precision).

Since different dosage forms and/or routes of administration may affect the bioavailability of the drug, a comparative bioavailability (bioequivalence) study may involve the comparison of different dosage forms (or formulations) of the same drug, generic drug product and the marketed (innovator) drug product of the same active ingredient, and different routes of administration.

1.3. PHARMACOKINETIC PARAMETERS

In a comparative bioavailability study in humans, following the administration of a drug, the blood or plasma concentration-time curve is often used to study the absorption and elimination of the drug. In general, the rate of absorption depends on the routes of administration of the drug. For example, the rate of absorption of an intravenous administration is much higher than that of an oral administration. Thus, the peak of the plasma or blood concentration-time curve for an intravenous administration will generally occur much earlier than that of an oral administration. The plasma or blood concentration-time curve can be characterized by taking blood samples immediately prior to and at various time points after drug administration. The profile of the plasma or blood concentration-time curve is then studied by means of several pharmacokinetic parameters such as area under the plasma or blood concentration-time curve (AUC), maximum concentration (C_{max}), and time to achieve maximum concentration (t_{max}). The measurements of these pharmacokinetic parameters can be derived either directly from the observed blood or plasma concentration-time curve which is independent of a model or obtained by fitting the observed concentrations to a one or multi-compartment pharmacokinetic model. In the following case, the determination of some pharmacokinetic parameters assumes first order absorption and elimination.

One of the primary pharmacokinetic parameters in a bioavailability study is the AUC. The AUC is often used to measure the extent of absorption or total amount of drug absorbed in the body. Several methods exist for estimating the AUC from zero time until time t, at which the last blood sample is taken. These methods include the interpolation using the trapezoidal rule, the Lagrange and spline methods, the use of a planimeter, the use of digital computers, and the physical method which compares the weight of a paper corresponding to the area under the experimental curve to the weight of a paper of known area. Among these methods, the method of interpolation appears to be the most commonly used method. Yeh and Kwan (1978) discussed the advantages and disadvantages of using the Lagrange and spline methods relative to the trapezoidal rule in the method of interpolation. For simplicity, we will only introduce the method of linear interpolation using the trapezoidal rule. Let $C_0, C_1, \ldots, C_k$ be the plasma or blood concentrations obtained at time $0, t_1, \ldots, t_k$, respectively. The AUC from zero to t_k, denoted by $AUC(0 - t_k)$, is obtained by

$$AUC(0 - t_k) = \sum_{i=1}^{k} \left(\frac{C_{i-1} + C_i}{2} \right) (t_i - t_{i-1}). \qquad (1.3.1)$$

It should be noted that the AUC should be calculated from zero to infinity not just to the time of the last blood sample as is so often done. The portion of the remaining area from t_k to infinity could be large if the blood level at t_k is substantial (Martinez and Jackson, 1991). The AUC from 0 to infinity, denoted by $AUC(0 - \infty)$, can be estimated as follows (Rowland and Tozer, 1980):

$$AUC(0 - \infty) = AUC(0 - t_k) + C_k/\lambda \qquad (1.3.2)$$

where C_k is the concentration at the last measured sample after drug administration and λ is the slope of the terminal portion of the log concentration-time curve multiplied by -2.303. The FDA regulation requires that sampling be continued through at least 3 half-lives of the active drug ingredient or therapeutic moiety, or its metabolites, measured in the blood or urine, or the decay of the acute pharmacologic effect so that the elimination will have been completed and any remaining area beyond time t_k is negligible. Note that a few missing values and/or unexpected observations in the plasma concentration-time curve within (t_1, t_k) will generally have little effect on the calculations of $AUC(0 - t_k)$ and

AUC($0 - \infty$). However, if there are many missing values and/or unexpected observations in the plasma concentration-time curve, especially at endpoints (i.e., t_1 and t_k), the bias of the estimate of AUC could be substantial.

In addition to the AUC, the absorption rate constant is usually studied during the absorption phase. Under the single compartment model, the absorption rate constant can be estimated based on the following equation using the method of residuals (Gibaldi and Perrier, 1982).

$$C_t = \frac{k_aFD_0}{V(k_a - k_e)} (e^{-k_et} - e^{-k_at}), \qquad (1.3.3)$$

where k_a and k_e are the absorption and elimination rate constants, respectively; D_0 is the dose administered; V is the volume of distribution, and F is the fraction of the dose that reaches the systemic circulation.

Based on the above formula, similarly, C_{max} and t_{max} can be obtained as follows:

$$t_{max} = \frac{2.303}{k_a - k_e} \log \left(\frac{k_a}{k_e}\right), \quad \text{and} \qquad (1.3.4)$$

$$C_{max} = \frac{k_aFD_0}{V(k_a - k_e)} (e^{-k_et_{max}} - e^{-k_at_{max}}). \qquad (1.3.5)$$

In practice, however, the estimates from a pharmacokinetic model usually are not used for the comparison of formulations. Thus, C_{max} is estimated directly from the observed concentrations. That is, $C_{max} = \max \{C_0, C_1, \ldots, C_k\}$. Similarly, t_{max} is estimated as the corresponding time point at which the C_{max} occurs.

During the elimination phase, the pharmacokinetic parameters which are often studied are the elimination half-life ($t_{1/2}$) and rate constant (k_e) (Chen and Pelsor, 1991). The plasma elimination half-life is the time taken for the plasma concentration to fall by one-half. Assume that the decline in plasma concentration is of first order, the $t_{1/2}$ can be obtained by considering

$$\log D = \log D_0 - \frac{k_et}{2.303}, \qquad (1.3.6)$$

where D is the amount of drug in the body. Thus, at $D = D_0/2$, i.e., $t = t_{1/2}$, we have

$$\log(1/2) = -\frac{k_et_{1/2}}{2.303}.$$

Hence,

$$t_{1/2} = 0.693/k_e,$$

where k_e is given by

$$k_e = (-2.303)\left(\frac{d \log D}{dt}\right).$$

The first order elimination half-life is independent of the amount of drug in the body. In practice, all the drug may be regarded as having been eliminated (about 97%) by 5 half-lives.

The above pharmacokinetic parameters are usually considered in a single dose trial. In practice, as is well known, drugs are most commonly prescribed to be taken on fixed-time-interval basis (i.e., multiple doses such as b.i.d., t.i.d., or q.i.d.). Dosing a drug several times a day can result in a different drug concentration profile than that produced by a single dose. If the dosing interval is less than the time required to eliminate the entire dose, the peak plasma level following the second and succeeding doses of a drug is always higher than the peak level after the first dose. This leads to drug accumulation in the body relative to the initial dose. For a multiple dose regimen, the amount of drug in the body is said to have reached a steady state level if the amount or average concentration of the drug in the body remains stable. The following pharmacokinetic parameters at steady state are usually studied:

$$C_{max} = \frac{D_0}{1 - (1/2)^\varepsilon},$$

$$C_{min} = \frac{D_0}{1 - (1/2)^\varepsilon}(1/2)^\varepsilon,$$

$$C_{av} = \frac{C_{max} - C_{min}}{\log(C_{max}/C_{min})}, \qquad (1.3.7)$$

$$\% \text{ fluctuation} = \left\{\frac{C_{min}}{C_{max}}\right\} \times 100\%,$$

where ε is the dosing interval τ divided by elimination half-lives. Note that τC_{av} is the area under the curve within a dosing interval at steady state, which is equal to that following a single dose. In a multiple dose study, however, how to choose or combine the information of several pairs of C_{max} and C_{min} from a subject is an interesting question. This certainly has some impact on statistical analysis of the data.

Example 1.3.1 To illustrate how to estimate AUC, C_{max}, t_{max}, $t_{1/2}$, and k_e from the observed concentrations, it is helpful to consider the following example. Table 1.3.1 lists the primidone concentrations ($\mu g/mL$) vs time points (hr) from a subject over a 32 hr period after administered a 250 mg tablet of a drug. The blood samples were drawn immediately prior to and at time points: 0.5, 1.0,

Table 1.3.1 Calculation of AUC Using the Trapezoidal Rule[a]

Blood sample i	t_i	C_i	$(C_i + C_{i-1})/2$	$t_i - t_{i-1}$	$(C_i + C_{i-1})(t_i - t_{i-1})/2$
1	0.0	0.0	—	—	—
2	0.5	0.0	0.00	0.5	0.00
3	1.0	2.8	1.40	0.5	0.70
4	1.5	4.4	3.60	0.5	1.80
5	2.0	4.4	4.40	0.5	2.20
6	3.0	4.7	4.55	1.0	4.55
7	4.0	4.1	4.40	1.0	4.40
8	6.0	4.0	4.05	2.0	8.10
9	8.0	3.6	3.80	2.0	7.60
10	12.0	3.0	3.30	4.0	13.20
11	16.0	2.5	2.75	4.0	11.00
12	24.0	2.0	2.25	8.0	18.00
13	32.0	1.6	1.80	8.0	14.40

[a]AUC(0 − 32) = 85.95.

1.5, 2, 3, 4, 6, 8, 12, 16, 24, and 32 hours. The plot of primidone concentration-time curve for the subject is exhibited in Figure 1.3.1. From Table 1.3.1, AUC(0 − 32) and C_{max} can be obtained as follows.

$$
\begin{aligned}
AUC(0 - 32) &= \sum_{i=1}^{12} \left[\frac{C_{i-1} + C_i}{2} \right] (t_i - t_{i-1}) \\
&= \frac{(0 + 0)}{2}(1 - 0.5) + \frac{(2.8 + 0)}{2}(1.5 - 1) + \cdots \\
&\quad + \frac{(1.6 + 2)}{2}(32 - 24) \\
&= 85.95 \ (\mu \cdot hr/mL) \\
C_{max} &= max(0, 0, 2.8, \ldots, 1.6) = 4.7 \ \mu g/mL.
\end{aligned}
$$

t_{max} is estimated as the corresponding time point at which C_{max} was achieved. Thus, $t_{max} = 3.0$ hr. For the estimation of the elimination rate k_e, the last seven concentrations during the elimination phase were used to fit a linear regression based on the log concentrations with a base 10 using the least squares method (see, e.g., Draper and Smith, 1981). The resultant regression line is given by

$$
\log_{10}(C_i) = 0.6713 - 0.01518 t_i.
$$

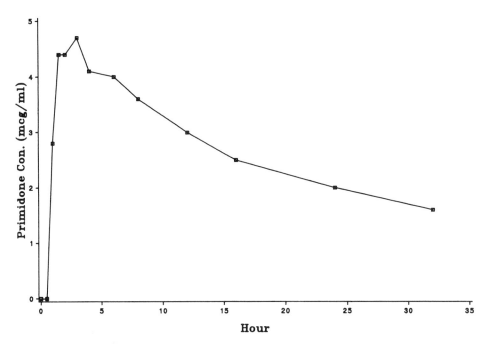

Figure 1.3.1   Primidone concentration-time curve.

Thus, the elimination rate is

$$k_e = (-2.303)(-0.01518) = 0.03496 \ (hr^{-1}).$$

Consequently, the elimination half-life is

$$t_{1/2} = 0.693/0.03496 = 19.8 \ (hr).$$

The AUC$(0 - \infty)$ can be obtained as

$$AUC(0 - \infty) = AUC(0 - 32) + C_{32}/0.03496$$
$$= 85.95 + 1.6/0.03496$$
$$= 131.72 \ (\mu g \cdot hr/mL).$$

In the above example, we selected the last seven concentrations during the elimination phase to calculate the elimination rate. In practice, the number of concentrations used may depend on the plasma concentration-time curve for each subject. This is an interesting statistical question which needs further attention.

1.4. CLINICALLY IMPORTANT DIFFERENCES

The definition of a clinically significant difference is important for the assessment of therapeutic equivalence in terms of efficacy, safety, and benefit/risk ratio. In bioavailability/bioequivalence studies, it is our intention to consider bioequivalence in terms of therapeutic equivalence. However, this ultimate assumption of bioequivalence can only be verified through rigorous prospective clinical trials which may relate bioavailability parameters with clinical endpoints through the data from blood concentrations and clinical efficacy/safety evaluations. In practice, such clinical trials are rarely carried out due to the following difficulties:

(i) Unlike healthy subjects who are often used for bioavailability/ bioequivalence studies, patients cannot be well controlled;

(ii) Patients are more heterogeneous in a wide variety of characteristics.

However, the ultimate obstacle lies in the estimation and translation of the differences in bioavailability into the therapeutic differences of interest.

Westlake (1979) pointed out that a statistically significant difference in the comparison of bioavailability between drug products does not necessarily imply that there is a clinically significant difference between drug products. For example, the AUC for the test product may exhibit an 80% bioavailability compared to the reference product. The 20% difference in AUC, which may be statistically significant, however, may not be of clinical significance in terms of therapeutic effect. In other words, although there is a 20% difference, both test and reference products can still reach the same therapeutic effect. Thus, they should be considered therapeutically equivalent. Generally, a set of bioequivalence limits, say (a, b) is given for the evaluation of clinical difference. If the difference (usually in %) in AUC between the test and reference products is within the limits, then there is no clinical difference or they are considered to be therapeutically equivalent. Bioequivalent limits for therapeutic equivalence generally depend upon the nature of the drug, patient population, and clinical endpoints (efficacy and

Table 1.4.1 Equivalence Limits for Binary Responses

Equivalence limits	Response rate for the reference drug
± 20%	50%–80%
± 15%	80%–90%
± 10%	90%–95%
± 5%	>95%

safety parameters) for the assessment of therapeutic effect. For example, for some drugs such as topical antifungals or vaginal antifungals which may not be absorbed in blood (Huque and Dubey, 1990), the FDA proposed some equivalent limits for some clinical endpoints (binary responses) such as cure rate as in Table 1.4.1. This table indicates that if the cure rate for the reference drug is greater than 95%, then a difference in cure rate within 5% is not considered a clinically important difference.

1.5. ASSESSMENT OF BIOEQUIVALENCE

The assessment of bioequivalence for different drug products is based on the following **Fundamental Bioequivalence Assumption**:

> When two drug products are equivalent in the rate and extent to which the active drug ingredient or therapeutic moiety is absorbed and becomes available at the site of drug action, it is assumed that they will be therapeutically equivalent.

The purpose of bioequivalence trials is to identify pharmaceutical equivalents or pharmaceutical alternatives that are intended to be used interchangeably for the same therapeutic effect (21 CFR, 320.50). Thus, bioequivalent drug products are therapeutic equivalents and can be used interchangeably.

Once the fundamental assumption and the purpose of bioequivalence trials are clearly defined and understood, the next question is what and how to assess bioequivalence. The essential pharmacokinetic parameters required in the FDA regulations for an in vivo bioavailability study are $AUC(0 - \infty)$, C_{max}, t_{max}, $t_{1/2}$, and k_e of the therapeutic moiety. As discussed in Section 1.3, these pharmacokinetic parameters can be derived either directly from the observed blood or plasma concentration-time curve or obtained by fitting the observed concentrations to a one or multi-compartment pharmacokinetic model. In general, the use of the observed $AUC(0 - \infty)$, C_{max}, or t_{max} from the blood or plasma concentration-time curve is preferred since they provide the essential information about the pharmacokinetic characteristics in assessment of bioequivalence, and are model-independent and easy to calculate. However, there are some drawbacks in these estimates. For example, the predetermined sampling time points are often too few to have reliable estimates on C_{max} and t_{max} in most bioavailability studies. Consequently, the distribution of the estimated t_{max} is not continuous but rather discrete. On the other hand, when a pharmacokinetic model is considered, the goodness of fit of the model should be performed by examining the residuals. In practice, it is almost impossible to fit the same theoretical model for each subject in the study. Moreover, the sampling time points are too few to provide reliable estimates for the pharmacokinetic parameters under the model, even though, theoretically, the assumed pharmacokinetic model may adequately describe the observed blood or plasma concentration-time curve.

The statistical answers to the questions of (i) How can the bioequivalence of two drug products be determined? and (ii) Can the two drug products be substituted for each other if they are judged bioequivalent? depend on whether the marginal distributions of the pharmacokinetic parameters of interest for the two drug products are equivalent. This concept is referred to as "population bioequivalence." Based on the fact that the distribution of some random variables (e.g., normal random variable) is uniquely determined by its moments, the equivalence between two distributions can be assessed through the moments of the marginal distributions of the test and reference formulations. The first two moments of the distribution reflect the average and the variability of the distribution. The comparison of the first moments of the distributions of the pharmacokinetic parameters (say, AUC$(0 - \infty)$) for the two drug products refers to the comparison of *average bioavailability,* while the comparison between the second moments refers to the *variability of bioavailability.* To provide a better

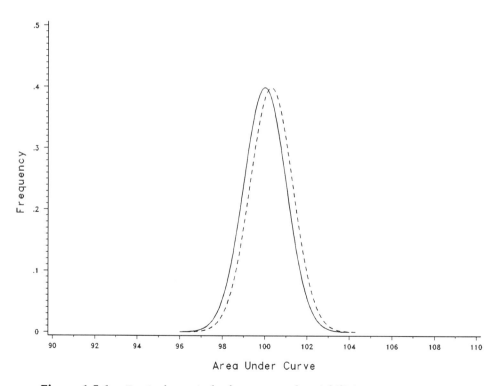

Figure 1.5.1   Equivalence in both means and variabilities.

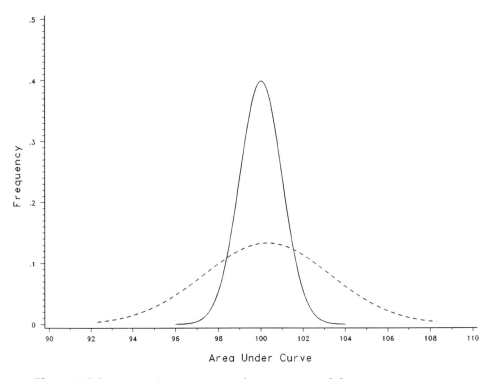

Figure 1.5.2   Equivalence in means but not in variabilities.

understanding of average bioavailability and variability in bioavailability, equivalence in averages and variabilities are illustrated in Figures 1.5.1–1.5.3. For example, if the distribution for AUC(0 − ∞) is normal and if the AUC(0 − ∞) of the two products are equivalent in both averages and variabilities, then the two drug products are bioequivalent. However, in general, equivalence in the first two moments does not guarantee equivalence between distributions. Currently the FDA regulations for bioequivalence of two drug products only require establishing bioequivalence in average bioavailability. This certainly does not ensure the ultimate goal that two drug products can be used interchangeably if they are only bioequivalent in average bioavailability. Some discussions on this issue can be found in Cornell (1980), Metzler and Huang (1983), and Liu (1991).

1.6. DECISION RULES AND REGULATORY ASPECTS

As discussed in previous sections, the association between bioequivalence limits and clinical difference is difficult to assess in practice. The following decision

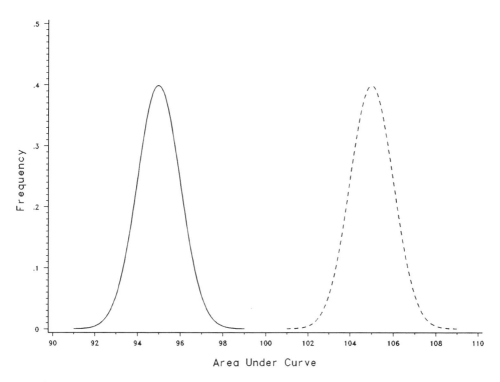

Figure 1.5.3   Equivalence in variabilities but not in means.

rules were proposed by the FDA between 1977 and 1980 (Purich, 1980) for testing the bioequivalence in terms of average bioavailability of specific drugs, such as anticonvulsants, carbonic anhydrase inhibitors and phenothiazines. Suppose AUC and C_{max} are the primary measures of the extent and rate of absorption. For each parameter, the following decision rules are applied.

The 75/75 Rule   Bioequivalence is claimed if at least 75% of individual subject ratios (relative individual bioavailability of the test formulation to the reference formulation) are within (75%, 125%) limits.

The 80/20 Rule   If the test average is not statistically significantly different from the reference average and if there is at least 80% power for detection of a 20% difference of the reference average, then bioequivalence is concluded.

The ±20 Rule   Bioequivalence is concluded if the average bioavailability of the test formulation is within ±20% of that of the reference formulation with a certain assurance.

For the 75/75 rule, although it possesses some advantages such as (i) it is easy to apply, (ii) it compares the relative bioavailability within each subject, and (iii) it removes the effect of heterogeneity of inter-subject variability from the comparison between the formulations, it is not viewed favorably by the FDA due to some undesirable statistical properties. In a simulation study, Haynes (1981) showed that the 75/75 rule is very sensitive for drugs that have large inter- or intra-subject variabilities even in the situation where the mean AUC's for the test and reference formulations are exactly the same. Metzler and Huang (1983), in another simulation study, also indicated that the 75/75 rule may reject as much as 56.3% of test products when the inter-subject variability is large. Thiyagarajan and Dobbins (1987) also discussed the use of the 75/75 rule for assessment of bioequivalence. Chow (1989) and Chow and Shao (1991) provided an analytic evaluation of the 75/75 rule relative to the ± 20 rule. The results suggest that the 75/75 rule will never be met when the intra-subject variability is large (say 20%) for any given true ratio of means. For small variability (say 10%), only 61.3% of individual subject ratios will fall within (75%, 125%) limits when the true ratio of means is within 80% and 120% limits. Anderson and Hauck (1990) discussed the 75/75 rule and considered the use of individual subject ratios for assessment of "individual bioequivalence."

The 80/20 rule, which often requires that a study be large enough to provide at least 80% chance of correctly detecting a 20% difference in average bioavailability, is based on the concept of testing a hypothesis of equality for a single variable rather than equivalence. In the past two decades, however, hypothesis testing for the evaluation of bioequivalence has been questioned and was not encouraged. The 80/20 rule is usually considered as a pre-study power calculation for sample size determination in the planning stage of study protocol.

The ± 20 rule, which allows a test formulation to exhibit up to a 20% variation in bioavailability in comparison to a reference formulation, is commonly employed for most drug products. Levy (1986), however, indicated that the ± 20 rule does not accommodate the impact that the 20% variation could have on the safety and efficacy of a specific drug. Another concern is the interchangeability of the formulations. As more generic products become available, the generic substitution for a brand name drug may involve the substitution of one generic product for another during a patient's therapy. Under the ± 20 rule, interchanging the generic products can lead to a more than 20% difference from one to another. For example, substitution of a bioequivalent product providing 120% of the reference for a bioequivalent product providing 80% of the reference would result in an increase in the relative dose of 50%. In this case, the toxicity and/or efficacy is significantly magnified when bioequivalent products varying by as much as 50% are interchanged one for another. Thus, it is suggested that individualized drug-by-drug bioequivalence criteria (i.e., the acceptable degree of variation in bioavailability) be developed by the FDA at the time a generic product becomes eligible for approval.

Based on the report by the Bioequivalence Task Force, the 75/75 rule is not required for the assessment of bioequivalence because it is not based on rigorous statistical tests. It appears that the ± 20 rule is acceptable to the FDA. The 80/20 rule is recommended as the secondary analysis which is often used as a supplement to the ± 20 rule. However, in many cases, the ± 20 rule and the 80/20 rule may result in inconsistent conclusions. That is, the bioequivalence is concluded based on the ± 20 rule, but the power for detecting a 20% difference is far below 80% or vice versa. The possible causes of the inconsistency between the two decision rules were discussed by Chow and Shao (1991).

It should be noted that bioequivalence determinations based on mean values do not account for the differences in inter- and/or intra-subject variabilities between formulations. Although the Pitman-Morgan test (Pitman, 1939; Morgan, 1939) is suggested for testing the equality of the variance between formulations, little or no attention in the literature has been given to address how much difference in variability including inter- and/or intra-subject variabilities would be of clinical significance. In general, a much larger sample size is required for testing a difference in variances than that of testing for a difference in averages. More details regarding the variability of bioavailability is given in Chapter 7.

1.7. STATISTICAL CONSIDERATIONS

In this section, some statistical considerations which may occur in the assessment of bioequivalence are summarized below.

1.7.1. AUC Calculation

As indicated in Section 1.3, among the pharmacokinetic parameters, AUC is the primary measure of the extent of absorption or the amount of drug absorbed in the body which is often used to assess bioequivalence between drug products. AUC is usually calculated using the trapezoidal rule based on the blood or plasma concentrations obtained at various sampling time points. In practice, a few missing values and/or unexpected observations may occur at some sampling time points due to laboratory error, data transcription error, or other causes unrelated to bioequivalence. Generally, missing values and/or unexpected observations between two end sampling time points have little effect on the comparison of bioavailability (Rodda, 1986). However, if many missing values and/or unexpected observations occur in the plasma concentration-time curve, especially at two end sampling time points, the bias of the estimated AUC could be substantial, and consequently may affect the comparison of bioavailability. Thus, how to justify the bias in the calculation of AUC is an important statistical issue. Furthermore, since the concentration at time zero (i.e., immediately prior to drug administration) may be different from subject to subject, whether or not the AUC should be adjusted from the baseline concentration is an interesting problem for both the clinician and biostatistician.

1.7.2. Model Selection and Normality Assumptions

Let μ_T and μ_R be the true averages for test and reference products, respectively. According to the ± 20 rule, the ratio of true averages (μ_T/μ_R) must be within (80%, 120%) with 90% assurance to claim bioequivalence. A typical approach is to construct a 90% confidence interval for μ_T/μ_R and compare it with (80%, 120%). If the constructed confidence interval is within (80%, 120%), then bioequivalence is concluded. To construct a 90% confidence interval for μ_T/μ_R, two statistical models, namely the raw data model (or additive model) and the log-transformed model (or multiplicative model), are often considered.

For the raw data model, an exact 90% confidence interval for $\mu_T - \mu_R$ is constructed based on the original data (raw data) and converted to be the confidence interval for μ_T/μ_R by dividing by the observed reference mean ($\overline{Y}_R$) (assuming that $\overline{Y}_R$ is the true μ_R). The constructed confidence interval, however, is not at the exact 90% confidence level because the method ignores the variability of $\overline{Y}_R$. Another method is to use the Fieller's theorem (Locke, 1984; Schuirmann, 1989) to construct an exact 90% confidence interval for μ_T/μ_R directly. This method is derived based on the ratio of sample means for test and reference products. However, the disadvantage of this method is that the distribution of the ratio of sample means is rather complicated and its moments may not exist (Hinkley, 1969). It is important to provide a further statistical evaluation of the above confidence intervals because the decision of bioequivalence is made based on whether or not the confidence interval is within 80% and 120% (Schuirmann, 1989).

The primary assumptions of the raw data model are normality assumptions. Since the AUC's, t_{max}, and C_{max} are positive quantities, the underlying distributions are, in fact, normal distribution truncated at 0. This is a valid argument against the raw data model. In addition, for many cases, the distribution of AUC is skewed. Thus, a log transformation on AUC is usually performed to remove the skewness. The log-transformed data is then analyzed using the raw data model which is equivalent to analyzing the raw data using the log-transformed model. Under the normality assumptions, the log-transformed model can provide an exact confidence interval for μ_T/μ_R (Mandallaz and Mau, 1981). Thus, compared to the raw data model, the FDA is in favor of the log-transformed model if transformation is required for the analysis of bioequivalence studies (see also, Attachment 5, report by the Bioequivalence Task Force, 1988).

The above method, based on either the raw data model or the log transformed model, are derived under the assumptions of normality or lognormality for between subject (inter-subject) and within subject (intra-subject) variabilities. One of the difficulties commonly encountered is whether or not the assumption of normality or lognormality is valid. It is suggested that the normality or lognormality assumptions be checked before an appropriate statistical model is used. The tests for normality or lognormality assumptions are critical for choosing an appropriate model. Unfortunately, thus far, there exist no convincing statistical tests for normality or lognormality assumptions for inter-subject and intra-subject variabilities in bioequivalence studies. However, Jones and Kenward (1989) recommended a method using studentized residuals, which are obtained under the model (they are approximately independent) for testing normality of an intra-subject variability based on the Shapiro-Wilk statistic (Shapiro and Wilk, 1965). A similar approach is also suggested for testing the normality of an inter-subject variability. Due to the difficulty of testing normality assumptions, in the past two decades, some research efforts were directed to the search for nonparametric alternatives (see, e.g., Koch, 1972; Cornell, 1980; Hauschke, Steinijans, and Dilletti, 1990).

1.7.3. Inter- and Intra-Subject Variabilities

Since individual subjects may differ widely in their responses to the drug, the knowledge of inter- and intra-subject variabilities may provide valuable information in the assessment of bioequivalence (Wagner, 1971). To remove the inter-subject variability from the comparison of bioavailability between drug products, a crossover design, which is the design of choice by many investigators and is acceptable to the FDA (21 CFR, 320.26 and 320.27), is often considered. The advantages of using a crossover design are

   (i)   Each subject can serve as his/her own control;
   (ii) The assessment of bioequivalence is based on the intra-subject variability;
  (iii) Fewer subjects are required to provide the desired degree of accuracy and power compared to other designs such as parallel design.

However, in a crossover design, the intra-subject variability may be confounded with some expected and unexpected variabilities such as lot-to-lot, product-to-product, and subject-by-product variabilities. These sources of variabilities are difficult to assess based on a crossover design or other current available designs (Ekbohm and Melander, 1989). Thus, appropriate statistical designs and/or methods are necessary for assessing these variabilities.

1.7.4. Interval Hypothesis and Two One-Sided Tests

As early as the 1970's, statisticians became aware that the usual hypotheses testing for equality was not appropriate for bioavailability studies (Metzler, 1974). The purpose of bioequivalence is to verify that two formulations are indeed bioequivalent. Thus, from a statistical viewpoint, it may be more appropriate to reverse the null hypothesis of bioequivalence and the alternative hypothesis of bioinequivalence. Let θ_1 and θ_2 be two known bioequivalence limits and θ be the parameter of interest. The hypotheses for assessment of bioequivalence are given below:

$$H_0: \theta \leq \theta_1 \quad \text{or} \quad \theta \geq \theta_2 \quad \text{vs} \quad H_a: \theta_1 < \theta < \theta_2,$$

which can be further decomposed into two one-sided hypotheses as:

$$H_{01}: \theta \leq \theta_1 \quad \text{vs} \quad H_{a1}: \theta_1 < \theta,$$

and

$$H_{02}: \theta \geq \theta_2 \quad \text{vs} \quad H_{a2}: \theta < \theta_2.$$

Because of hypothesis of bioequivalence in H_a is expressed as an interval, it is then referred to as the interval hypothesis. The test procedures for the average bioavailability based on the interval hypothesis were proposed by Schuirmann (1981, 1987) and Anderson and Hauck (1983). The distribution of the observed test statistic proposed by Anderson and Hauck can be approximated by a central t-distribution. Schuirmann's procedure uses two one-sided tests for assessment of equivalence in average bioavailability. In this approach, two p values are obtained to evaluate whether the bioavailability of the test product is not too low for one side (H_{01} vs H_{a1}) and whether the bioavailability is not too high for the other side (H_{02} vs H_{a2}). However, it is not clear what the overall p value is for H_0 vs H_a since for any given θ_1 and θ_2 and the observed statistic for H_{01} vs H_{a1}, the p value for H_{02} vs H_{a2} is not a random variable but a fixed known quantity. In addition, the above two approaches suffer from the fact that under the normality assumption there is no unconditional uniformly most powerful unbiased (UMPU) (nor invariant) test (Lehmann, 1959; Kendall and Stuart, 1965). In other words, there always exists procedures with greater power for the same hypotheses under certain conditions. Alternatively, several nonparametric procedures have been proposed (Hauschke et al., 1990; Liu, 1991). However, there is little or no information available regarding the relative efficiency of the nonparametric procedures to the parametric methods.

1.7.5. Outlier Detection

As indicated in the report by the Bioequivalence Task Force, the detection and treatment of outlying data in bioequivalence studies are important issues because the results and decisions of bioequivalence could be totally different by including or excluding the outlying data in the analysis. Recently, several tests have been proposed for the detection of outlying data (Chow and Tse, 1990a; Liu and Weng, 1991; Lin and Tsong, 1990). However, additional research and the development of some robust procedures are needed in this area.

1.7.6. Exchangeability

As indicated in Section 1.5, one of the objectives in bioavailability/ bioequivalence studies is to examine whether the test and reference products can be used interchangeably. For this purpose, equivalence must be established at least in both average and variability. In this case, there are two sets of bioequivalence hypotheses to be tested, one for average bioavailability and the other for variability of bioavailability. However, the following question arises. Do we test the two sets of hypotheses separately, each at the α level or do we test these two sets of hypotheses jointly at the α level? For the latter case, it may be necessary to develop a combined test statistic from two independent test statistics (one from the average and the other from the variability).

1.7.7. Other Issues

Several issues regarding the assessment of bioequivalence have been recently discussed. These include the possible use of individual subject ratios as a preliminary test for bioequivalence (Peace, 1986; Anderson and Hauck, 1990), the use of the ratio of medians as an alternative measure for bioequivalence (Metzler and Huang, 1983; Chinchilli and Durham, 1989), and the justification of significant level when a three- (or higher) way crossover design is used. Some of these issues will be discussed further in Chapter 10.

1.8. AIMS AND STRUCTURE OF THE BOOK

This is intended to be the first book entirely devoted to the design and analysis of bioavailability studies. It covers all of the statistical issues that may occur in the various stages of design and data analysis in bioavailability studies. It is our goal to provide a useful desk reference and state-of-the art examination of this area to scientists engaged in pharmaceutical research, those in government regulatory agencies who have to make decisions on the bioequivalence between drug products, and to biostatisticians who provide the statistical support for bioavailability studies and related clinical projects. More importantly, we would like to provide graduate students in pharmacokinetics, clinical pharmacology, biopharmaceutics, and biostatistics an advanced textbook in bioavailability studies. We hope that this book can serve as a bridge among the pharmaceutical industry, government regulatory agencies, and academia.

In this chapter, the history, definition, decision rules, and some statistical considerations for bioavailability studies have been discussed. In the next chapter, some basic considerations regarding the concerns of the investigator, monitor, and biostatistician for the designs of bioavailability studies will be discussed. We will then introduce some designs which are currently available for bioavailability studies. The relative advantages of a crossover design which is acceptable to the FDA will also be extensively discussed in this chapter. In Chapter 3, statistical inference for a variety of effects from a standard 2×2 crossover design will be discussed. Statistical methods currently available for the assessment of bioequivalence will be provided in Chapter 4. The nonparametric methods including bootstrap resampling procedure will also be extensively explored in this chapter. These methods will be compared in terms of power and relative efficiency in Chapter 5. Sample size determination will also be included in this chapter. The log-transformed model and the approach using individual subject ratios will be given in Chapter 6. In addition to the examination of intra-subject variability and inter-subject variability, the assessment of bioequivalence using the variability of bioavailability will be explored in Chapter 7. In Chapter 8, some tests for normality assumptions and procedures for detection of outliers will be derived. Chapter 9 provides statistical methods for assessing bioequivalence under a higher-order crossover design for two formulations. Assessment of bioequivalence for more than two formulations are outlined in Chapter 10. Chapter 11 gives an introduction for assessment of bioequivalence based upon clinical endpoints such as response data and time to onset of a therapeutic response when plasma concentrations are negligible. In Chapter 12, some related problems in bioavailability studies such as dose proportionality studies, drug interaction studies, steady state analyses for multiple doses, and population pharmacokinetics are given.

2

Designs of Bioavailability Studies

2.1. INTRODUCTION

Before a clinical trial is conducted, a protocol which details the conduct of the trial is usually developed. A thoughtful and well-organized protocol includes study objective(s), study design, patient selection criteria, dosing schedules, and statistical methods. Unlike clinical trials, bioavailability studies are often conducted with healthy volunteers. Thus, the choice of the designs and the statistical methods for the analysis of data become two important aspects in planning a bioavailability study. These two aspects are closely related to each other since the method of analysis depends on the design employed. Generally meaningful conclusions can only be drawn based on data collected from a valid scientific design using appropriate statistical methods. General considerations that one should consider when planning a bioavailability study include:

(i)   What is to be studied or what is the study objective(s)?
(ii) How are the data to be collected or what design is to be employed?
(iii) How is the data to be analyzed or what statistical methods are to be used?

In this chapter, our efforts will be directed to the determination of study objectives and the selection of an appropriate design for a bioavailability study. We intend to explore and compare some basic designs which are currently available for bioavailability studies. Some specific designs which are used for different purposes under various circumstances will be discussed further in Chapter 9. Unless otherwise specified, throughout this book, for the sake of convenience, we will restrict our attention to the comparison of different formulations of the same drug product. The comparison of different drug products of the same active ingredient and different ways of administration can be treated similarly.

The choice of the design primarily depends on the variability in the observations. For example, as indicated in Section 1.7.3, the individual subjects may differ very widely in their responses to the drug products. Thus, one major source of variability arises from differences between subjects. As a result, a criterion for choosing an appropriate design is whether or not the selected design can identify and isolate the inter-subject variability in data analysis. Any design which can remove this variation from the comparison between formulations would be appropriate. Such a design is generally more efficient than a design which cannot account for the inter-subject variability. In this chapter, we will introduce several designs which are often considered for bioavailability/bioequivalence studies. These designs include the complete randomized designs (or the parallel designs), the randomized block designs, the crossover designs, the Latin square designs, and the (balanced) incomplete block designs. These designs, which may remove the expected variability from the comparison of bioavailability between formulations, may be useful depending on the parameters to be evaluated, the characteristics of the drug and/or the medical restrictions.

The remainder of this chapter is organized as follows. In the next section, objectives for some studies related to bioavailability such as bioequivalence studies, dose proportionality studies, and steady state analyses are discussed. In Section 2.3, we provide some design considerations when planning a bioavailability study. In Section 2.4, a brief description of a parallel design is given. An extensive discussion on crossover designs is presented in Section 2.5. Balanced incomplete block designs are introduced in Section 2.6. Some factors for choosing an appropriate design for bioavailability studies will be discussed in Section 2.7.

2.2. THE STUDY OBJECTIVE

In clinical trials, a description of the general aims of the study is a useful preliminary which helps to explain why the study is considered worthwhile (Pocock, 1983). The statement of study objective(s) is a concise and precise definition of prespecified hypotheses or parameters concerning drug products which are to be examined or estimated. In clinical trials, a clear statement of study objectives not only ensures that the investigator adhere to the hypotheses at the time of analysis and interpretation of results, but also enables statisticians to select an appropriate design and statistical methods for data analysis.

In the following, some examples of study objectives and corresponding hypotheses or parameters of interest in bioavailability and related studies are given.

2.2.1. Bioequivalence Studies

One of the objectives of a bioequivalence study is to compare bioavailability between two formulations (a test and a reference formulation) of a drug product and to determine bioequivalence in terms of the rate and extent of absorption. The primary hypothesis is whether the difference in average bioavailability between a test and reference product is within $\pm 20\%$ of the reference mean with certain assurance. To achieve this objective, a parallel design or a crossover design is often considered. Several statistical methods are available for the evaluation of the hypothesis.

2.2.2. Dose Proportionality Studies

For a dose proportionality (or dose linearity) study, the objective is to evaluate whether the relationship between dose level and a pharmacokinetic parameter (such as AUC) is linear over a given dose range. The results may provide useful information in determining dose levels at which the minimum concentration for therapeutic effect and toxic concentration will be achieved. The hypothesis of interest is that there is a linear relationship between dose level and AUC. Several statistical tests for the hypothesis of dose proportionality are available for both serial blood collection and single time point blood collection. More details regarding dose proportionality studies will be discussed further in Chapter 12.

2.2.3. Steady State Studies

For a steady state study, a comparison of the blood (or plasma) concentration is made after steady state is achieved (generally, after multiple dosing). The objective of such a study is to determine whether a steady state was reached and when it was reached. This may be evaluated by testing the hypothesis that there is no difference in concentrations at the end of each dosing interval. In Chapter 12, more details regarding a steady state analysis will be given.

2.2.4. Variability and Interchangeability

Since the determination of bioequivalence may not adequately characterize different types of variation that can occur both within a given individual as well as among different individuals, an appropriate design may be considered to provide information on the inter-subject and the intra-subject variabilities and the interchangeability of one formulation for another. The objective of such design is to estimate the inter-subject and intra-subject variabilities and provide statistical inference on both the variability and the interchangeability. This issue will be examined in Chapter 7.

Popock (1983) indicated that, in clinical trials, the study objectives are built on more expansive descriptions of patient selection criteria, treatment schedules, and the methods of patient evaluation. Although a precise and detailed explanation of these issues can help to ensure that an unbiased assessment of the study objectives is achieved, a valid scientific design with appropriate statistical methods for the analysis of data is the key to carrying out the study objectives.

2.3. BASIC DESIGN CONSIDERATIONS

In the *Federal Register* (Vol. 42, No. 5, Sect. 320.25(b)), the FDA (1977) indicated that a basic design for an in vivo bioavailability study is determined by the following:

   (i)   The scientific questions to be answered;
  (ii)   The nature of the reference material and the dosage form to be tested;
 (iii)   The availability of analytical methods;
  (iv)   Benefit-risk considerations in regard to human testing.

Consideration of the reference dosage form is critical. For example, a suspension may not be an appropriate reference material because of high variability in bioavailability of the suspension dosage form. In many instances, a suspension of a poorly soluble active drug ingredient may be more poorly absorbed than a well-formulated tablet.

The availability of the analytical method which is used to measure acute pharmacological effect or concentration of the active drug ingredient, therapeutic moiety, or metabolites is important. The FDA requires that the analytical method used in bioavailability studies be of sufficient accuracy, sensitivity and reproducibility to discriminate between inequivalent products. The requirement implies that a product of known poor bioavailability must be compared against the reference product to determine whether the method can detect differences between the two products.

Finally, in practice, most bioavailability studies are conducted with healthy normal subjects. Bioavailability studies conducted on critically ill patients may not be appropriate and contrary to the best medical practice unless there is a definitive benefit to the patients. For example, a bioavailability study with kanamycin in patients with stable renal disease would permit dosage adjustments based on renal creatinine clearance and serum kanamycin levels.

In addition to these basic design considerations, some specific considerations when planning a design for a bioavailability study are given below.

2.3.1. Experimental Design

The *Federal Register* (Vol. 42, No. 5, Sects. 320.26(b) and 320.27(b), 1977) indicated that a bioavailability study (single-dose or multiple-dose) should be crossover in design, unless a parallel or other designs is more appropriate for valid scientific reasons. For a parallel design, each subject receives one and only one formulation in random fashion, while for a crossover design each subject receives more than one formulation at different time periods. In practice, subjects account for a large source of variability in plasma or blood drug concentrations. Thus, an appropriate design should allow estimation and removal of the inter-subject variability from drug comparison. More details regarding the parallel, crossover, and other designs are discussed in the following sections.

2.3.2. Randomization

Valid statistical inferences are usually drawn based on the assumption that the errors in observations are independently distributed random variables. Randomization usually ensures this assumption valid. The randomization schedules depend on the design selected. For example, for a parallel design comparing two formulations of a drug product, the subjects are assigned to receive each formulation at random. For a crossover design, each subject is a block which represents a restriction on complete randomization because the formulations are randomized within the subject. An example of randomization for a standard 2×2 crossover design is given in Section 2.5.4.

2.3.3. Sampling Time Intervals

For the estimation of the rate and extent of absorption, although the sampling time intervals for both the test and reference formulations need not be the same, it is preferred that sampling time intervals be identical to assure true equivalence. Blood or plasma samples should be collected at the time prior to dosing and over an interval of sufficient time (e.g., 3–5 half-lives of the drug active ingredient or therapeutic moiety) to accurately determine the individual terminal disposition curve.

2.3.4. Drug Elimination Period

For a single-dose study, the terminal drug elimination period should allow at least three half-lives of the active drug ingredient or therapeutic moiety, or its metabolite, either measured in the blood or as the decay of the acute pharmacological effect. For a multiple dose study, the drug elimination period should allow at least 5 half-lives.

2.3.5.   Number of Subjects

For a bioavailability study, usually 18–24 healthy normal subjects are used. To detect a clinically important difference (e.g., 20%), a pre-study power calculation is often performed to determine the number of subjects needed for detection of such difference with a desired probability (e.g., 80%). The issue of power and sample size determination will be discussed further in Chapter 5.

2.4.   THE PARALLEL DESIGN

A parallel design is a complete randomized design in which each subject receives one and only one formulation of a drug in a random fashion. The simplest parallel design is the two-group parallel design which compares two formulations of a drug. Each group usually contains the same number of subjects. An example of a two-group parallel design is illustrated in Table 2.4.1.

For phase II and phase III clinical trials, the parallel design probably is the most frequently used design. However, it may not be an appropriate design for bioavailability/bioequivalence studies. This is because the variability in observations (e.g., AUC) consists of the inter-subject and intra-subject variabilities and the assessment of bioequivalence between formulations is usually made based on the intra-subject variability. A parallel design, however, is not able to identify and separate these two sources of variations because each subject in the parallel design usually receives the same drug during the entire course of study. Although the equivalence in average bioavailability between formulations can still be

Table 2.4.1   Two-Group Parallel Design

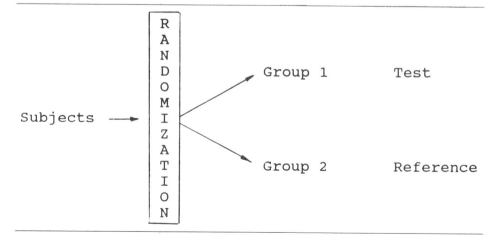

established through this design, the comparison is made based on the inter-subject and intra-subject variabilities. As a result, for a fixed number of subjects, the parallel design would, in general, provide a less precise statistical inference for the difference in bioavailability between formulations than that of a crossover design.

Although the parallel design is not widely used for bioavailability studies due to the incapability of identifying and removing the inter-subject variability from the comparison between formulations, there are some rare occasions in which a parallel design may be more appropriate than a crossover design. For example, for generic topical antifungals bioequivalence study, the FDA requires a three-arm parallel design (i.e., test, reference, and vehicle control). If the drug is known to have a very long half-life, it is not desirable to adapt a crossover design. In a crossover design, a sufficient length of washout is necessary to eliminate the possible carry-over effects and consequently, the study may take considerable time. This, in turn, may increase the number of dropouts and make the completion of a study difficult. In addition, if the study is to be conducted with very ill patients, a parallel design is usually recommended so that the study can be completed quickly. As a result, a parallel design may be considered as an alternative to a crossover design if (i) the inter-subject variability is relatively small compared to the intra-subject variability, (ii) the drug is potentially toxic and/or has a very long elimination half-life, (iii) the population of interest consists of very ill patients, and (iv) the cost for increasing the number of subjects is much less than that of adding an additional treatment period.

2.5. THE CROSSOVER DESIGN

2.5.1. Introduction

A crossover design is a modified randomized block design in which each block receives more than one formulation of a drug at different time periods. A block may be a subject or a group of subjects. Subjects in each block receive a different sequence of formulations. A crossover design is called a complete crossover design if each sequence contains each of the formulations. For a crossover design, it is not necessary that the number of formulations in each sequence be greater than or equal to the number of formulations to be compared. We shall refer to a crossover design as a g × p crossover design if there are g sequences of formulations which are administered at p different time periods. For bioavaila-bility/bioequivalence studies, the crossover design is viewed favorably by the FDA due the following advantages:

(i) Each subject serves as his/her own control. It allows a within subject comparison between formulations.

    (ii)   It removes the inter-subject variability from the comparison between formulations.

    (iii) With a proper randomization of subjects to the sequence of formulation administrations, it provides the best unbiased estimates for the differences (or ratios) between formulations.

The use of crossover designs for clinical trials has been greatly discussed in the literature. See, for example, Brown (1980), Huitson et al. (1982), and Jones and Kenward (1989).

In the following, we introduce several different types of crossover designs which are often used in bioavailability studies. The relative advantages and drawbacks of these designs will also be discussed.

2.5.2. Washout and Carry-Over Effects

It is helpful to introduce the concepts of washout and carry-over effects (or residual effects) in a crossover design since the presence of carry-over effects usually has an impact on statistical inference of bioavailability between formulations.

The washout period is defined as the rest period between two treatment periods for which the effect of one formulation administered at one treatment period does not carry over to the next. In a crossover design, the washout period should be long enough for the formulation effects to wear off so that there is no carry-over effect from one treatment period to the next. The washout period depends on the nature of the drug. A suitable washout period should be long enough to return any relevant changes which influence bioavailability to baseline (usually, at least three times the blood/plasma elimination half-life of the active ingredient, therapeutic moiety, or its metabolite, or the decay of the acute pharmacological effect).

If a drug has a long half-life or if the washout period between treatment periods is too short, the effect of the drug might persist after the end of dosing period. In this case, it is necessary to distinguish the difference between the *direct* drug effect and the carry-over effects. The direct drug effect is the effect that a drug product has during the period in which the drug is administered, while the carry-over effect is the drug effect that persists after the end of the dosing period. Carry-over effects which last only one treatment period are called first-order carry-over effects. A drug is said to have c-order carry-over effects if the carry-over effects last up to c treatment periods. In bioavailability/bioequivalence studies, however, it is unlikely that a drug effect will carry over more than one treatment period since a sufficient length of washout is usually considered. In this book, therefore, we will only consider the first-order carry-over effects if they are present.

2.5.3. Statistical Model and Linear Contrast

In a crossover design, since the direct drug effect may be confounded with any carry-over effects, it is important to remove the carry-over effects from the comparison if possible. To account for these carry-over effects, the following statistical model is usually considered. Let Y_{ijk} be the response (e.g., AUC) of the ith subject in the kth sequence at the jth period.

$$Y_{ijk} = \mu + S_{ik} + P_j + F_{(j,k)} + C_{(j-1,k)} + e_{ijk}, \tag{2.5.1}$$

where

> μ = the overall mean;
> S_{ik} = the random effect of the ith subject in the kth sequence, where $i = 1, 2, \ldots, n_k$ and $k = 1, 2, \ldots, g$;
> P_j = the fixed effect of the jth period, where $j = 1, \ldots, p$ and $\Sigma_j P_j = 0$;
> $F_{(j,k)}$ = the direct fixed effect of the formulation in the kth sequence which is administered at the jth period, and $\Sigma F_{(j,k)} = 0$;
> $C_{(j-1,k)}$ = the fixed first order carry-over effect of the formulation in the kth sequence which is administered at the $(j - 1)$th period, where $C_{(0,k)} = 0$, and $\Sigma C_{(j-1,k)} = 0$;
> e_{ijk} = the (within subject) random error in observing Y_{ijk}.

It is assumed that $\{S_{ik}\}$ are independently and identically distributed (i.i.d.) with mean 0 and variance σ_s^2 and $\{e_{ijk}\}$ are independently distributed with mean 0 and variances σ_t^2, where $t = 1, 2, \ldots, L$ (the number of formulations to be compared). $\{S_{ik}\}$ and $\{e_{ijk}\}$ are assumed mutually independent. The estimate of σ_s^2 is usually used to explain the inter-subject variability, while the estimates of σ_t^2 are used to assess the intra-subject variabilities for the tth formulation.

Let $\overline{Y}_{\cdot 1k}, \overline{Y}_{\cdot 2k}, \ldots, \overline{Y}_{\cdot pk}$ be the observed means for periods in the kth sequence. That is,

$$\overline{Y}_{\cdot jk} = \frac{1}{n_k} \sum_{i=1}^{n_k} Y_{ijk}, \qquad j = 1, \ldots, p \quad \text{and} \quad k = 1, \ldots, g. \tag{2.5.2}$$

Under the normality assumptions, the carry-over effects and other fixed effects such as the direct drug effect and the period effect can be estimated based on these gp means since there are $(gp - 1)$ degrees of freedom (df) among these gp means which can be decomposed as follows (Jones and Kenward, 1989):

$$(gp - 1) = (p - 1) + (g - 1) + (p - 1)(g - 1),$$

where $(p - 1)$ df are attributed to the period effect, $(g - 1)$ df are assigned to the sequence effect, and $(p - 1)(g - 1)$ are associated with the sequence-by-period interaction. The $(p - 1)(g - 1)$ df are of particular interest because they preserve the information related to the direct drug effect and the carry-over effects. For example, for a 2×2 crossover design, there are 3 df associated with four sequence-by-period means: one for the sequence effect, one for the period effect, and one for the sequence-by-period interaction which is, in fact, the direct drug effect when there are no carry-over effects.

A within-subject linear contrast for the kth sequence is defined as a linear combination of $\overline{Y}_{.1k}, \overline{Y}_{.2k}, \ldots,$ and $\overline{Y}_{.pk}$. That is,

$$l = c_1\overline{Y}_{.1k} + c_2\overline{Y}_{.2k} + \cdots + c_p\overline{Y}_{.pk},$$

where $\Sigma_j c_j = 0$.

Two linear combinations of $\overline{Y}_{.jk}, j = 1, 2, \ldots, p$ are said to be orthogonal if the sum of the cross product of the coefficients of the two contrasts is 0. In other words, let

$$l_1 = \sum_{j=1}^{p} c_{1j}\overline{Y}_{.jk} \quad \text{and} \quad l_2 = \sum_{j=1}^{p} c_{2j}\overline{Y}_{.jk}$$

be two linear contrasts, then l_1 and l_2 are orthogonal if

$$\sum_{j=1}^{p} c_{1j}c_{2j} = 0.$$

It can be seen that the variance of l only involves the intra-subject variabilities $\sigma_t^2, t = 1, 2, \ldots, L$. Thus, statistical inferences for the fixed effects such as the period effects, the direct drug effects, and the carry-over effects can be made based on within subject variabilities using appropriate linear contrasts of these gp means.

2.5.4. Crossover Designs for Two Formulations

In this section, we will focus on the assessment of bioequivalence between a test formulation (T) and a reference (or standard) formulation (R) of a drug product. The most commonly used statistical design for comparing two formulations of a drug probably is a two-period, two-sequence, crossover design. We shall refer to this design as a standard 2×2 crossover design. For a standard 2×2 crossover design, each subject is randomly assigned to either sequence RT or sequence TR at two dosing periods. In other words, subjects within RT (TR) receive formulation R(T) at the first dosing period and formulation T(R) at the second dosing period. The dosing periods, of course, are separated by a washout period of sufficient length for the drug received in the first period to be completely metabolized and/or excreted from the body. This design appears to be preferred by the FDA Division of Biopharmaceutics. An example of a

Table 2.5.1 2 × 2 Crossover Designs

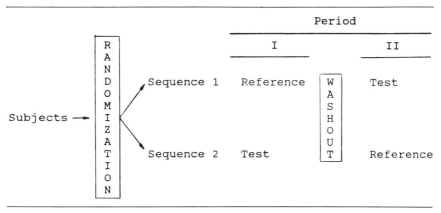

2 × 2 crossover design is illustrated in Table 2.5.1. Although the crossover design is a variant of the Latin square design, the number of the formulations in a crossover design does not necessarily have to be equal to the number of periods. One example is a 2 × 3 crossover design for comparing two formulations as illustrated in Table 2.5.2. In this design, there are two formulations but three periods. Subjects in each sequence receive one of the formulations twice at two different periods. The design of this kind is known as the higher-order crossover design which will be discussed in details in Chapter 9.

Randomization for a standard 2 × 2 crossover design can be carried out by using either a table of random numbers or a SAS procedure, PROC PLAN (SAS,*

Table 2.5.2 2 × 3 Crossover Designs

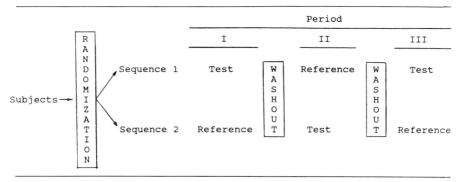

*Registered trademark SAS Institute, Cary, North Carolina

1990). For example, suppose a standard 2×2 crossover design is to be conducted with 24 healthy volunteers to assess bioequivalence between a test formulation and a reference formulation of a drug product. Since there are two sequences of formulations (RT and TR), 12 subjects are to be assigned to each of the two sequences. In other words, one group will receive the first sequence of formulations (RT) and the other group will receive the second sequence of formulations (TR). Thus, we first generate a set of random numbers from 1 to 24 using PROC PLAN, which is given below:

    16, 19, 20, 11, 4, 24, 1, 12, 5, 23, 15, 6,
    17, 2, 10, 14, 18, 13, 21, 3, 7, 8, 22, 9.

Then, subjects are sequentially assigned a number from 1 through 24. Subjects with numbers in the first half of the above random order are assigned to the first sequence (RT) and the rest are assigned to the second sequence (TR) (see Table 2.5.3). In practice, a randomization code for more than the total number of subjects planned is usually prepared to account for the possible replacement of dropouts.

For a standard 2×2 crossover design, from model (2.5.1), the two responses for the ith subject in each sequence are given below:

$$\text{Sequence 1} \quad Y_{i11} = \mu + S_{i1} + P_1 + F_1 + e_{i11}$$
$$Y_{i21} = \mu + S_{i1} + P_2 + F_2 + C_1 + e_{i21} \qquad (2.5.3)$$
$$\text{Sequence 2} \quad Y_{i12} = \mu + S_{i2} + P_1 + F_2 + e_{i12}$$
$$Y_{i22} = \mu + S_{i2} + P_2 + F_1 + C_2 + e_{i22}$$

where $P_1 + P_2 = 0$, $F_1 + F_2 = 0$, and $C_1 + C_2 = 0$.

For each subject, a pair of observations are observed at period 1 and period 2. Thus, we may consider a bivariate random vector (i.e., (period 1, period 2)) as follows.

$$\mathbf{Y}_{ik} = (Y_{i1k}, Y_{i2k}), \quad i = 1, 2, \ldots, n_k \quad \text{and} \quad k = 1, 2. \qquad (2.5.4)$$

Then, $\mathbf{Y}_{ik}$ are independently distributed with the following mean vector and covariance matrix:

$$\text{Sequence 1} \quad \boldsymbol{\alpha}_1 = \begin{bmatrix} \mu + P_1 + F_1 \\ \mu + P_2 + F_2 + C_1 \end{bmatrix},$$
$$\Sigma_1 = \begin{bmatrix} \sigma_1^2 + \sigma_s^2 & \sigma_s^2 \\ \sigma_s^2 & \sigma_2^2 + \sigma_s^2 \end{bmatrix}$$

$$\text{Sequence 2} \quad \boldsymbol{\alpha}_2 = \begin{bmatrix} \mu + P_1 + F_2 \\ \mu + P_2 + F_1 + C_2 \end{bmatrix},$$
$$\Sigma_2 = \begin{bmatrix} \sigma_2^2 + \sigma_s^2 & \sigma_s^2 \\ \sigma_s^2 & \sigma_1^2 + \sigma_s^2 \end{bmatrix} \qquad (2.5.5)$$

Table 2.5.3   Randomization Code for a
Standard 2 × 2 Crossover Design with 24
Subjects

Subject	Sequence	Formulations
1	1	RT
2	2	TR
3	2	TR
4	1	RT
5	1	RT
6	1	RT
7	2	TR
8	2	TR
9	2	TR
10	2	TR
11	1	RT
12	1	RT
13	2	TR
14	2	TR
15	1	RT
16	1	RT
17	2	TR
18	2	TR
19	1	RT
20	1	RT
21	2	TR
22	2	TR
23	1	RT
24	1	RT

It can be seen that the intra-subject variabilities are different between formulations. If, however, $\sigma_1^2 = \sigma_2^2 = \sigma_e^2$, then $\Sigma_1 = \Sigma_2 = \Sigma$, where

$$\Sigma = \begin{bmatrix} \sigma_e^2 + \sigma_s^2 & \sigma_s^2 \\ \sigma_s^2 & \sigma_e^2 + \sigma_s^2 \end{bmatrix}. \tag{2.5.6}$$

When the carry-over effects are present (i.e., $C_1 \neq 0$ and $C_2 \neq 0$), a standard 2 × 2 crossover design may not be desirable since it may not provide estimates for some fixed effects. For example, as indicated in the previous subsection, there is only one degree of freedom which is attributed to the sequence effect. The sequence effect, which cannot be estimated separately, is confounded (or aliased) with any carry-over effects. If the carry-over effects are unequal (i.e., $C_1 \neq C_2 \neq 0$), then there exists no unbiased estimate for the direct drug effect from both periods. In addition, the carry-over effects cannot be precisely estimated because it can only be evaluated based on the between subject comparison. Furthermore, the intra-subject variabilities σ_1^2 and σ_2^2 cannot be estimated independently and directly from the observed data because each subject receives either the test formulation or the reference formulation only once during the study. In other words, there is no replicates for each formulation within each subject.

To overcome the above undesirable properties, a higher-order crossover design may be useful. A higher-order crossover design is defined as a crossover design in which either the number of periods is greater than the number of formulations to be compared or the number of sequences is greater than the number of

Table 2.5.4 Optimal Crossover Designs for Two Formulations

Design A		
	Period	
Sequence	I	II
1	T	T
2	R	R
3	R	T
4	T	R

Design B		
	Period	
Sequence	I	II	III
1	T	R	R
2	R	T	T

Design C			
	Period		
Sequence	I	II	III	IV
1	T	T	R	R
2	R	R	T	T
3	T	R	R	T
4	R	T	T	R

Table 2.5.5 Variances for Designs A, B, and C in Multiples of $\hat{\sigma}_e^2/n$

Design	$V(\hat{C}|F)$	$V(\hat{F}|C)$	$V(\hat{F})$
S^a	—^{b}	—^{c}	1.0000
A	4.0000	2.0000	1.0000
B	1.0000	0.7500	0.7500
C	0.3636	0.2500	0.2500

^{a}S is a standard 2×2 crossover design.
$^b V(\hat{C}|F) = 4 (2\hat{\sigma}_s^2 + \hat{\sigma}_e^2)/n$.
^{c}The direct drug effect is not estimable in the presence of the carry-over effects.

formulations to be compared. There are a number of higher-order crossover designs available in the literature (Kershner and Federer, 1981; Laska, Meinser, and Kushner, 1983; Laska and Meinser, 1985; Jones and Kenward, 1989). These designs, however, have their own advantages and disadvantages. An in-depth discussion can be found in Jones and Kenward (1989).

In the following, we shall discuss three commonly used higher-order crossover designs, which possess some optimal statistical properties, for comparing two formulations. We shall refer to these three designs as design A, design B, and design C, respectively. Designs A, B, and C are given in Table 2.5.4. In each of the three designs, the estimates of the direct drug effect and carry-over effects are obtained based on the within subject linear contrasts. As a result, statistical inferences for direct drug effect and carry-over effects are mainly based on the intra-subject variability. For the comparisons of these three designs with the standard 2×2 crossover design, it is helpful to use the following notations. The direct drug effect after adjustment for the carry-over effects is denoted by F|C. Then, F simply refers to the unadjusted direct drug effect. Also the variance of the estimate of F|C (i.e., $\hat{F}$|C) is denoted by $V(\hat{F}|C)$. Table 2.5.5 gives the variances (in the multiples of σ_e^2/n) of the direct drug effect and carry-over effects for the three designs and the standard 2×2 crossover design (Jones and Kenward, 1989). The variances of designs A, B, C are derived under the assumptions that (i) $n_k = n$ for all k; (ii) $\sigma_1^2 = \sigma_2^2 = \sigma_e^2$; and (iii) there is no direct drug by carry-over interaction. For designs B and C, the variances of the direct drug effect adjusted for the carry-over effects are the same as the unadjusted direct drug effect (i.e., no carry-over effects). This is because the direct drug effect and carry-over effects for designs B and C are estimated by the linear contrasts which are orthogonal to each other. Note that an orthogonality of linear contrasts for the direct drug effect and carry-over effects implies that their covariance is zero. In other words, the estimators of the direct drug effect and carry-over effects in designs B and C are not correlated (or independent).

Design A is also known as Balaam's design (Balaam, 1968). It is an optimal design in the class of the crossover designs with two periods and two formulations. This design is formed by adding two more sequences (sequences 1 and 2) to the standard 2×2 crossover design (sequences 3 and 4). These two augmented sequences are TT and RR. With additional information provided by the two augmented sequences, not only can the carry-over effects be estimated using the within subject contrasts but the intra-subject variability for both test and reference formulations can also be obtained because there are replicates for each formulation within each subject.

Design B is an optimal design in the class of the crossover designs with two sequences, three periods, and two formulations. It can be obtained by adding an additional period to the standard 2×2 crossover designs. The treatments administered in the third period are the same as those in the second period. This type of designs is also known as the extended-period or extra-period designs. Note that this design is made of a pair of dual sequences TRR and RTT. Two sequences whose treatments are mirror images of each other are said to be a pair of dual sequences. As pointed out by Jones and Kenward (1989), the only crossover designs worth considering are those which are made up of dual sequences. Compared to the standard 2×2 crossover design, the variance for the direct drug effect is reduced by 25%. For the carry-over effects, the variance is reduced by about 75% as compared to the Balaam's design. In addition, the intra-subject variability can be estimated based on the data collected from periods 2 and 3.

Design C is an optimal design in the class of the crossover designs with four sequences, four periods, and two formulations. It is also made up of 2 pairs of dual sequences (TTRR, RRTT) and (TRRT, RTTR). Note that the first two periods of design C are the same as those in Balaam's design and the last two periods are the mirror image of the first two periods. This design is much more complicated than designs A and B, though it produces the maximum in variance reduction for both the direct drug effect and the carry-over effects among the designs considered.

2.5.5. Crossover Designs for Three or More Formulations

The crossover designs for comparing three or more formulations are much more complicated than those for comparing two formulations. For simplicity, in this section, we will restrict our attention to those designs in which the number of periods equals the number of formulations to be compared. In the next section, the designs for comparing a large number of formulations with a small number of treatment periods will be discussed.

For comparing three formulations of a drug, there are a total of three possible pairwise comparisons between formulations: formulation 1 vs formulation 2, formulation 1 vs formulation 3, and formulation 2 vs formulation 3. It is desirable to estimate these pairwise differences between formulations with the same degree of precision. In other words, it is desirable to have equal variances for each pairwise difference between formulations, i.e., $V(\hat{F}_i - \hat{F}_j) = v\sigma_e^2$, where v is a constant and σ_e^2 is the intra-subject variability. Designs with this property are known as variance-balanced designs. It should be noted that, in practice, v may vary from design to design. Thus, an ideal design is one with the smallest v such that all pairwise differences between formulations can be estimated with the same and possibly best precision. However, to achieve this goal, the design must be balanced. A design is said to be balanced if it satisfies the following conditions (Jones and Kenward, 1989):

(i)   Each formulation occurs only once with each subject;
(ii) Each formulation occurs the same number of times in each period;
(iii) The number of subjects who receive formulation i in some period followed by formulation j in the next period is the same for all i ≠ j.

Under the constraint of the number of periods (p) being equal to the number of formulations (t), balance can be achieved by using a complete set of "Orthogonal Latin Squares" (John, 1971; Jones and Kenward, 1989). However, if p = t, a complete set of orthogonal Latin squares consists of t(t − 1) sequences except for t = 6. Some examples of orthogonal Latin squares with t = 3 and t = 4 are presented in Table 2.5.6. As a result, when the number of formulations to be compared is large, more sequences and consequently more subjects are required. This, however, may not be of practical use.

A more practical design has been proposed by Williams (1949). We shall refer to this as a Williams design. A Williams design possesses balance property and requires fewer sequences and periods. The algorithm for constructing a Williams design with t periods and t formulations is summarized in the following numerical steps (Jones and Kenward, 1989):

(1) Number the formulations from 1, 2, . . . , t;

(2) Start with a t × t standard Latin square. In this square, the formulations in the ith row are given by i, i + 1, . . . , t, 1, 2, . . . , i − 1;

(3) Get a mirror image of the standard Latin square.

(4) Interlace each row of the standard Latin square with the corresponding mirror image to get a t × 2t arrangement.

(5) Slice the 2 × 2t arrangement down to the middle to yield two t × t squares. The columns of each t × t squares correspond to the periods and the rows are the sequences. The number within the square are the formulations.

Table 2.5.6 Orthogonal Latin Squares for t = 3 and 4

Three formulations (t = 3)

Sequence	Period		
	I	II	III
1	R^a	T_1	T_2
2	T_1	T_2	R
3	T_2	R	T_1
4	R	T_2	T_1
5	T_1	R	T_2
6	T_2	T_1	R

Four formulations (t = 4)

Sequence	Period			
	I	II	III	IV
1	R^a	T_1	T_2	T_3
2	T_1	R	T_3	T_2
3	T_2	T_3	R	T_1
4	T_3	T_2	T_1	R
5	R	T_3	T_1	T_2
6	T_1	T_2	R	T_3
7	T_2	T_1	T_3	R
8	T_3	R	T_2	T_1
9	R	T_2	T_3	T_1
10	T_1	T_3	T_2	R
11	T_2	R	T_1	T_3
12	T_3	T_1	R	T_2

[a]R is the reference formulation and T_1, T_2, and T_3 are the test formulations 1, 2, and 3, respectively.

(6) If t is even, choose any one of the two t × t squares. If t is odd, use both squares.

In the following, to illustrate the use of the above algorithm as an example, we will construct a Williams design with t = 4 (one reference and three test formulations) by following the above steps.

(1) Denote the reference formulation by 1, and test formulations 1, 2, and 3 by 2, 3, and 4.

(2) A 4 × 4 standard Latin square is given below

$$
\begin{array}{cccc}
1 & 2 & 3 & 4 \\
2 & 3 & 4 & 1 \\
3 & 4 & 1 & 2 \\
4 & 1 & 2 & 3
\end{array}
$$

(3) The minor image of the 4 × 4 standard Latin square is then given by

$$
\begin{array}{cccc}
4 & 3 & 2 & 1 \\
1 & 4 & 3 & 2 \\
2 & 1 & 4 & 3 \\
3 & 2 & 1 & 4
\end{array}
$$

(4) The 4 × 8 arrangement after interlacing the 4 × 4 standard Latin square with its mirror image is

$$
\begin{array}{cccc|cccc}
1 & 4 & 2 & 3 & 3 & 2 & 4 & 1 \\
2 & 1 & 3 & 4 & 4 & 3 & 1 & 2 \\
3 & 2 & 4 & 1 & 1 & 4 & 2 & 3 \\
4 & 3 & 1 & 2 & 2 & 1 & 3 & 4
\end{array}
$$

(5) The two 4 × 4 squares obtained by slicing the above 4 × 8 arrangement are

			Period		
Square	Sequence	I	II	III	IV
1	1	1	4	2	3
	2	2	1	3	4
	3	3	2	4	1
	4	4	3	1	2
2	1	3	2	4	1
	2	4	3	1	2
	3	1	4	2	3
	4	2	1	3	4

(6) Because t = 4, we can choose either square 1 or square 2. The resultant Williams design from square 1 is given in Table 2.5.7 by replacing 1, 2, 3, 4 with R, T_1, T_2, and T_3.

Table 2.5.7   Williams Designs for t = 3, 4
and 5

	Three formulations (t = 3) Period		
Sequence	I	II	III
1	R^a	T_2	T_1
2	T_1	R	T_2
3	T_2	T_1	R
4	T_1	T_2	R
5	T_2	R	T_1
6	R	T_1	T_2

	Four formulations (t = 4) Period			
Sequence	I	II	III	IV
1	R	T_3	T_1	T_2
2	T_1	R	T_2	T_3
3	T_2	T_1	T_3	R
4	T_3	T_2	R	T_1

	Five formulations (t = 5) Period				
Sequence	I	II	III	IV	V
1	R	T_4	T_1	T_3	T_2
2	T_1	R	T_2	T_4	T_3
3	T_2	T_1	T_3	R	T_4
4	T_3	T_2	T_4	T_1	R
5	T_4	T_3	R	T_2	T_1
6	T_2	T_3	T_1	T_4	R
7	T_3	T_4	T_2	R	T_1
8	T_4	R	T_3	T_1	T_2
9	R	T_1	T_4	T_2	T_3
10	T_1	T_2	R	T_3	T_4

[a]R is the reference formulation and T_1, T_2, T_3, and T_4
are test formulations 1, 2, 3, and 4, respectively.

From the above example, it can be seen that a Williams design requires only 4 sequences to achieve the property of "variance-balanced," while a complete set of 4 × 4 orthogonal Latin squares requires 12 sequences. A Williams design with t = 3 and 5 constructed using the above algorithm are also given in Table 2.5.7.

2.6. THE BALANCED INCOMPLETE BLOCK DESIGN

When comparing three or more formulations of a drug product, a complete crossover design may not be of practical interest for the following reasons (Westlake, 1973):

(i) If the number of formulations to be compared is large, the study may be too time consuming since t formulations require t − 1 washout periods.

(ii) It may not be desirable to draw many blood samples for each subject due to medical concerns.

(iii) Moreover, a subject is more likely to drop out when he/she is required to return frequently for tests.

These considerations suggest that one should keep the number of formulations that a subject receives as small as possible when planning a bioavailability study. In this case, a randomized incomplete block design may be useful. An incomplete block design is a randomized block design in which not all formulations are present in every block. A block is called incomplete if the number of formulations in the block is less than the number of formulations to be compared. It should be noted that for an incomplete block design the blocks and formulations are not orthogonal to each other. That is, the block effects and formulation effects may not be estimated separately.

When an incomplete block design is used, it is recommended that the formulations in each block be randomly assigned in a *balanced* way so that the design will possess some optimal statistical properties. We shall refer to such a design as a balanced incomplete block design. A balanced incomplete block design is an incomplete block design in which any two formulations appear together an equal number of times. The advantages of using a balanced incomplete block design rather than an incomplete design are given below.

(i) The difference between the effects of any two formulations can always be estimated with the same degree of precision.

(ii) The analysis is simple in spite of the nonorthogonality provided that the balance is preserved.

(iii) Unbiased estimates of formulation effects are available.

Suppose that there are t formulations to be compared and each subject can only receive exactly p formulations (t > p). A balanced incomplete block design may be constructed by taking $C(t,p)$, the combinations of p out of t formulations, and assigning a different combination of formulations to each subject. However, to minimize the period effect, it is preferable to assign the formulations in such a way that the design is balanced over period (i.e., each formulation appears

Table 2.6.1 Balanced Incomplete Block Designs for t = 4 with p = 2 and 3

I. Each sequence receives two formulations (p = 2)

Sequence[a]	Period I	Period II
1	R[b]	T_1
2	T_1	T_2
3	T_2	T_3
4	T_3	R
5	R	T_2
6	T_1	T_3
7	T_3	T_1
8	T_2	R
9	R	T_3
10	T_3	T_2
11	T_2	T_1
12	T_1	R

II. Each sequence receives three formulations (p = 3)

Sequence	Period I	Period II	Period III
1	T_1	T_2	T_3
2	T_2	T_3	R
3	T_3	R	T_1
4	R	T_1	T_2

[a]A sequence (or block) may represent a subject or a group of homogeneous subjects.

[b]R is the reference formulation and T_1, T_2, and T_3 are test formulations 1, 2, and 3, respectively.

Table 2.6.2 Balanced Incomplete Block Designs for t = 5 with p = 2, 3, and 4

I. Each sequence receives two and three formulations

	p = 2			p = 3		
	Period				Period	
Sequence[a]	I	II	Sequence	I	II	III
1	R[b]	T_1	1	T_2	T_3	T_4
2	T_1	T_2	2	T_3	T_4	R
3	T_2	T_3	3	T_4	R	T_1
4	T_3	T_4	4	R	T_1	T_2
5	T_4	R	5	T_1	T_2	T_3
6	R	T_2	6	T_1	T_3	T_4
7	T_2	T_4	7	T_3	R	T_1
8	T_4	T_1	8	R	T_2	T_3
9	T_1	T_3	9	T_2	T_4	R
10	T_3	R	10	T_4	T_1	T_2

II. Each sequence receives four formulations (p = 4)

	Period			
Sequence	I	II	III	IV
1	T_1	T_2	T_3	T_4
2	T_2	T_3	T_4	R
3	T_3	T_4	R	T_1
4	T_4	R	T_1	T_2
5	R	T_1	T_2	T_3

[a]A sequence (or block) may represent a subject or a group of homogeneous subjects.
[b]R is the reference formulation and T_1, T_2, T_3, and T_4 are test formulations 1, 2, 3, and 4, respectively.

the same number of times in each period). In general, if the number of formulations is even (i.e., t = 2n) and p = 2, the number of blocks (sequences) required is g = 2n(2n − 1). On the other hand, if the number of formulations is odd (i.e., t = 2n + 1) and p = 2, then g = (2n + 1)n. Some examples for balanced incomplete block design are given in Table 2.6.1 and 2.6.2. Table 2.6.1 gives examples for p = 2 and 3 when four formulations (t = 4) are to be compared. For p = 2, the first six blocks are required for a balanced incomplete block design. However, to ensure the balance over period, an additional six blocks (7 through 12) are needed. For t = 5, Table 2.6.2 lists examples for a balanced incomplete block design with p = 2, 3, and 4. A balanced incomplete block design for p = 3 is the complementary part of that balanced incomplete block design for p = 2. The design for p = 4 can be constructed by deleting each formulation in turn to obtain five blocks successively.

For t > 5, several methods for constructing balanced incomplete block designs are available. Among these, the easiest way probably is the method of cyclic substitution. For this method to work, we first choose an appropriate initial block. The other blocks can be obtained successively by changing formulations A to B, B to C, . . . , and so on in each block. For example, for t = 6 and p = 3, if we start with (A,B,D), then the second block is (B,C,E) and the third block is (C,D,F) and so on.

Note that a balanced incomplete block design is in fact a special case of variance-balanced design which will be discussed in Chapter 10. For an incomplete block design, balance may be achieved with fewer than C(t,p) blocks. Such designs are known as partially balanced incomplete block designs. The analysis of these designs, however, are complicated and hence of little practical interest. More details on balanced incomplete block designs and partially balanced incomplete block designs can be found in Fisher and Yates (1953), Bose, Clatworthy, and Shrikhande (1954), Cochran and Cox (1957), and John (1971).

2.7. THE SELECTION OF DESIGN

In previous sections, we briefly discussed three basic statistical designs, the parallel design, the crossover design, and the balanced incomplete block design for bioavailability/bioequivalence studies. Each of these has its own advantages and drawbacks under different circumstances. How to select an appropriate design when planning a bioavailability study is an important question. The answer to this question depends upon many factors which are summarized below.

(1) The number of formulations to be compared;
(2) The characteristics of the drug and its disposition;
(3) The study objective;
(4) The availability of subjects;
(5) The inter- and intra-subject variabilities;
(6) The duration of the study or the number of periods allowed;
(7) The cost of adding a subject relative to that of adding one period.
(8) Dropout rates.

For example, if the intra-subject variability is the same as or larger than the inter-subject variability, the inference on the difference in bioavailability would be the same regardless of which design is used. Actually, a crossover design in this situation would be a poor choice, since blocking results in the loss of some degrees of freedom and will actually lead to a wider confidence interval on the difference between formulations.

If a bioavailability/bioequivalence study compares more than three formulations, a crossover design may not be appropriate. The reasons, as indicated in Section 2.6, are (i) it may be too time consuming to complete the study since a washout is required between treatment periods, (ii) it may not be desirable to draw many blood samples for each subject due to medical concerns, and (iii) too many periods may increase the number of dropouts. In this case, a balanced incomplete block design is preferred. However, if we compare several test formulations with a reference formulation, the within-subject comparison is not reliable since the subjects in some sequences may not receive the reference formulation.

If the drug has a very long half-life, or it possesses a potential toxicity, or bioequivalence must be established by clinical endpoint because some drugs do not work through systemic absorption, then a parallel design may be a possible choice. With this design, the study avoids a possible cumulative toxicity due to the carry-over effects from one treatment period to the next. In addition, the study can be completed quickly. However, the drawback is that the comparison is made based on the inter-subject variability. If the inter-subject variability is large relative to the intra-subject variability, the statistical inference on the difference in bioavailability between formulations is not reliable. Even if the inter-subject variability is relatively small, a parallel design may still require more subjects in order to reach the same degree of precision achieved by a crossover design.

In practice, a crossover design, which can remove the inter-subject variability from the comparison of average bioavailability between formulations, is often considered to be the design of choice if the number of formulations to be compared is small, say no more than three. If the drug has a very short half-life (i.e., there may not be carry-over effects if the length of washout is long enough to eliminate the residual effects), a crossover design may be useful for the assessment of the intra-subject variability provided that the cost for adding one period is comparable to that of adding a subject.

In summary, to choose an appropriate design for a bioavailability study is an important issue in the development of a study protocol. The selected design may affect the data analysis, the interpretation of the results, and the determination of bioequivalence between formulations. Thus, all factors listed above should be carefully evaluated before an appropriate design is chosen.

Statistical Inferences for Effects from a Standard 2 × 2 Crossover Design

3.1. INTRODUCTION

In the previous chapter, several useful designs for assessing bioequivalence in a variety of situations were discussed. Among these designs, the standard 2 × 2 crossover design, as outlined below, appears to be the most common design for assessing bioequivalence between two formulations (a test formulation T and a reference formulation R) of a drug product.

Sequence	Period I	Period II
1 (RT)	Reference Formulation	Test Formulation
	Data: Y_{i11}	Data: Y_{i21}
2 (TR)	Test Formulation	Reference Formulation
	Data: Y_{i12}	Data: Y_{i22}

Each subject is randomly assigned to either sequence 1 (RT) or sequence 2 (TR). Subjects within sequence RT (TR) receive formulation R (T) during the first dosing period and formulation T (R) during the second period. Dosing periods are usually separated by a washout period of at least three times the half-life of the active drug ingredient or therapeutic moiety.

The general model (2.5.1) can be used to describe the above standard 2 × 2 crossover design as follows:

$$Y_{ijk} = \mu + S_{ik} + P_j + F_{(j,k)} + C_{(j-1,k)} + e_{ijk}, \qquad (3.1.1)$$

where i(subject) = 1,2, . . . ,n_k and j(period), k(sequence) = 1,2. $F_{(j,k)}$ is the direct fixed effect of the formulation administered at period j in sequence k. In a standard 2 × 2 crossover design, there are only two formulations. Thus, since the formulation administered at the first period in the first sequence is the reference formulation, then

$$F_{(j,k)} = \begin{cases} F_R & \text{if } k = j \\ F_T & \text{if } k \neq j, \end{cases} \quad k = 1,2; \quad j = 1,2. \qquad (3.1.2)$$

$C_{(j-1,k)}$ is the residual effect carried over from the $(j - 1)$th period to the jth period in sequence k. For a standard 2 × 2 crossover design, the carry-over effects can only occur at the second period. We will denote the carry-over effect of the reference formulation from the first period to the second period at sequence 1 by C_R. Thus,

$$C_{(j-1,k)} = \begin{cases} C_R & \text{if } k = 1, \quad j = 2 \\ C_T & \text{if } k = 2, \quad j = 2. \end{cases} \qquad (3.1.3)$$

It can be seen that the above model includes fixed effects such as the period effect, the direct drug effect, and the carry-over effects. For each subject, the fixed effects which occur at each period in each sequence are summarized as follows.

Sequence	Period I	Period II
1 (RT)	$\mu_{11} = \mu + P_1 + F_R$	$\mu_{21} = \mu + P_2 + F_T + C_R$
2 (TR)	$\mu_{12} = \mu + P_1 + F_T$	$\mu_{22} = \mu + P_2 + F_R + C_T$

where $\mu_{jk} = E(Y_{ijk})$, $P_1 + P_2 = 0$, $F_R + F_T = 0$, and $C_R + C_T = 0$.

For the comparison of bioavailability between formulations, it is desirable to estimate and separate these effects from the direct drug effect (or formulation effect). In practice, for a bioavailability/bioequivalence study, it is usually assumed that (i) there is no period effect, and (ii) there are no carry-over effects. This is because (i) a well-conducted study can eliminate the possible period effect and (ii) the washout period of sufficient length can be chosen to insure that there are no residual effects from previous dosing period to the next dosing period. In many cases, however, the period effect and/or carry-over effects may still be present. The presence of the carry-over effects can certainly increase the complexity of statistical analyses for the assessment of bioequivalence between formulations. Thus, it is of interest to perform some preliminary tests for the presence of the period effect and/or the carry-over effects before the comparison of bioavailability between formulations is made.

In this chapter, statistical inferences on these effects will be reviewed under model (3.1.1) with the following assumptions:

(i) $\{S_{ik}\}$ are i.i.d. normal with mean 0 and variance σ_s^2;
(ii) $\{e_{ijk}\}$ are i.i.d. normal with mean 0 and variance σ_e^2;
(iii) $\{S_{ik}\}$ and $\{e_{ijk}\}$ are mutually independent. (3.1.4)

These assumptions are much stronger than those specified in model (2.5.1) which does not impose normality assumptions on $\{S_{ik}\}$ and $\{e_{ijk}\}$ and which allows the intra-subject variability to vary from formulation to formulation. Under these assumptions, statistical inferences such as estimation, confidence interval, and hypotheses testing for the fixed effects can be derived based on two-sample t statistics (Hills and Armitage, 1979; Jones and Kenward, 1989).

In Sections 3.2 through 3.4, statistical inferences for the carry-over effects, the direct drug effect, and the period effect will be obtained based on two-sample t statistics. The method of the analysis of variance for a general crossover design will be presented in Section 3.5. In Section 3.6, an example will be given to illustrate the use of the derived statistical methods.

3.2. THE CARRY-OVER EFFECTS

For the assessment of carry-over effects, it is helpful to consider the following subject totals for each sequence.

$$U_{ik} = Y_{i1k} + Y_{i2k}, \quad i = 1,2, \ldots ,n_k; \quad k = 1,2. \quad (3.2.1)$$

The expected value and variance for U_{ik} are given by

$$E(U_{ik}) = \begin{cases} 2\mu + C_R & \text{for subjects in sequence 1} \\ 2\mu + C_T & \text{for subjects in sequence 2.} \end{cases} \quad (3.2.2)$$

$$\sigma_u^2 = V(U_{ik}) = 2(2\sigma_s^2 + \sigma_e^2) \quad \text{for all subjects.} \quad (3.2.3)$$

Let $C = C_T - C_R$. Then, C can be used to assess the carry-over effects. Under the constraint of $C_R + C_T = 0$, carry-over effects are equal for the two formulations, i.e., $C = 0$ if and only if $C_R = C_T = 0$. Therefore, a test for no carry-over effects is equivalent to a test for equal carry-over effects. When there are no carry-over effects, the direct drug effect (i.e., $F = F_T - F_R$) can be estimated based on the data from both periods. However, there exists no unbiased estimator for the direct drug effect if unequal carry-over effects are present. Thus, it is of interest to examine whether or not the unequal carry-over effects are present. The unequal carry-over effects can be determined by testing the following hypotheses:

$$H_0: C = 0 \text{ (or } C_R = C_T) \quad vs \quad H_a: C \neq 0 \text{ (or } C_R \neq C_T). \tag{3.2.4}$$

The rejection of the null hypothesis leads to the conclusion of the presence of unequal carry-over effects between formulations. To draw a statistical inference on C, it is useful to consider the following sample mean of the subject totals for each sequence:

$$\overline{U}_{\cdot k} = \frac{1}{n_k} \sum_{i=1}^{n_k} U_{ik}, \quad k = 1,2. \tag{3.2.5}$$

$\overline{U}_{\cdot 1}$ and $\overline{U}_{\cdot 2}$ are two independent random samples from normal populations with equal variances. Thus, statistical inference on C can be made based on a two-sample t statistic.

First, C can be estimated by the difference in sample means of the subject totals for the two sequences. That is,

$$\hat{C} = \overline{U}_{\cdot 2} - \overline{U}_{\cdot 1}$$
$$= (\overline{Y}_{\cdot 12} + \overline{Y}_{\cdot 22}) - (\overline{Y}_{\cdot 11} + \overline{Y}_{\cdot 21}). \tag{3.2.6}$$

Under the assumptions (i)–(iii) as specified in the previous section, $\hat{C}$ is normally distributed with mean C and variance $V(\hat{C})$ which is given by

$$V(\hat{C}) = 2(2\sigma_s^2 + \sigma_e^2)\left(\frac{1}{n_1} + \frac{1}{n_2}\right)$$
$$= \sigma_u^2\left(\frac{1}{n_1} + \frac{1}{n_2}\right). \tag{3.2.7}$$

The variance $V(\hat{C})$ can be estimated by replacing σ_u^2 with $\hat{\sigma}_u^2$, the pooled sample variance of the subject totals from the two sequences. That is,

$$\hat{V}(\hat{C}) = \hat{\sigma}_u^2\left(\frac{1}{n_1} + \frac{1}{n_2}\right), \tag{3.2.8}$$

where

$$\hat{\sigma}_u^2 = \frac{1}{(n_1 + n_2 - 2)} \sum_{k=1}^{2} \sum_{i=1}^{n_k} (U_{ik} - \overline{U}_{\cdot k})^2.$$

Note that $\hat{C}$ is the minimum variance unbiased estimator (MVUE) for C and $\hat{\sigma}_u^2$ is an unbiased estimator for σ_u^2. The MVUE is the unbiased estimator with the smallest variance among all unbiased estimators. Furthermore, $(n_1 + n_2 - 2)\hat{\sigma}_u^2$ is distributed as

$$\sigma_u^2 \chi^2(n_1 + n_2 - 2),$$

where $\chi^2(n_1 + n_2 - 2)$ is a chi-square random variable with $n_1 + n_2 - 2$ degrees of freedom, which is independent of $\hat{C}$. Thus, under H_0 in (3.2.4),

$$T_c = \frac{\hat{C}}{\hat{\sigma}_u \sqrt{\dfrac{1}{n_1} + \dfrac{1}{n_2}}} \tag{3.2.9}$$

has a student central t distribution with $n_1 + n_2 - 2$ degrees of freedom. As a result, we would reject the null hypothesis of H_0: $C_R = C_T$ and in favor of H_a: $C_R \neq C_T$ at the α level of significance if

$$|T_c| > t(\alpha/2, n_1 + n_2 - 2), \tag{3.2.10}$$

where $t(\alpha/2, n_1 + n_2 - 2)$ is the upper $\alpha/2$ critical value of a t distribution with $n_1 + n_2 - 2$ degrees of freedom. In other words, we reject the hypothesis of no carry-over effects or equal carry-over effects and conclude that there are unequal carry-over effects if $|T_c| > t(\alpha/2, n_1 + n_2 - 2)$.

Since the test statistic T_c involves the estimate of $\sigma_u^2 = 2(2\sigma_s^2 + \sigma_e^2)$ which includes the inter-subject and the intra-subject variabilities, it may have little power when the inter-subject variability is relatively larger than the intra-subject variability. This is because in most bioavailability/bioequivalence studies, the sample size is chosen based on a pre-study power calculation on the direct drug effect which only involves the intra-subject variability. To increase the test power, however, Grizzle (1965) suggested testing the null hypothesis at the $\alpha = 10\%$ level instead of the traditional 5% level.

Based on the t statistic, a $(1 - \alpha) \times 100\%$ confidence interval for C can be obtained as follows:

$$\hat{C} \pm t(\alpha/2, n_1 + n_2 - 2)\,\hat{\sigma}_u \sqrt{\frac{1}{n_1} + \frac{1}{n_2}}. \tag{3.2.11}$$

If the confidence interval contains 0, then we are in favor of (or fail to reject) the null hypothesis of no or equal carry-over effects for the two formulations. If the confidence interval does not include 0, we conclude that there are unequal carry-over effects between the two formulations.

3.3. THE DIRECT DRUG EFFECT

It is helpful to start with the period differences for each subject within each sequence which are defined as follows:

$$d_{ik} = \tfrac{1}{2}(Y_{i2k} - Y_{i1k}), \qquad i = 1, 2, \ldots, n_k; \quad k = 1, 2. \tag{3.3.1}$$

The expected value and variance of the period differences are given by

$$E(d_{ik}) = \begin{cases} \frac{1}{2} [(P_2 - P_1) + (F_T - F_R) + C_R] & \text{for subjects in sequence 1} \\ \frac{1}{2} [(P_2 - P_1) + (F_R - F_T) + C_T] & \text{for subjects in sequence 2} \end{cases}$$

$$\tag{3.3.2}$$

$$V(d_{ik}) = \sigma_d^2 = \sigma_e^2/2. \tag{3.3.3}$$

It can be seen that the variance of the period differences only involves the intra-subject variability which reflects the usefulness of the crossover design in comparing the direct drug effects. However, the expected value of d_{ik} consists of the period effects and the carry-over effects.

Denote the period effect and the direct drug effect by $P = P_2 - P_1$ and $F = F_T - F_R$, respectively. To draw statistical inference on F, consider the sample means of the period differences for each sequence. That is,

$$\bar{d}_{\cdot k} = \frac{1}{n_k} \sum_{i=1}^{n_k} d_{ik}, \qquad k = 1,2. \tag{3.3.4}$$

The difference between sequences (i.e., $\bar{d}_{\cdot 1} - \bar{d}_{\cdot 2}$) is clearly not an unbiased estimator of F unless there are no unequal carry-over effects (i.e., $C_R = C_T$) since

$$\begin{aligned} E(\bar{d}_{\cdot 1} - \bar{d}_{\cdot 2}) &= (F_T - F_R) + (C_R - C_T)/2 \\ &= F - C/2, \end{aligned} \tag{3.3.5}$$

where $C = C_T - C_R$.

As a result, if $C_R \neq C_T$, there exists no unbiased estimator for F based on the data from both periods. On the other hand, if $C_R = C_T$, then

$$\begin{aligned} \hat{F} &= \bar{d}_{\cdot 1} - \bar{d}_{\cdot 2} \\ &= \frac{1}{2} [(\bar{Y}_{\cdot 21} - \bar{Y}_{\cdot 11}) - (\bar{Y}_{\cdot 22} - \bar{Y}_{\cdot 12})] \\ &= \bar{Y}_T - \bar{Y}_R \end{aligned} \tag{3.3.6}$$

is the MVUE of F, where

$$\bar{Y}_R = \frac{1}{2} (\bar{Y}_{\cdot 11} + \bar{Y}_{\cdot 22}) \quad \text{and} \quad \bar{Y}_T = \frac{1}{2} (\bar{Y}_{\cdot 21} + \bar{Y}_{\cdot 12}). \tag{3.3.7}$$

Note that $\bar{Y}_R$ and $\bar{Y}_T$ are the so-called least squares (LS) means for the reference and test formulation, respectively. $\hat{F} = \bar{Y}_T - \bar{Y}_R$ is a linear contrast of the sequence by period means.

In practice, F is often estimated by the difference between the direct sample means for the two formulations. That is,

$$\hat{F}^* = \bar{Y}_T^* - \bar{Y}_R^*,$$

where

$$\overline{Y}_R^* = \frac{1}{n_1 + n_2} \left\{ \sum_{i=1}^{n_1} Y_{i11} + \sum_{i=1}^{n_2} Y_{i22} \right\},$$

and

$$\overline{Y}_T^* = \frac{1}{n_1 + n_2} \left\{ \sum_{i=1}^{n_1} Y_{i21} + \sum_{i=1}^{n_2} Y_{i12} \right\}.$$

When $C_R = C_T$, we have

$$E(\overline{Y}_R^*) = \frac{1}{n_1 + n_2} [(n_1 + n_2)\mu + (n_1 + n_2)F_R + n_1P_1 + n_2P_2]$$

$$E(\overline{Y}_T^*) = \frac{1}{n_1 + n_2} [(n_1 + n_2)\mu + (n_1 + n_2)F_T + n_1P_2 + n_2P_1].$$

Hence,

$$E[\overline{Y}_T^* - \overline{Y}_R^*] = (F_T - F_R) + \frac{1}{n_1 + n_2} [(n_2 - n_1)P_1 + (n_1 - n_2)P_2].$$

Therefore, the difference between the direct sample means for the two formulations, $\hat{F}^*$, is not an unbiased estimator for F unless $n_1 = n_2$.

Under assumptions (i)–(iii) as specified in (3.1.4), the difference between the LS means for the two formulations, $\hat{F}$, is normally distributed with mean F and variance

$$V(\hat{F}) = \sigma_d^2 \left(\frac{1}{n_1} + \frac{1}{n_2} \right). \tag{3.3.8}$$

Since $\{d_{i1}\}\, i = 1, \ldots, n_1$ and $\{d_{i2}\}, i = 1, \ldots, n_2$ are two independent samples from normal populations with equal variances when no unequal carry-over effects are present, a test for the direct drug effect can be obtained based on a two-sample t statistic as follows

$$T_d = \frac{\hat{F}}{\hat{\sigma}_d \sqrt{\dfrac{1}{n_1} + \dfrac{1}{n_2}}}, \tag{3.3.9}$$

where $\hat{\sigma}_d^2$ is the pooled sample variance of period differences from both sequences and is an unbiased estimator of σ_d^2. It is given by

$$\hat{\sigma}_d^2 = \frac{1}{n_1 + n_2 - 2} \sum_{k=1}^{2} \sum_{i=1}^{n_k} (d_{ik} - \overline{d}_{.k})^2. \tag{3.3.10}$$

Since $(n_1 + n_2 - 2)\hat{\sigma}_d^2$ is distributed as $\sigma_d^2 \chi^2(n_1 + n_2 - 2)$, T_d has a central student t distribution with $n_1 + n_2 - 2$ degrees of freedom. A $(1 - \alpha) \times 100\%$ confidence interval for F can then be obtained as follows:

$$\hat{F} \pm t(\alpha/2, n_1 + n_2 - 2)\hat{\sigma}_d \sqrt{\frac{1}{n_1} + \frac{1}{n_2}}. \tag{3.3.11}$$

The presence of the direct drug effect can be examined by testing the hypotheses:

$$H_0: F_R = F_T \quad vs \quad H_a: F_R \neq F_T \tag{3.3.12}$$

We reject H_0 if

$$|T_d| > t(\alpha/2, n_1 + n_2 - 2). \tag{3.3.13}$$

Note that the above testing procedure is for the equality of the direct drug effects, not for equivalence of direct drug effects which will be discussed in the next chapter.

As mentioned earlier, $\hat{F}$ is not an unbiased estimator for F in the presence of unequal carry-over effects (i.e., $C_R \neq C_T$). However, an unbiased estimator of F can still be obtained from the data from the first period at the expense of precision. Let $\overline{Y}_{.11}$ and $\overline{Y}_{.12}$ be sample means of the two sequences at the first period. Then,

$$\begin{aligned} E(\overline{Y}_{.12} - \overline{Y}_{.11}) &= (\mu + P_1 + F_T) - (\mu + P_1 + F_R) \\ &= F_T - F_R \\ &= F. \end{aligned}$$

Denote $\overline{Y}_{.12} - \overline{Y}_{.11}$ by $\hat{F}|C$. Thus, $\hat{F}|C$ is an unbiased estimator of F in the presence of unequal carry-over effects. The variance of $\hat{F}|C$ is given by

$$V(\hat{F}|C) = (\sigma_s^2 + \sigma_e^2)\left(\frac{1}{n_1} + \frac{1}{n_2}\right). \tag{3.3.14}$$

Note that

$$V(\hat{F}|C) - V(\hat{F}) = \left(\sigma_s^2 + \frac{\sigma_e^2}{2}\right)\left(\frac{1}{n_1} + \frac{1}{n_2}\right). \tag{3.3.15}$$

Hence, in the presence of unequal carry-over effects, an unbiased estimator for F can only be obtained using the data at the first period at the expense of losing the precision by at least 50% even when $\sigma_s^2 = 0$. Thus, in practice, it is extremely important to have a sufficient length of washout period between dosing periods to eliminate the residual effects from a previous dosing period before initiating the next dosing period. In the presence of unequal carry-over effects, however, a $(1 - \alpha) \times 100\%$ confidence interval for F and a test statistic for the hypothesis of no direct drug effect can also be obtained based on a two-sample t statistic using the data from the first period.

First, an unbiased estimator of $V(\hat{F}|C)$ is

$$\hat{V}(\hat{F}|C) = S_f^2 \left(\frac{1}{n_1} + \frac{1}{n_2} \right), \tag{3.3.16}$$

where

$$S_f^2 = \frac{1}{n_1 + n_2 - 2} \sum_{k=1}^{2} \sum_{i=1}^{n_k} (Y_{i1k} - \overline{Y}_{\cdot 1k})^2. \tag{3.3.17}$$

Note that although S_f^2 is an unbiased estimator of $\sigma_e^2 + \sigma_s^2$, individual estimates for σ_e^2 and σ_s^2 are not available based on the data from the first period only. A $(1 - \alpha) \times 100\%$ confidence interval for F in the presence of unequal carry-over effects is then given by

$$\hat{F}|C \pm t(\alpha/2, n_1 + n_2 - 2)S_f \sqrt{\frac{1}{n_1} + \frac{1}{n_2}}. \tag{3.3.18}$$

The null hypothesis of no direct drug effect is rejected if

$$\left| \frac{\hat{F}|C}{S_f \sqrt{\frac{1}{n_1} + \frac{1}{n_2}}} \right| > t(\alpha/2, n_1 + n_2 - 2). \tag{3.3.19}$$

In practice, in the presence of unequal carry-over effects, the data in the first period are analyzed to assess the bioequivalence between formulations in bio-vailability studies. However, one should be aware of the following consequences:

(i)   There is little power for detection of a clinically significant difference due to the increase in variability; and

(ii) The sacrifice of the information in the second period negates the benefit of a crossover design which removes the inter-subject variability from the comparison between formulations.

3.4. THE PERIOD EFFECT

Define the crossover differences as follows

$$O_{ik} = \begin{cases} d_{ik} & \text{for subjects in sequence 1} \\ -d_{ik} & \text{for subjects in sequence 2.} \end{cases} \tag{3.4.1}$$

The expected value and variance of the crossover differences are

$$E(O_{ik}) = \begin{cases} \frac{1}{2}[(P_2 - P_1) + (F_T - F_R) + C_R] & \text{for subjects in sequence 1} \\ \frac{1}{2}[(P_1 - P_2) + (F_T - F_R) - C_T] & \text{for subjects in sequence 2.} \end{cases}$$

(3.4.2)

and $V(O_{ik}) = \sigma_d^2 = \sigma_e^2/2$, respectively.

Let $\overline{O}_{\cdot 1}$ and $\overline{O}_{\cdot 2}$ be the sample means of the crossover differences in sequences 1 and 2. Then,

$$\overline{O}_{\cdot k} = \begin{cases} \overline{d}_{\cdot 1} & \text{for } k = 1 \\ -\overline{d}_{\cdot 2} & \text{for } k = 2. \end{cases}$$

(3.4.3)

An unbiased estimator for the period effect P can then be obtained as

$$\begin{aligned} \hat{P} &= \overline{O}_{\cdot 1} - \overline{O}_{\cdot 2} \\ &= \frac{1}{2}[(\overline{Y}_{\cdot 21} - \overline{Y}_{\cdot 11}) - (\overline{Y}_{\cdot 12} - \overline{Y}_{\cdot 22})]. \end{aligned}$$

(3.4.4)

Since $C_R + C_T = 0$, $\hat{P}$ is the minimum variance unbiased estimator for P regardless of the presence of unequal carry-over effects. A $(1 - \alpha) \times 100\%$ confidence interval for P is given as follows:

$$\hat{P} \pm t(\alpha/2, n_1 + n_2 - 2)\hat{\sigma}_d\sqrt{\frac{1}{n_1} + \frac{1}{n_2}}.$$

(3.4.5)

We reject the null hypothesis of no period effect, i.e.,

$$H_0: P_1 = P_2 \quad \text{vs} \quad H_a: P_1 \neq P_2,$$

(3.4.6)

if

$$|T_o| > t(\alpha/2, n_1 + n_2 - 2),$$

(3.4.7)

where

$$T_o = \frac{\hat{P}}{\hat{\sigma}_d\sqrt{\frac{1}{n_1} + \frac{1}{n_2}}}.$$

(3.4.8)

The statistical inferences for the carry-over effects, the direct drug effect, and the period effect for a standard 2×2 crossover design are summarized in Table 3.4.1.

Table 3.4.1. Statistical Inferences for Fixed Effects in a Standard 2 × 2 Crossover Design

Effect	Unequal carry-over effects	MVUE[a]	$(1 - \alpha) \times 100\%$ C.I.	Test statistic
Carry-over	—	$\hat{C} = \bar{U}_{\cdot 2} - \bar{U}_{\cdot 1} = (\bar{Y}_{\cdot 11} + \bar{Y}_{\cdot 21}) - (\bar{Y}_{\cdot 12} + \bar{Y}_{\cdot 22})$	$\hat{C} \pm t(\alpha/2, n_1 + n_2 - 2)\hat{\sigma}_u \sqrt{\dfrac{1}{n_1} + \dfrac{1}{n_2}}$	$T_C = \dfrac{\hat{C}}{\hat{\sigma}_u \sqrt{\dfrac{1}{n_1} + \dfrac{1}{n_2}}}$
Direct drug	No	$\hat{F} = \bar{d}_{\cdot 1} - \bar{d}_{\cdot 2} = \frac{1}{2}[(\bar{Y}_{\cdot 21} - \bar{Y}_{\cdot 11}) - (\bar{Y}_{\cdot 22} - \bar{Y}_{\cdot 12})]$	$\hat{F} \pm t(\alpha/2, n_1 + n_2 - 2)\hat{\sigma}_d \sqrt{\dfrac{1}{n_1} + \dfrac{1}{n_2}}$	$T_d = \dfrac{\hat{F}}{\hat{\sigma}_d \sqrt{\dfrac{1}{n_1} + \dfrac{1}{n_2}}}$
Direct drug	Yes	$\hat{F}|C = \bar{Y}_{\cdot 12} - \bar{Y}_{\cdot 11}$	$\hat{F}|C \pm t(\alpha/2, n_1 + n_2 - 2)S_f \sqrt{\dfrac{1}{n_1} + \dfrac{1}{n_2}}$	$T_f = \dfrac{\hat{F}|C}{S_f \sqrt{\dfrac{1}{n_1} + \dfrac{1}{n_2}}}$
Period	—	$\hat{P} = \bar{O}_{\cdot 1} - \bar{O}_{\cdot 2} = \frac{1}{2}[(\bar{Y}_{\cdot 21} - \bar{Y}_{\cdot 11}) - (\bar{Y}_{\cdot 12} - \bar{Y}_{\cdot 22})]$	$\hat{P} \pm t(\alpha/2, n_1 + n_2 - 2)\hat{\sigma}_d \sqrt{\dfrac{1}{n_1} + \dfrac{1}{n_2}}$	$T_o = \dfrac{\hat{P}}{\hat{\sigma}_d \sqrt{\dfrac{1}{n_1} + \dfrac{1}{n_2}}}$

[a]MVUE = Minimum Variance Unbiased Estimate.

3.5. THE ANALYSIS OF VARIANCE

In previous sections, statistical inferences for the fixed effects in model (3.1.1) for a standard 2×2 crossover design were derived based on two-sample t statistics. In this section, the method of the analysis of variance which is often considered for a general situation will be introduced. The two-sample t statistic is shown to be equivalent to a special case of the method of analysis of variance.

The concept of the analysis of variance is to study the variability in the observed data by partitioning the total sum of squares (SS) of the observations into components of the fixed effects and the random errors. For example, for the standard 2×2 crossover design, we would partition the total sum of squares of the $2(n_1 + n_2)$ observations into components for the carry-over effects, the period effect, the direct drug effect, and the error. Let $\overline{Y}...$ be the grand mean of all observations. Then the total corrected sum of squares is given by

$$SS_{Total} = \sum_{k=1}^{2} \sum_{j=1}^{2} \sum_{i=1}^{n_k} (Y_{ijk} - \overline{Y}...)^2 \tag{3.5.1}$$

$$= \sum_{k=1}^{2} \sum_{j=1}^{2} \sum_{i=1}^{n_k} (Y_{ijk} - \overline{Y}_{i \cdot k} + \overline{Y}_{i \cdot k} - \overline{Y}...)^2$$

$$= \sum_{k=1}^{2} \sum_{j=1}^{2} \sum_{i=1}^{n_k} (Y_{ijk} - \overline{Y}_{i \cdot k})^2 + 2 \sum_{k=1}^{2} \sum_{i=1}^{n_k} (\overline{Y}_{i \cdot k} - \overline{Y}...)^2$$

$$= SS_{Within} + SS_{Between},$$

where

$$\overline{Y}_{i \cdot k} = \tfrac{1}{2} \sum_{j=1}^{2} Y_{ijk},$$

and $SS_{Between}$ is the sum of squares due to subjects (i.e., between subjects) and SS_{Within} is the sum of squares for the within subjects. Since there are $2(n_1 + n_2)$ observations, SS_{Total} has $2(n_1 + n_2) - 1$ degrees of freedom. There are $n_1 + n_2$ subjects in both sequences. Thus, $SS_{Between}$ and SS_{Within} have $n_1 + n_2 - 1$ and $n_1 + n_2$ degrees of freedom, respectively. The $SS_{Between}$ can be further partitioned into two components, one for the carry-over effects and the other for the inter-subject error. That is,

$$SS_{Between} = SS_{Carry} + SS_{Inter}, \tag{3.5.2}$$

where

$$SS_{Carry} = \frac{2n_1 n_2}{n_1 + n_2} [(\overline{Y}_{\cdot 12} + \overline{Y}_{\cdot 22}) - (\overline{Y}_{\cdot 11} + \overline{Y}_{\cdot 21})]^2, \quad \text{and}$$

$$SS_{Inter} = \sum_{k=1}^{2} \sum_{i=1}^{n_k} \frac{Y_{i \cdot k}^2}{2} - \sum_{k=1}^{2} \frac{Y_{\cdot \cdot k}^2}{2n_k},$$

where $Y_{i \cdot k}$ and $Y_{\cdot \cdot k}$ are the sum of Y_{jkk} over the corresponding indices. SS_{Carry} and SS_{Inter} have 1 and $n_1 + n_2 - 2$ degrees of freedom, respectively. Each sum of squares divided by its degrees of freedom is a mean squares (MS). The expected value of the mean squares for SS_{Carry} and SS_{Inter} can be shown to be

$$E(MS_{Carry}) = \frac{2n_1 n_2}{n_1 + n_2} (C_T - C_R)^2 + 2\sigma_s^2 + \sigma_e^2 \qquad (3.5.3)$$

$$E(MS_{Inter}) = 2\sigma_s^2 + \sigma_e^2. \qquad (3.5.4)$$

Therefore, to test the hypotheses in (3.2.4) (i.e., the equality of the carry-over effects), we would use the test statistic

$$F_c = \frac{MS_{Carry}}{MS_{Inter}}, \qquad (3.5.5)$$

which follows an F distribution with degrees of freedom 1 and $n_1 + n_2 - 2$ if the null hypothesis in (3.2.4) is true. We reject H_0 if

$$F_c > F(\alpha, 1, n_1 + n_2 - 2),$$

where $F(\alpha, 1, n_1 + n_2 - 2)$ is the upper α percentile of the F distribution with degrees of freedom 1 and $n_1 + n_2 - 2$. Note that an F distribution with degrees of freedom 1 and ν is equal to the square of a t distribution with degrees of freedom ν. Thus, the above test statistic is equivalent to test statistic T_c in (3.2.10) since $F_c = T_c^2$.

Similarly, the SS_{Within} can be further decomposed into three components: a SS for the direct drug effect, a SS for the period effect, and a SS for the intra-subject residuals. That is,

$$SS_{Within} = SS_{drug} + SS_{Period} + SS_{Intra}, \qquad (3.5.6)$$

where

$$SS_{Drug} = \frac{2n_1 n_2}{n_1 + n_2} \{ \tfrac{1}{2} [(\overline{Y}_{\cdot 21} - \overline{Y}_{\cdot 11}) - (\overline{Y}_{\cdot 22} - \overline{Y}_{\cdot 12})] \}^2$$

$$SS_{Period} = \frac{2n_1 n_2}{n_1 + n_2} \{ \tfrac{1}{2} [(\overline{Y}_{\cdot 21} - \overline{Y}_{\cdot 11}) - (\overline{Y}_{\cdot 12} - \overline{Y}_{\cdot 22})] \}^2,$$

$$SS_{Intra} = \sum_{k=1}^{2} \sum_{j=1}^{2} \sum_{i=1}^{n_k} Y_{ijk}^2 - \sum_{k=1}^{2} \sum_{i=1}^{n_k} \frac{Y_{i \cdot k}^2}{2} - \sum_{k=1}^{2} \sum_{j=1}^{2} \frac{Y_{\cdot jk}^2}{n_k} + \sum_{k=1}^{2} \frac{Y_{\cdot \cdot k}^2}{2n_k}.$$

There is one degree of freedom for each of SS_{Drug} and SS_{Period} and $n_1 + n_2 - 2$ degrees of freedom for SS_{Intra}. The expected values for their mean squares are given below:

$$E(MS_{Drug}) = \frac{2n_1 n_2}{n_1 + n_2} \left[(F_T - F_R) + \frac{C_R - C_T}{2} \right]^2 + \sigma_e^2, \qquad (3.5.7)$$

$$E(MS_{Period}) = \frac{2n_1n_2}{n_1 + n_2} (P_2 - P_1)^2 + \sigma_e^2, \tag{3.5.8}$$

$$E(MS_{Intra}) = \sigma_e^2. \tag{3.5.9}$$

Note that $MS_{Intra} = 2\hat{\sigma}_d^2$, where $\hat{\sigma}_d^2$ is the pooled sample variance of period differences. When $C_R = C_T$, the null hypothesis in (3.3.12) of no direct drug effect can be tested using the following statistic

$$F_d = \frac{MS_{Drug}}{MS_{Intra}}, \tag{3.5.10}$$

which is distributed as an F distribution with degrees of freedom 1 and $n_1 + n_2 - 2$. We reject the null hypothesis if

$$F_d > F(\alpha, 1, n_1 + n_2 - 2).$$

Test statistic F_d is equivalent to test statistic T_d in (3.3.9) since $F_d = T_d^2$.

For testing the null hypothesis (3.4.6) of no period effect, we may consider the following test statistic

$$F_p = \frac{MS_{Period}}{MS_{Intra}}, \tag{3.5.11}$$

which follows an F distribution with degrees of freedom 1 and $n_1 + n_2 - 2$. The null hypothesis is then rejected if

$$F_p > F(\alpha, 1, n_1 + n_2 - 2).$$

It can be verified that $F_p = T_o^2$. Thus, test statistic F_p is equivalent to test statistic T_o in (3.4.8).

For a general crossover design, the method of the analysis of variance is useful in deriving statistical inferences for the fixed effects in model (2.5.1) under some normality assumptions. It can be seen that for a standard 2 × 2 crossover design, a two-sample t statistic is equivalent to a special case of the method of the analysis of variance. The analysis of variance table for the standard 2 × 2 crossover design is summarized in Table 3.5.1.

From Table 3.5.1, in addition, a test for the hypotheses of the presence of the inter-subject variability, i.e.,

$$H_0: \sigma_s^2 = 0 \quad \text{vs} \quad H_a: \sigma_s^2 > 0, \tag{3.5.12}$$

can also be obtained by considering

$$F_v = \frac{MS_{Inter}}{MS_{Intra}}, \tag{3.5.13}$$

where F_v is distributed as an F distribution with degrees of freedom $n_1 + n_2 -$

Table 3.5.1 Analysis of Variance Table for a Standard 2 × 2 Crossover Design

Source of variation	df	SS	MS = SS/df	E(MS)	F
Inter-subjects					
Carry-over	1	SS_{Carry}	SS_{Carry}	$\dfrac{2n_1n_2}{n_1 + n_2}(C_T - C_R)^2 + 2\sigma_s^2 + \sigma_e^2$	$F_C = MS_{Carry}/MS_{Inter}$
Residuals	$n_1 + n_2 - 2$	SS_{Inter}	$SS_{Inter}/n_1 + n_2 - 2$	$2\sigma_s^2 + \sigma_e^2$	$F_V = MS_{Inter}/MS_{Intra}$
Intra-subjects					
Direct drug	1	SS_{Drug}	SS_{Drug}	$\dfrac{2n_1n_2}{n_1 + n_2}\left[(F_T - F_R) + \dfrac{C_R - C_T}{2}\right]^2 + \sigma_e^2$	$F_d^* = MS_{Drug}/MS_{Intra}$ [a]
Period	1	SS_{Period}	SS_{Period}	$\dfrac{2n_1n_2}{n_1 + n_2}(P_2 - P_1)^2 + \sigma_e^2$	$F_p = MS_{Period}/MS_{Intra}$
Residuals	$n_1 + n_2 - 2$	SS_{Intra}	$SS_{Intra}/n_1 + n_2 - 2$	σ_e^2	
Total	$2(n_1 + n_2) - 1$	SS_{Total}			

[a] F_d^* valid only if $C_R = C_T$.

2 and $n_1 + n_2 - 2$ under H_0. Thus, we reject the null hypothesis of no inter-subject variability if

$$F_v > F(\alpha, n_1 + n_2 - 2, n_1 + n_2 - 2).$$

Statistical inferences for the inter-subject variability (σ_s^2) and the intra-subject variability (σ_e^2) will be discussed in Chapter 7.

It should be noted that a standard 2×2 crossover design can only provide estimates and tests for the period effect, the direct drug effect, and the carry-over effects. It does not provide any inference on the interactions among these effects and the interactions between fixed and random effects. To draw some statistical inference on the interactions of interest (e.g., the subject-by-formulation, the formulation-by-period, and the sequence-by-period interactions), a higher-order crossover design is necessary. However, it is important to determine which interaction term is to be examined before an appropriate design is chosen. For example, to test the subject-by-formulation interaction in comparing two formulations, each subject must receive each of the test and reference formulations twice. Consequently, a four-period design will have to be used.

3.6. AN EXAMPLE

Example 3.6.1 To illustrate the above statistical inferences for the fixed effects in model (3.1.1) for a standard 2×2 crossover design, let's consider the following example concerning the comparison of bioavailability between two formulations of a drug product. This study was conducted with 24 healthy volunteers. During each dosing period, each subject was administered either five 50 mg tablets (test formulation) or five ml of an oral suspension (50 mg/ml) (reference formulation). Blood samples were obtained at 0 hours prior to dosing and at various times after dosing. AUC values from 0 to 32 hours, given in Table 3.6.1, were calculated using the trapezoidal method.

For a preliminary examination of the data, the plots of subject profiles for each sequence and sequence-by-period means are useful and presented in Figures 3.6.1–3.6.3. Figures 3.6.1 and 3.6.2 indicate that the variability in AUC at the second sequence seems larger than that at the first sequence. Moreover, some drastic changes in AUC in each sequence were observed. In Figure 3.6.3, 1T and 2T (1R and 2R) are the sample means for test (reference) formulation in sequences 1 and 2, which are given as follows.

Sequence	Period I	Period II	Sequence mean
1	$1R = \bar{Y}_{\cdot 11} = 85.82$	$1T = \bar{Y}_{\cdot 21} = 81.80$	$\bar{Y}_{\cdot\cdot 1} = 83.81$
2	$2T = \bar{Y}_{\cdot 12} = 78.74$	$2R = \bar{Y}_{\cdot 22} = 79.30$	$\bar{Y}_{\cdot\cdot 2} = 79.02$
Period mean	$\bar{Y}_{\cdot 1\cdot} = 82.28$	$\bar{Y}_{\cdot 2\cdot} = 80.55$	$\bar{Y}_{\cdot\cdot\cdot} = 81.42$

Table 3.6.1   AUC(0 − 32) for Test and Reference Formulations

Seq.	Subject number	Period I	Period II	Subject total	P.D.[a]	C.D.[b]
1						
RT	1	74.675	73.675	148.350	− 1.000	− 1.000
RT	4	96.400	93.250	189.650	− 3.150	− 3.150
RT	5	101.950	102.125	204.075	0.175	0.175
RT	6	79.050	69.450	148.500	− 9.600	− 9.600
RT	11	79.050	69.025	148.075	− 10.025	− 10.025
RT	12	85.950	68.700	154.650	− 17.250	− 17.250
RT	15	69.725	59.425	129.150	− 10.300	− 10.300
RT	16	86.275	76.125	162.400	− 10.150	− 10.150
RT	19	112.675	114.875	227.550	2.200	2.200
RT	20	99.525	116.250	215.775	16.725	16.725
RT	23	89.425	64.175	153.600	− 25.250	− 25.250
RT	24	55.175	74.575	129.750	19.400	19.400
2						
TR	2	74.825	37.350	112.175	− 37.475	37.475
TR	3	86.875	51.925	138.800	− 34.950	34.950
TR	7	81.675	72.175	153.850	− 9.500	9.500
TR	8	92.700	77.500	170.200	− 15.200	15.200
TR	9	50.450	71.875	122.325	21.425	− 21.425
TR	10	66.125	94.025	160.150	27.900	− 27.900
TR	13	122.450	124.975	247.425	2.525	− 2.525
TR	14	99.075	85.225	184.300	− 13.850	13.850
TR	17	86.350	95.925	182.275	9.575	− 9.575
TR	18	49.925	67.100	117.025	17.175	− 17.175
TR	21	42.700	59.425	102.125	16.725	− 16.725
TR	22	91.725	114.050	205.775	22.325	− 22.325

[a]P.D. = 2 × (Period Difference).
[b]C.D. = 2 × (Crossover Difference).

These results indicate that the mean AUC's for both test and reference formulations in sequence 1 are higher than those in sequence 2. In particular, the mean AUC of the reference formulation in sequence 1 is about 8.2% higher than that in sequence 2. As mentioned earlier, for a 2 × 2 crossover design, the sequence-by-period interaction represents the direct drug effect if there is no carry-over effects. Thus, a preliminary test for the presence of carry-over effects is necessarily performed before the assessment of bioequivalence between for-

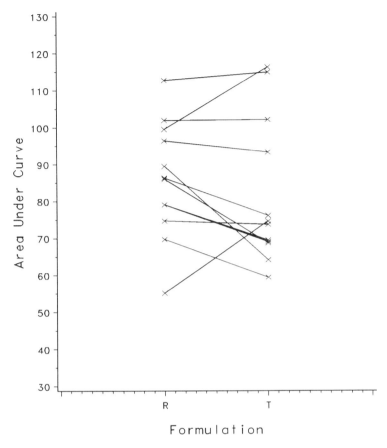

Figure 3.6.1 Subject profiles sequence = 1.

mulations is made. We now carry out the statistical inferences for the fixed effects discussed in previous sections as follows.

The Carry-Over Effects

Since $\overline{U}_{\cdot 1} = 167.63$, $\overline{U}_{\cdot 2} = 158.04$, and $\hat{\sigma}_u^2 = 1473.77$, the test statistic is given by

$$T_c = \frac{158.04 - 167.63}{\sqrt{1473.77 \left(\frac{1}{12} + \frac{1}{12}\right)}} = -0.6120.$$

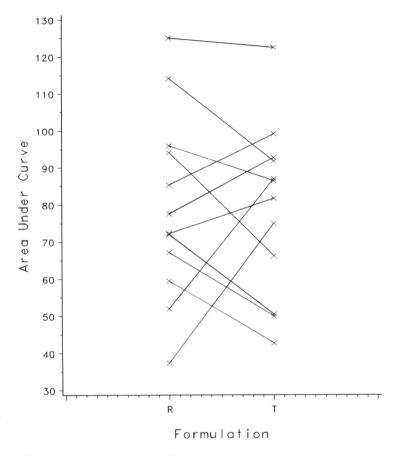

Figure 3.6.2   Subject profiles sequence = 2.

Thus, $|T_c| = 0.612 < t(0.025,22) = 2.074$. Hence, we fail to reject the hypothesis of no unequal carry-over effects. The observed p value is 0.5468 which also indicates that there is little evidence for the presence of unequal carry-over effects. The test result suggests that it may be appropriate to use the data from both periods for inference on the direct drug effect.

The Direct Drug Effect

From Table 3.6.1, it can be verified that

$$\bar{d}_{.1} = -2.0094, \qquad \bar{d}_{.2} = 0.2781, \quad \text{and} \quad \hat{\sigma}_d^2 = 83.623.$$

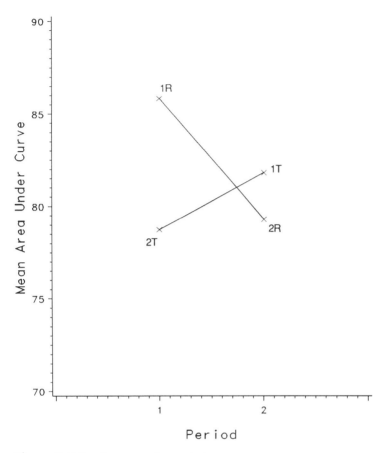

Figure 3.6.3   Sequence-by-period means.

Thus, the point estimate of F is

$$\hat{F} = \bar{d}_{\cdot 1} - \bar{d}_{\cdot 2} = -2.29,$$

with a 95% confidence interval

$$\hat{F} \pm t(0.025,22)\hat{\sigma}_d\sqrt{\frac{1}{n_1} + \frac{1}{n_2}}$$
$$= -2.29 \pm (2.074)(9.1446)\sqrt{0.1667}$$
$$= (-10.03, 5.46).$$

As indicated by the 95% confidence interval, no significant direct drug effect was detected. It should be noted that the null hypothesis of the equality between formulations does not imply the bioequivalence between formulations. The assessment of bioequivalence between formulations according to the FDA requirement will be discussed extensively in the next chapter.

The Period Effect

From Table 3.6.1, the crossover differences can be obtained as follows:

$$\overline{O}_{.1} = -2.0094 \quad \text{and} \quad \overline{O}_{.2} = -0.2781,$$

Thus, the t statistic is given by

$$T_o = \frac{0.2781 - 2.0094}{\sqrt{(83.623)(0.1667)}} = -0.4637.$$

Hence, we fail to reject the hypothesis of no period effect since $|T_o| = 0.4637 < t(0.025,22) = 2.074$ (p-value $= 0.6474$).

The results of statistical inferences on the fixed effects obtained from two-sample t statistics are summarized in Table 3.6.2.

The Analysis of Variance

For the data given in Table 3.6.1, the method of the analysis of variance was also performed. The results are given in Table 3.6.3. It can be easily verified that the results from the method of the analysis of variance are equivalent to the results obtained using two-sample t statistics.

Test for Inter-Subject Variability

It can be seen from the ANOVA table given in Table 3.6.3 that $MS_{inter} = 736.885$ and $MS_{Intra} = 167.246$. Thus, the F statistic for $H_0: \sigma_s^2$ vs $H_a: \sigma_s^2 > 0$ is

Table 3.6.2   Statistical Inferences for the Fixed Effects

Effect	MVUE	Variance estimate	95% C.I.[a]	T	P-Value
Carry-over	−9.59	245.63	(−42.10, 22.91)	−0.612	0.5468
Direct drug	−2.29	13.97	(−10.03, 5.46)	−0.613	0.5463
Period	−1.73	13.97	(−9.47, 6.01)	−0.464	0.6474

[a]95% C.I. = 95% confidence interval.

Table 3.6.3 Analysis of Variance Table for Data in Table 3.6.1

Source of variation	df	SS	MS	F	P-Value
Inter-subjects					
Carry-over	1	276.00	276.00	0.37	0.5468
Residuals	22	16211.49	736.89	4.41	0.0005
Intra-subjects					
Direct drug	1	62.79	62.79	0.38	0.5463
Period	1	35.97	35.97	0.22	0.6474
Residuals	22	3679.43	167.25		
Total	47	20265.68			

$$F_v = \frac{736.885}{167.246} = 4.41,$$

which is greater than $F(0.05, 22, 22) = 2.12$. The observed p value is 0.0005 which favors the presence of inter-subject variability.

Statistical Methods for Average Bioavailability

4.1. INTRODUCTION

In the previous chapter, statistical tests for the presence of the fixed effects from a crossover design were reviewed. These tests are often used as a preliminary analysis of the data before the assessment of bioequivalence between formulations is made. Failing to reject the null hypothesis of the equality of formulation effects, however, does not imply the bioequivalence between formulations. In this chapter, we will introduce a number of statistical methods based on raw data (or untransformed data), which are derived under a crossover design, for the assessment of bioequivalence between formulations in average bioavailability. Statistical methods for transformed data will be discussed in later chapters.

In practice, it is usually assumed that there are no carry-over effects since a washout period of sufficient length can be chosen to completely eliminate the residual effects from one dosing period to the next. In this case, model (2.5.1) for a general crossover design reduces to:

$$Y_{ijk} = \mu + S_{ik} + F_{(j,k)} + P_j + e_{ijk}, \tag{4.1.1}$$

where Y_{ijk}, μ, S_{ik}, $F_{(j,k)}$, P_j and e_{ijk} were defined in (2.5.1). In (4.1.1), the term $F_{(j,k)}$ for formulation is determined when the sequence and period are determined. In some cases, however, the reduced model is sometimes expressed as follows:

$$Y_{ijk} = \mu + S_{ik} + F_j + P_{(j,k)} + e_{ijk}, \tag{4.1.2}$$

where Y_{ijk} is the response of the ith subject in the kth sequence for the jth formulation, in which $i = 1,2, \ldots ,n_k, j = 1,2, \ldots ,J$, and $k = 1,2, \ldots ,g$; μ is the overall mean; F_j is the fixed effect for the jth formulation; $P_{(j,k)}$ is the fixed period effect at which the jth formulation in the kth sequence is administered; S_{ik} and e_{ijk} are defined as in (2.5.1). It is also assumed that $\{S_{ik}\}$ and $\{e_{ijk}\}$ are mutually independently distributed with mean 0 and variances σ_s^2 and σ_e^2, respectively. It should be noted that the above two models are equivalent but different representations of the responses. In this chapter, however, unless otherwise stated, all methods for assessing bioequivalence in average bioavailability will be derived under model (4.1.1) for a standard 2×2 crossover design comparing a test formulation with a reference formulation.

To claim bioequivalence in average bioavailability, as indicated in Section 1.6, the ± 20 rule requires that the ratio of the two true formulation means μ_T/μ_R for AUC and C_{max} be within (80%, 120%) limits (or the difference $\mu_T - \mu_R$ is within $\pm 20\%$ of μ_R). The FDA requires that the bioequivalence be concluded (i.e., $\mu_T/\mu_R \in$ (80%, 120%)) with 90% assurance. Along this line, several methods have been proposed in the past two decades. These methods include the confidence interval approach, the method of interval hypotheses testing, the Bayesian approach and nonparametric methods. For the confidence interval approach, for example, we may construct a confidence interval for μ_T/μ_R (Westlake, 1976) and compare the constructed confidence interval with (80%, 120%) limits. If the constructed confidence interval falls within the limits, then the two formulations are considered bioequivalent. Alternatively, we may construct an exact confidence interval for μ_T/μ_R based on Fieller's theorem (Locke, 1984) or construct a confidence region for (μ_R,μ_T) (Chow, 1990; Chow and Shao, 1990) and compare it with the acceptance region $\mathcal{R}$ which is bounded by two straight lines $\mu_T = 0.8\mu_R$ and $\mu_T = 1.2\mu_R$. For the method of interval hypotheses testing, Schuirmann (1981, 1987) proposed an approach using two one-sided tests. In this approach, two p-values are obtained to evaluate whether the bioavailability of the test formulation is not too low for one side and not too high for the other side. Instead of testing whether or not the true ratio of means is within (80%, 120%), Anderson and Hauck (1983) suggested a test which rejects the null hypothesis of bioinequivalence in favor of bioequivalence for a small p value. Rodda and Davis (1980) first introduced the idea of Bayesian analysis for assessing bioequivalence. Mandallaz and Mau (1981) proposed a Bayesian method for the ratio assuming that (i) there are no carry-over effects; (ii) the subject effects are fixed and (iii) the number of subjects in each sequence are the same. Grieve (1985) considered a Bayesian approach using a prior distribution which takes into account the inter-subject variability. These methods are mainly based on the normality or lognormality assumptions. Other Bayesian methods were also considered in the literature. See, for example, Selwyn, Dempster and Hall (1981), Racine-Poon et al. (1986), and Racine-Poon et al. (1987). Recently, due to the difficulty of testing these assumptions, the search for appropriate nonparametric methods has received tremendous attention (Cornell, 1980; Steinijans and Diletti, 1983, 1985). Several nonparametric methods have been proposed. These methods include the Wilcoxon-Mann-Whitney two one-sided tests procedure (Hauschke, Steinijans, and Diletti, 1990; Liu, 1991) and the bootstrap resampling procedure (Chow, Cheng, and Shao, 1990).

As indicated in Chapter 3, when the numbers of subjects in each sequence are different, the difference in least squares means is the MVUE of the direct drug effect, but the difference in raw (direct sample) means is not. In practice, the number of randomized subjects in each sequence is often different from the number of subjects who complete the study in each sequence. Thus, to have a correct and valid statistical inference, all statistical methods for the assessment of bioequivalence discussed in this chapter are based on least squares means. Note that if the number of subjects in each sequence is the same, then the least squares means are the same as the raw means. For the sake of convenience, the least squares (LS) means for the test and reference formulations, denoted by $\overline{Y}_T$ and $\overline{Y}_R$, are referred to as the (observed) test and reference means, respectively. It is easily seen that

$$E(\overline{Y}_R) = E[(\overline{Y}_{.11} + \overline{Y}_{.22})/2] = \mu + F_R, \quad \text{and}$$

$$E(\overline{Y}_T) = E[(\overline{Y}_{.21} + \overline{Y}_{.12})/2] = \mu + F_T.$$

Let $\mu_R = \mu + F_R$ and $\mu_T = \mu + F_T$. Then μ_R and μ_T are indeed the unknown true reference and test formulation means. Note that μ_R and μ_T do not contain any nuisance fixed effects such as period effects. Therefore, the difference or the ratio of μ_T and μ_R only involves the direct formulation effects. Thus, in this chapter, a number of methods using LS means for statistical inferences of either $\mu_T - \mu_R$ or μ_T/μ_R will be introduced for the assessment of bioequivalence in average bioavailability.

In the following sections, more details regarding the confidence interval approach, the method of interval hypotheses testing, the Bayesian approach and some nonparametric methods are given. The example presented in Section 3.6 will be used to illustrate these methods. In the last section, other possible alternatives for the assessment of bioequivalence will be discussed. A brief discussion is also included in this section. Some SAS programs for the methods introduced in this chapter are included in Appendix B.

4.2. THE CONFIDENCE INTERVAL APPROACH

In Chapter 3, a test for the null hypothesis of the equality of the two formulations of a drug product was derived under a standard 2×2 crossover design. Such a test for assessing bioequivalence, which has been criticized by many researchers (see, e.g., Westlake, 1972; Metzler, 1974; Dunnett and Gent, 1977), is not an appropriate statistical method. Alternatively, Westlake (1976) and Metzler (1974) indicated that the method of confidence interval is an appropriate method for assessing bioequivalence. Based on the confidence interval, Westlake (1981) suggested the following action for decision-making:

If a $(1 - 2\alpha) \times 100\%$ confidence interval for the difference $(\mu_T - \mu_R)$ or the ratio (μ_T/μ_R) is within the acceptance limits as recommended by the regulatory agency, then accept the test formulation (i.e., the test formulation is bioequivalent to the reference formulation); otherwise reject it.

If the ± 20 rule, as indicated in Section 1.6, is adopted, then α is usually chosen to be 0.05 and the acceptance limits (or equivalence limits) are either

(i) $\pm 20\%$ of the observed average bioavailability for the reference formulation if the 90% confidence interval for $\mu_T - \mu_R$ is used, or

(ii) 80% and 120% if the 90% confidence interval for μ_T/μ_R is used.

Thus, we may conclude bioequivalence if either the 90% confidence interval for the difference in average bioavailability of the two formulations is within $\pm 20\%$ of the observed reference mean or the 90% confidence interval for the ratio of average bioavailability of the two formulations is within 80% and 120% limits. Note that the two decision rules are equivalent when μ_R is known. Although the confidence interval for the difference can be converted to be a confidence interval for the ratio, they may lead to different conclusions of bioequivalence when $\overline{Y}_R$ is assumed to be the true μ_R.

Based on the above action of decision-making, several methods for constructing a 90% confidence interval for $\mu_T - \mu_R$ and/or μ_T/μ_R have been proposed under a raw data (or untransformed data) model. These methods include (i) the classical confidence interval which is also known as the shortest confidence interval; (ii) Westlake's symmetric confidence interval; (iii) confidence interval for μ_T/μ_R based on Fieller's theorem; and (iv) Chow and Shao's joint confidence region for (μ_R, μ_T).

In the following, these methods will be derived under the standard 2×2 crossover design assuming that there are no carry-over effects. For a general crossover design with more than two formulations, the methods can be treated similarly.

4.2.1. The Classical (Shortest) Confidence Interval

Let $\overline{Y}_T$ and $\overline{Y}_R$ be the respective least squares means for the test and reference formulations, which can be obtained from the sequence-by-period means. The classical (or shortest) $(1 - 2\alpha) \times 100\%$ confidence interval can then be obtained based on the following t statistic

$$T = \frac{(\overline{Y}_T - \overline{Y}_R) - (\mu_T - \mu_R)}{\hat{\sigma}_d \sqrt{\dfrac{1}{n_1} + \dfrac{1}{n_2}}}, \tag{4.2.1}$$

where n_1 and n_2 are the numbers of subjects in sequences 1 and 2, respectively, and $\hat{\sigma}_d$ is given in Section 3.3. Under normality assumptions, T follows a central student t distribution with degrees of freedom $n_1 + n_2 - 2$. Thus, the classical $(1 - 2\alpha) \times 100\%$ confidence interval for $\mu_T - \mu_R$ can be obtained as follows:

$$L_1 = (\overline{Y}_T - \overline{Y}_R) - t(\alpha, n_1 + n_2 - 2)\hat{\sigma}_d \sqrt{\frac{1}{n_1} + \frac{1}{n_2}},$$

$$U_1 = (\overline{Y}_T - \overline{Y}_R) + t(\alpha, n_1 + n_2 - 2)\hat{\sigma}_d \sqrt{\frac{1}{n_1} + \frac{1}{n_2}}. \tag{4.2.2}$$

The above confidence interval for $\mu_T - \mu_R$ can be converted into a $(1 - 2\alpha)$ $\times 100\%$ approximate confidence interval for μ_T/μ_R by dividing by $\overline{Y}_R$ (assuming that $\overline{Y}_R$ is the true reference mean μ_R). That is,

$$L_2 = (L_1/\overline{Y}_R + 1) \times 100\%,$$

$$U_2 = (U_1/\overline{Y}_R + 1) \times 100\%. \tag{4.2.3}$$

Let θ_L and θ_U be the respective lower and upper equivalence limits for the difference. Also, let δ_L and δ_U be the respective lower and upper equivalence limits for the ratio. Then, we conclude bioequivalence if

$$(L_1, U_1) \in (\theta_L, \theta_U)$$

$$\text{or} \quad (L_2, U_2) \in (\delta_L, \delta_U), \tag{4.2.4}$$

where $\theta_L = -0.2\mu_R$, $\theta_U = 0.2\mu_R$; $\delta_L = 80\%$, $\delta_U = 120\%$ for the ± 20 rule.

The use of confidence interval as a tool for assessing bioequivalence based on (4.2.4) seems intuitively appealing. There is, however, a discrepancy between the concept of the confidence interval and the action of decision-making. A $(1 - 2\alpha) \times 100\%$ confidence interval for $\mu_T - \mu_R$ is a random interval and its associate confidence limits are, in fact, random variables. The fundamental concept of a $(1 - 2\alpha) \times 100\%$ confidence interval for $\mu_T - \mu_R$ is that if the same study can be repeatedly carried out for a large number of times, say B, then $(1 - 2\alpha) \times 100\%$ times of the B constructed random intervals will cover $\mu_T - \mu_R$ (Bickel and Doksum, 1977). In other words, in the long run, a $(1 - 2\alpha) \times 100\%$ confidence interval will have at least a $1 - 2\alpha$ chance to cover the true mean difference (or ratio) since, under the normality assumptions, we have

$$P\{\mu_T - \mu_R \in (L_1, U_1)\} = 1 - 2\alpha.$$

However, the $(1 - 2\alpha) \times 100\%$ confidence interval does not guarantee that, in the long run, the chance of the $(1 - 2\alpha) \times 100\%$ confidence interval being within the equivalence limits is at least $1 - 2\alpha$, i.e., the probability

$$P\{(L_1, U_1) \in (\theta_L, \theta_U)\}$$

is not necessarily greater than or equal to $1 - 2\alpha$. To demonstrate this, as an example, a small simulation study was conducted. A total of 1000 sets of AUC values were generated from the statistical model for a standard 2×2 crossover design under normality assumptions. For simplicity, we assumed there were no period and carry-over effects. The true test and reference formulation means were both chosen to be 100 with $n_1 = n_2 = 9, 12$. Thus, the equivalence limits for the difference are from -20 to 20. The following intra-subject variabilities were considered: $\sigma_e^2 = 400, 900$, and 1600 (or CV $= 20\%, 30\%$, and 40%), where CV is defined as $(\sigma_e/\mu_R) \times 100\%$. For each random sample, (L_1, U_1) was calculated. The results are summarized in Table 4.2.1. The results indicate that, for $n_1 = n_2 = 9$, in the long run, 76.8% of confidence intervals (L_1, U_1) will be within the equivalence limits when CV is 20%. However, only 24.7% and 2.1% of confidence intervals will be within the equivalence limits for CV $= 30\%$ and 40%, respectively. For $n_1 = n_2 = 12$, a similar pattern was observed. Thus, when there is a large CV (or large variability), the confidence interval approach for assessing bioequivalence in average bioavailability may not have the desired level of assurance as required by the FDA. In other words, the probability of correctly concluding bioequivalence may not be of the desired level. In this case, it is suggested that a parametric bootstrap random samples (Efron, 1982) be simulated based on the observed values (i.e., assuming these values are the true population values) to evaluate the performance of (L_1, U_1) in the long run (i.e., to determine the level of assurance of (L_1, U_1)). Note that the probability of correctly concluding bioequivalence for (L_1, U_1) and (L_2, U_2) is the same when μ_R is known.

Table 4.2.1   Summary of Simulation Results

Sample size	Reference mean	Test mean	CV(%)	% of C.I.s containing 0	% of C.I.s within $(-20,20)$
$n_1 = n_2$ $= 9$	100	100	20	90.3	76.8
			30	91.3	24.7
			40	90.4	2.1
$n_1 = n_2$ $= 12$	100	100	20	90.7	91.5
			30	89.2	43.9
			40	88.6	7.5

Example 4.2.1   To illustrate the use of the classical (shortest) confidence intervals for $\mu_T - \mu_R$ (4.2.2) and μ_T/μ_R (4.2.3), consider the example presented in Section 3.6. Based on the AUC values from Table 3.6.1, we have

$$\overline{Y}_T = 80.272;$$
$$\overline{Y}_R = 82.559;$$
$$t(0.05, 22) = 1.717; \quad \text{and}$$
$$\hat{\sigma}_d = 9.145.$$

Thus the equivalence limits for $\mu_T - \mu_R$ are given by -16.51 and 16.51. The confidence interval (4.2.2) for $\mu_T - \mu_R$ can be obtained as follows:

$$(80.272 - 82.559) \pm (1.717)(9.145)\sqrt{0.167}.$$

Thus, $(L_1, U_1) = (-8.698, 4.123)$ and the corresponding approximate 90% confidence interval for μ_T/μ_R, (L_2, U_2), is given by $(89.46\%, 104.99\%)$. Since both (L_1, U_1) and (L_2, U_2) satisfy (4.2.4), we conclude that the test formulation is bioequivalent to the reference formulation.

The finite samples performance of (L_1, U_1) was also evaluated through a small simulation study. A total of 1000 bootstrap samples were generated from the standard 2×2 crossover model (4.1.1) under normality assumptions with $n_1 = n_2 = 12$, $\mu_R = 82.599$, $\mu_T = 80.272$, and $\sigma_e = 12.933$ (i.e., $\sigma_d = 9.145$). For each bootstrap sample, (L_1, U_1) was calculated. The results indicate that, as expected, 90.6% of the 1000 confidence intervals cover $\mu_T - \mu_R = -2.287$. On the other hand, 98% of these intervals are within the equivalence limits $(-16.51, 16.51)$, where $16.51 = (0.2)(82.559)$. The average lower and upper limits for these intervals are -8.68 and 4.00 which are close to $(-8.698, 4.123)$ obtained from the observed data. Thus, there is more than 95% assurance that the observed confidence interval will be within the equivalence limits $(-16.51, 16.51)$ in the long run. This certainly supports the conclusion of bioequivalence between the two formulations.

Suppose that the observed test and reference means remain the same as 80.272 and 82.559, respectively, but the observed intra-subject variability $\hat{\sigma}_e$ increases to 27.659 (or $\sigma_d = 19.558$). It can be verified that $(L_1, U_1) = (-15.989, 11.413)$ are totally within the equivalence limits $(-16.51, 16.51)$. Hence, according to (4.2.4), the test formulation is also bioequivalent to the reference formulation. However, based on a simulation study of 1000 bootstrap samples with $\mu_R = 82.559$, $\mu_T = 80.272$, $\sigma_e = 27.659$, and $n_1 = n_2 = 12$, it was found that 89.6% of the 1000 90% confidence intervals cover $\mu_T - \mu_R = -2.287$ with the average lower and upper limits being -15.832 and 11.250, but, only 26.7% of these 1000 random intervals are within the equivalence limits $(-16.51, 16.51)$. Therefore, the simulation result contradicts the decision made based on the observed 90% confidence interval for $\mu_T - \mu_R$. This example reveals the importance of the intra-subject variability in the assessment of bio-equivalence. Thus, it is suggested that a simulation study be conducted to evaluate the finite sample performance of (L_1, U_1) before a decision regarding bioequivalence is made.

4.2.2. Westlake's Symmetric Confidence Interval

It is apparent from (4.2.2) that the classical $(1 - 2\alpha) \times 100\%$ confidence interval for $\mu_T - \mu_R$ is symmetric about $\overline{Y}_T - \overline{Y}_R$ and not symmetric about 0. Also, it can be seen from (4.2.3) that the confidence interval for μ_T/μ_R is symmetric about $\overline{Y}_T/\overline{Y}_R$, and not about unity. Basically, the classical confidence interval is derived from a two-sample t statistic in (4.2.1) as follows

$$|T| < k \quad \text{or} \quad -k < T < k,$$

where k is the upper α percentile of a central t distribution with degrees of freedom $n_1 + n_2 - 2$. In general, a $(1 - 2\alpha) \times 100\%$ confidence interval for the difference $\mu_T - \mu_R$ can be expressed as

$$k_2 < T < k_1, \tag{4.2.5}$$

where k_1 and k_2 are chosen so that the probability from k_2 to k_1 based on a central t distribution with $n_1 + n_2 - 2$ degrees of freedom is $1 - 2\alpha$, i.e.,

$$\int_{k_2}^{k_1} T \, dt = 1 - 2\alpha.$$

When $k_2 = -k_1$, (4.2.5) reduces to the classical confidence interval (4.2.2).

As the equivalence limits are usually given in a symmetric form (e.g., -20% to 20%), Westlake (1976) suggested that the confidence interval be adjusted to be symmetric about 0 for the difference (or about unity for the ratio). For the difference, it is desirable to construct a confidence interval as follows

$$-\Delta < \mu_T - \mu_R < \Delta. \tag{4.2.6}$$

This is, however, equivalent to constructing a confidence interval for the test formulation (μ_T) which is symmetric about the reference mean (μ_R), i.e.,

$$\mu_R - \Delta < \mu_T < \mu_R + \Delta. \tag{4.2.7}$$

The inequality in (4.2.5) can be rearranged as:

$$\mu_R + k_2\hat{\sigma}_d\sqrt{\frac{1}{n_1} + \frac{1}{n_2}} - (\overline{Y}_R - \overline{Y}_T) < \mu_T$$

$$< \mu_R + k_1\hat{\sigma}_d\sqrt{\frac{1}{n_1} + \frac{1}{n_2}} - (\overline{Y}_R - \overline{Y}_T).$$

Hence,

$$\Delta = k_1 \hat{\sigma}_d \sqrt{\frac{1}{n_1} + \frac{1}{n_2}} - (\overline{Y}_R - \overline{Y}_T)$$

$$= -k_2 \hat{\sigma}_d \sqrt{\frac{1}{n_1} + \frac{1}{n_2}} + (\overline{Y}_R - \overline{Y}_T).$$

This implies that

$$(k_1 + k_2)\hat{\sigma}_d \sqrt{\frac{1}{n_1} + \frac{1}{n_2}} = 2(\overline{Y}_R - \overline{Y}_T). \qquad (4.2.8)$$

The test formulation is concluded to be bioequivalent to the reference formulation according to the ± 20 rule if $|\Delta| < 0.2\mu_R$. To determine the values of k_1 and k_2, an iterative method is needed to solve the equation

$$\int_{k_2}^{k_1} T \, dt = 1 - 2\alpha$$

for k_1 and k_2 under constraint (4.2.8). This can be done using some statistical software such as SAS. In Appendix B, a SAS program for determination of k_1 and k_2 is given.

Example 4.2.2   For the example in Section 3.6, we can apply the SAS program to determine k_1 and k_2. After several iterations, k_1 and k_2 in (4.2.8) can be found to be 2.599 and -1.373, respectively. This gives Δ a value of 7.413. Thus, Westlake's symmetric confidence interval for the true test formulation mean are within 7.413 (or 8.98%) of the true reference mean. Hence, we conclude that the test formulation is bioequivalent to the reference formulation according to the ± 20 rule.

Let p_w be the coverage probability of Westlake's symmetric confidence interval. It was shown (Westlake, 1976) that

$$\lim_{|\mu_T - \mu_R| \mapsto \infty} p_w = 1 - 2\alpha \leqslant p_w \leqslant \lim_{|\mu_T - \mu_R| \mapsto 0} p_w = 1. \qquad (4.2.9)$$

Thus, Westlake's symmetric confidence interval approach is somewhat conservative in the determination of bioequivalence since it has at least $1 - 2\alpha$ coverage probability.

The concept of using a symmetric confidence interval for assessing bioequivalence has been discussed and criticized by many researchers since being introduced by Westlake (1976). For examples, see Mantel (1977), Kirkwood (1981), Mandallaz and Mau (1981), Shirley (1976). Steinijans and Diletti (1983), and Wijnand and Timmer (1983). Among these criticisms, the following are probably the most common:

(i) Since Westlake's symmetric confidence interval is symmetric about μ_R rather than $\overline{Y}_T - \overline{Y}_R$, the interval is, in fact, shifted away from the direction in which the sample difference was observed;

(ii) The tail probabilities associated with Westlake's symmetric confidence interval is not symmetric. Westlake's symmetric confidence interval moves from a two-sided to a one-sided approach as $\mu_T - \mu_R$ and/or σ_e^2 increases.

Metzler (1988) argued that Westlake's symmetric confidence interval should be used as a tool for decision-making rather than for estimation or testing. It, however, should be recognized that the assessment of bioequivalence depends on the results from valid statistical inference such as estimation and/or interval hypotheses testing, which should not be separated from the decision-making.

4.2.3. Confidence Interval Based on Fieller's Theorem

It can be seen that both the classical confidence interval and Westlake's symmetric confidence interval for the ratio μ_T/μ_R do not take into account the variability of $\overline{Y}_R$ and the correlation between $\overline{Y}_R$ and $\overline{Y}_T - \overline{Y}_R$. To account for the variability of $\overline{Y}_R$, Locke (1984) proposed an approach for constructing a $(1 - 2\alpha) \times 100\%$ confidence interval for μ_T/μ_R based on Fieller's theorem (Fieller, 1954). This method has become very attractive because (i) it provides an exact $(1 - 2\alpha) \times 100\%$ confidence interval for μ_T/μ_R, and (ii) it does take into account the variability of $\overline{Y}_R$. Recall that in Chapter 2, the responses for the ith subject in sequence 1 at periods I and II, (Y_{i11}, Y_{i21}), had a bivariate normal distribution with mean vector α_1 and covariance Σ_1, where

$$\alpha_1 = \begin{bmatrix} \mu + F_R + P_1 \\ \mu + F_T + P_2 \end{bmatrix} \quad \text{and} \quad \Sigma_1 = \begin{bmatrix} \sigma_R^2 + \sigma_s^2 & \sigma_s^2 \\ \sigma_s^2 & \sigma_T^2 + \sigma_s^2 \end{bmatrix}.$$

Similarly, (Y_{i12}, Y_{i22}) are i.i.d. normal with mean vector α_2 and covariance Σ_2, where

$$\alpha_2 = \begin{bmatrix} \mu + F_T + P_1 \\ \mu + F_R + P_2 \end{bmatrix} \quad \text{and} \quad \Sigma_2 = \begin{bmatrix} \sigma_T^2 + \sigma_s^2 & \sigma_s^2 \\ \sigma_s^2 & \sigma_R^2 + \sigma_s^2 \end{bmatrix}.$$

Let $\delta = \mu_T/\mu_R = (\mu + F_T)/(\mu + F_R)$. Define

$$U_{ik}^* = \begin{cases} \frac{1}{2}(Y_{i21} - \delta Y_{i11}), & i = 1, 2, \ldots, n_1, \quad k = 1 \\ \frac{1}{2}(Y_{i12} - \delta Y_{i22}), & i = 1, 2, \ldots, n_2, \quad k = 2. \end{cases} \quad (4.2.10)$$

Then U_{ii}^* are i.i.d. normal with mean $(P_2 - \delta P_1)/2$ and variance $\sigma_\delta^2/4$, where

$$\sigma_\delta^2 = (\sigma_T^2 + \sigma_s^2) - 2\delta\sigma_s^2 + \delta^2(\sigma_R^2 + \sigma_s^2), \quad \text{for } i = 1,2, \ldots, n_1,$$

and U_{i2}^* are i.i.d. normal with mean $(P_1 - \delta P_2)/2$ and variance $\sigma_\delta^2/4$, for $i = 1, 2, \ldots, n_2$. $\{U_{i1}^*, i = 1, 2, \ldots, n_1\}$ and $\{U_{i2}^*, i = 1, 2, \ldots, n_2\}$ are two independent samples from normal distributions with equal variances. Thus, a $(1 - 2\alpha) \times 100\%$ confidence interval for δ can be obtained based on a two-sample t statistic. First, let $\overline{U}_{\cdot k}^*$ be the sample means of U_{ik}^* for $k = 1, 2$ and S_u^2 be the pooled sample variance of U_{ik}^*, i.e.,

$$\overline{U}_{\cdot k}^* = \frac{1}{n_k} \sum_{i=1}^{n_k} U_{ik}^*, \quad \text{and} \tag{4.2.11}$$

$$S_u^2 = \frac{1}{n_1 + n_2 - 2} \sum_{k=1}^{2} \sum_{i=1}^{n_k} (U_{ik} - \overline{U}_{\cdot k})^2. \tag{4.2.12}$$

Note that $(\overline{U}_{\cdot 1}^* + \overline{U}_{\cdot 2}^*)$ is normally distributed with mean 0 and variance $\omega \sigma_\delta^2$, where

$$\omega = \frac{1}{4}\left(\frac{1}{n_1} + \frac{1}{n_2}\right).$$

Therefore, the statistic

$$T = \frac{\overline{U}_{\cdot 1}^* + \overline{U}_{\cdot 2}^*}{\sqrt{\omega S_u^2}}$$

has a central t distribution with $n_1 + n_2 - 2$ degrees of freedom. However, $\overline{U}_{\cdot 1}^* + \overline{U}_{\cdot 2}^*$ can be expressed in terms of the least squares means for the test and reference formulations as follows:

$$
\begin{aligned}
\overline{U}_{\cdot 1}^* + \overline{U}_{\cdot 2}^* &= \tfrac{1}{2}(\overline{Y}_{\cdot 21} - \delta\,\overline{Y}_{\cdot 11}) + \tfrac{1}{2}(\overline{Y}_{\cdot 12} - \delta\,\overline{Y}_{\cdot 22}) \\
&= \tfrac{1}{2}(\overline{Y}_{\cdot 21} + \overline{Y}_{\cdot 12}) - \delta \tfrac{1}{2}(\overline{Y}_{\cdot 11} + \overline{Y}_{\cdot 22}) \\
&= \overline{Y}_T - \delta\overline{Y}_R.
\end{aligned}
\tag{4.2.13}
$$

Similarly, it can be verified that

$$S_u^2 = S_{TT}^2 - 2\delta S_{TR} + \delta^2 S_{RR}^2,$$

where

$$S_{RR}^2 = \frac{1}{n_1 + n_2 - 2}\left[\sum_{i=1}^{n_1}(Y_{i11} - \overline{Y}_{\cdot 11})^2 + \sum_{i=1}^{n_2}(Y_{i22} - \overline{Y}_{\cdot 22})^2\right],$$

$$S_{TT}^2 = \frac{1}{n_1 + n_2 - 2}\left[\sum_{i=1}^{n_1}(Y_{i21} - \overline{Y}_{\cdot 21})^2 + \sum_{i=1}^{n_2}(Y_{i12} - \overline{Y}_{\cdot 12})^2\right],$$

$$S_{TR} = \frac{1}{n_1 + n_2 - 2} \left[\sum_{i=1}^{n_1} (Y_{i11} - \overline{Y}_{\cdot 11})(Y_{i21} - \overline{Y}_{\cdot 21}) \right.$$

$$\left. + \sum_{i=1}^{n_2} (Y_{i12} - \overline{Y}_{\cdot 12})(Y_{i22} - \overline{Y}_{\cdot 22}) \right]$$

Thus, T can be rewritten as

$$T = \frac{\overline{Y}_T - \delta \overline{Y}_R}{\sqrt{\omega(S_{TT}^2 - 2\delta S_{TR} + \delta^2 S_{RR}^2)}}. \tag{4.2.14}$$

Hence, a $(1 - 2\alpha) \times 100\%$ confidence interval for δ can be obtained as follows:

$$\{\delta | T^2 \le t^2(\alpha, n_1 + n_2 - 2)\}. \tag{4.2.15}$$

The lower and upper limits of the $(1 - 2\alpha) \times 100\%$ confidence interval for δ, if they exist, are the two roots of the following quadratic equation

$$(\overline{Y}_T - \delta \overline{Y}_R)^2 - t^2(\alpha, n_1 + n_2 - 2)\omega(S_{TT}^2 - 2\delta S_{TR} + \delta^2 S_{RR}^2), \tag{4.2.16}$$

which are given by

$$L_3 = \frac{1}{1 - G} \left[\left\{ \frac{\overline{Y}_T}{\overline{Y}_R} - G \frac{S_{TR}}{S_{RR}^2} \right\} - \left\{ t(\alpha, n_1 + n_2 - 2) \frac{\sqrt{\omega S_{RR}^2}}{\overline{Y}_R} G^* \right\} \right]$$

$$U_3 = \frac{1}{1 - G} \left[\left\{ \frac{\overline{Y}_T}{\overline{Y}_R} - G \frac{S_{TR}}{S_{RR}^2} \right\} + \left\{ t(\alpha, n_1 + n_2 - 2) \frac{\sqrt{\omega S_{RR}^2}}{\overline{Y}_R} G^* \right\} \right], \tag{4.2.17}$$

where

$$G = \{t(\alpha, n_1 + n_2 - 2)\}^2 \left[\frac{\omega S_{RR}^2}{\overline{Y}_R^2} \right],$$

and

$$G^{*2} = \left\{ \frac{\overline{Y}_T}{\overline{Y}_R} \right\}^2 + \frac{S_{TT}^2}{S_{RR}^2} (1 - G) + \frac{S_{TR}}{S_{RR}^2} \left\{ G \frac{S_{TR}}{S_{RR}^2} - 2 \frac{\overline{Y}_T}{\overline{Y}_R} \right\}.$$

As indicated by Fieller (1954), Kendall and Stuart (1961), and Locke (1984), the solutions to the quadratic equation (4.2.16) may not result in an actual interval. The conditions for both L_3 and U_3 to be finite positive real numbers are

(a) $\dfrac{\overline{Y}_R}{\sqrt{\omega S_{RR}^2}} > t(\alpha, n_1 + n_2 - 2)$, and

(b) $\dfrac{\overline{Y}_T}{\sqrt{\omega S_{TT}^2}} > t(\alpha, n_1 + n_2 - 2)$. \hfill (4.2.18)

In other words, it requires that the least squares means for both the test and the reference formulations be statistically significantly greater than 0 at the α level of significance. Condition (a) in (4.2.18) implies that if the variability of the reference formulation is large enough for (a) to be false, then there may not exist an interval for δ or the interval may contain imaginary numbers. On the other hand, the interval may contain negative values if the variability of the test formulation is sufficiently large for condition (b) to be false even when the reference formulation satisfies condition (a). Thus, in practice, we conclude bioequivalence if

    (1)   Both conditions (a) and (b) of (4.2.18) hold, and
    (2) $L_3 > 80\%$ and $U_3 < 120\%$;

Otherwise, we are in favor of bioinequivalence.

The exact $(1 - 2\alpha) \times 100\%$ confidence interval for μ_T/μ_R based on Fieller's theorem takes into account not only the variability of $\overline{Y}_R$ but also the inter-subject variability as well. In addition, it was derived under a very mild assumption which only requires the normality of the two responses observed on the same subject. Hence, even when $\sigma_T^2 \neq \sigma_R^2$, the exact $(1 - 2\alpha) \times 100\%$ confidence interval for μ_T/μ_R by Fieller's theorem is still valid.

Example 4.2.3   For the example in Section 3.6, it can be verified that both conditions (a) and (b) of (4.2.18) are satisfied, i.e.,

$$\frac{\overline{Y}_R}{\sqrt{\omega S_{RR}^2}} = 19.27 \quad \text{and} \quad \frac{\overline{Y}_T}{\sqrt{\omega S_{TT}^2}} = 18.73,$$

which are both greater than $t(0.05,22) = 1.717$. Thus, Fieller's theorem will give a positive real interval. It can be found that $G = 0.003$ and $G^* = 0.735$ which leads to a confidence interval for the ratio of $(89.78\%, 105.19\%)$. Thus, we conclude the two formulations are bioequivalent.

Schuirmann (1989) considered a special case of (4.2.15) by assuming that $\sigma_R^2 = \sigma_T^2$ and $\sigma_s^2 = 0$ (i.e., the subject effects are assumed to be fixed). Schuirmann (1989) refers to such an approach as the fixed Fieller's confidence interval approach. In this case, the lower and upper limits of $(1 - 2\alpha) \times 100\%$ confidence interval for δ become

$$L_4 = \frac{1}{1 - G_1}\left[\frac{\overline{Y}_T}{\overline{Y}_R} - t(\alpha, n_1 + n_2 - 2)\right.$$
$$\left. \frac{\sqrt{\omega MS_{Intra}}}{\overline{Y}_R}\sqrt{\left(\frac{\overline{Y}_T}{\overline{Y}_R}\right)^2 + (1 - G_1)}\right]$$

$$U_4 = \frac{1}{1 - G_1} \left[\frac{\overline{Y}_T}{\overline{Y}_R} + t(\alpha, n_1 + n_2 - 2) \right.$$

$$\left. \frac{\sqrt{\omega MS_{Intra}}}{\overline{Y}_R} \sqrt{\left(\frac{\overline{Y}_T}{\overline{Y}_R}\right)^2 + (1 - G_1)} \right],$$

$$(4.2.19)$$

where

$$G_1 = [t(\alpha, n_1 + n_2 - 2)]^2 \left(\frac{\omega MS_{Intra}}{\overline{Y}_R^2}\right).$$

It can be verified that there exists no interval for the fixed Fieller's confidence interval unless $G_1 < 1$.

For the calculation of the lower and upper bound of Fieller's confidence interval, a SAS program is developed which is given in Appendix B.1. The SAS program also includes the calculation of the above fixed Fieller's confidence interval.

Yee (1986) also attempted to account for the inter-subject variability for construction of a confidence interval for the ratio based on the Fieller's theorem through a linear combination of the inter-subject and intra-subject mean squared errors by a Satterthwaite approximation. His approach, however, is not an exact procedure but an approximation since the procedure involves the substitution of an estimate of the intra-subject correlation $\sigma_s^2/(\sigma_s^2 + \sigma_e^2)$ (Schiurmann, 1989).

4.2.4. Chow and Shao's Joint Confidence Region

Instead of constructing $(1 - 2\alpha) \times 100\%$ confidence intervals for $\mu_T - \mu_R$ and/or μ_T/μ_R, Chow and Shao (1990) proposed an alternative approach for assessing bioequivalence by constructing an exact $(1 - 2\alpha) \times 100\%$ confidence region for (μ_R, μ_T). If the constructed confidence region is within the following bioequivalence region:

$$\{(\mu_R, \mu_T) \mid \delta_L < \mu_T/\mu_R < \delta_U\},$$

then we conclude that the two formulations are bioequivalent. In the following, the method will be derived under model (4.1.2). Since S_{ik} consists of the fixed effect of the kth sequence and the random effect of the ith subject in the kth sequence, model (4.1.2) can be rewritten as

$$Y_{ijk} = \mu + S_{ik} + F_j + P_{(j,k)} + e_{ijk}$$
$$= \mu + G_k + S_{i(k)} + F_j + P_{(j,k)} + e_{ijk}, \qquad (4.2.20)$$

where Y_{ijk}, μ, F_j, $P_{(j,k)}$ and e_{ijk} are defined as before; G_k is the fixed effect of the kth sequence; $S_{i(k)}$ is the random effect of the ith subject nested in the kth sequence. Thus,

$$\mu_R = \mu + F_R \quad \text{and} \quad \mu_T = \mu + F_T.$$

Under the normality assumptions,

$$\mathbf{X}_{ik} = \begin{bmatrix} Y_{i1k} \\ Y_{i2k} \end{bmatrix} \sim N(\boldsymbol{\mu}_k, \Sigma)$$

and $\mathbf{X}_{ik}$, $i = 1, 2, \ldots, n_k$, $k = 1, 2$, are mutually independent, where $\Sigma > 0$ is the covariance matrix and

$$\boldsymbol{\mu}_1 = \begin{bmatrix} \mu_R + P_1 + G_1 \\ \mu_T + P_2 + G_1 \end{bmatrix} \quad \text{and} \quad \boldsymbol{\mu}_2 = \begin{bmatrix} \mu_R + P_2 + G_2 \\ \mu_T + P_1 + G_2 \end{bmatrix}. \tag{4.2.21}$$

Chow and Shao (1990) suggested testing the presence of the joint period and sequence effects as the first step for the assessment of bioequivalence. That is, the hypotheses

$$H_0: \boldsymbol{\mu}_1 = \boldsymbol{\mu}_2 \quad \text{vs} \quad H_a: \boldsymbol{\mu}_1 \neq \boldsymbol{\mu}_2 \tag{4.2.22}$$

are tested using the following test statistic

$$T = \frac{n_1 n_2}{2(n_1 + n_2)} (n_1 + n_2 - 3) (\overline{\mathbf{X}}_{\cdot 1} - \overline{\mathbf{X}}_{\cdot 2})' (S_1 + S_2)^{-1} (\overline{\mathbf{X}}_{\cdot 1} - \overline{\mathbf{X}}_{\cdot 2}),$$

where

$$\overline{\mathbf{X}}_{\cdot k} = \frac{1}{n_k} \sum_{i=1}^{n_k} \mathbf{X}_{ik},$$

$$S_k = \sum_{i=1}^{n_k} (\mathbf{X}_{ik} - \overline{\mathbf{X}}_{\cdot k}) (\mathbf{X}_{ik} - \overline{\mathbf{X}}_{\cdot k})', \qquad k = 1, 2.$$

We reject H_0 in (4.2.22) at the α level of significance if

$$T \geq F(1 - \alpha, 2, n_1 + n_2 - 3),$$

where $F(1 - \alpha, 2, n_1 + n_2 - 3)$ is the $1 - \alpha$ quantile of the F distribution with 2 and $n_1 + n_2 - 3$ degrees of freedom.

If we fail to reject H_0 in (4.2.22), then $\mathbf{X}_{ik}$, $i = 1, 2, \ldots, n_k$, $k = 1, 2$ have the same distribution. Let

$$S = \sum_{k=1}^{2} \sum_{i=1}^{n_k} (\mathbf{X}_{ik} - \overline{\mathbf{X}}_{\cdot\cdot})(\mathbf{X}_{ik} - \overline{\mathbf{X}}_{\cdot\cdot})',$$

where

$$\overline{X}_{..} = \frac{1}{n_1 + n_2} \sum_{k=1}^{2} \sum_{i=1}^{n_k} X_{ik} = \begin{bmatrix} \overline{Y}_R^* \\ \overline{Y}_T^* \end{bmatrix},$$

where $\overline{Y}_R^*$ and $\overline{Y}_T^*$ are the direct sample means for the reference and the test formulation, respectively. Then, S and $\overline{X}_{..}$ are independent and

$$T_1 = \frac{(n_1 + n_2)(n_1 + n_2 - 2)}{2} (\overline{X}_{..} - \mu)' S^{-1}(\overline{X}_{..} - \mu)$$

has an F distribution with degrees of freedom 2 and $n_1 + n_2 - 2$ and $\mu = (\mu_R, \mu_T)'$. Then, an exact $(1 - 2\alpha) \times 100\%$ confidence region for μ is

$$\{(\mu_R, \mu_T) \mid T_1 < F(1 - 2\alpha, 2, n_1 + n_2 - 2)\}. \tag{4.2.23}$$

Note that the confidence region is the interior of an ellipse on a 2×2 plane. Bioequivalence is claimed if and only if this ellipse is within the region bounded by two lines $\mu_T = \delta_L \mu_R$ and $\mu_T = \delta_U \mu_R$ (see Figure 4.2.1). When bioequivalence is claimed, we have $1 - 2\alpha$ assurance that the true μ_T/μ_R is within (δ_L, δ_U).

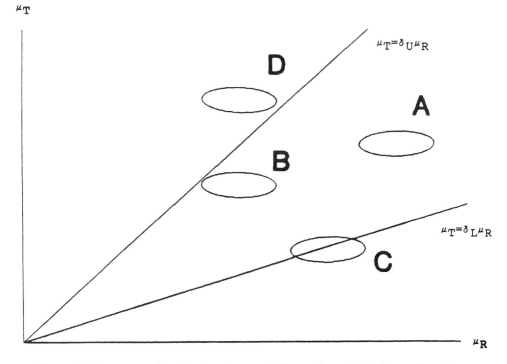

Figure 4.2.1 Note that the locations of ellipses A and B indicate that the two formulations are bioequivalent, while locations C and D indicate that the two formulations are not bioequivalent.

When Y_{ijk} are observed and δ_L and δ_U are given, we can check whether the confidence region (4.2.23) is within $\mu_T = \delta_L\mu_R$ and $\mu_T = \delta_U\mu_R$ as follows. Let

$$S_{pq} = \sum_{k=1}^{2} \sum_{i=1}^{n_k} (Y_{ipk} - \overline{Y}_{.p.})(Y_{iqk} - \overline{Y}_{.q.}), \qquad p,q = 1,2.$$

Then

$$S = \begin{bmatrix} S_{11} & S_{12} \\ S_{12} & S_{22} \end{bmatrix} \quad \text{and} \quad S^{-1} = \frac{1}{S_{11}S_{22} - S_{12}S_{12}} \begin{bmatrix} S_{22} & -S_{12} \\ -S_{12} & S_{11} \end{bmatrix}.$$

Thus, (4.2.23) is equivalent to

$$S_{22}(\overline{Y}_R^* - \mu_R)^2 + S_{11}(\overline{Y}_T^* - \mu_T)^2 - 2S_{12}(\overline{Y}_R^* - \mu_R)(\overline{Y}_T^* - \mu_T)$$
$$< \frac{2(S_{11}S_{22} - S_{12}S_{12})}{(n_1 + n_2)(n_1 + n_2 - 2)} F(1 - 2\alpha, 2, n_1 + n_2 - 2) \equiv d.$$

Therefore, the two formulations are bioequivalent, i.e., the ellipse defined in (4.2.23) is within the region bounded by $\mu_T = \delta_L\mu_R$ and $\mu_T = \delta_U\mu_R$ if and only if the following conditions hold simultaneously:

(1) $\delta_L < \overline{Y}_T^*/\overline{Y}_R^* < \delta_U$;
(2) The following equations have at most one solution:

$$\begin{cases} \mu_T = \delta_U\mu_R \\ S_{22}(\overline{Y}_R^* - \mu_R)^2 + S_{11}(\overline{Y}_T^* - \mu_T)^2 \\ - 2S_{12}(\overline{Y}_R^* - \mu_R)(\overline{Y}_T^* - \mu_T) = d; \end{cases}$$

(3) The following equations have at most one solution:

$$\begin{cases} \mu_T = \delta_L\mu_R \\ S_{22}(\overline{Y}_R^* - \mu_R)^2 + S_{11}(\overline{Y}_T^* - \mu_T)^2 \\ - 2S_{12}(\overline{Y}_R^* - \mu_R)(\overline{Y}_T^* - \mu_T) = d. \end{cases}$$

Conditions (1) through (3) are equivalent to the following:

(1′) $\delta_L < \overline{Y}_T^*/\overline{Y}_R^* < \delta_U$;
(2′) $(S_{22}\overline{Y}_R^* + \delta_U S_{11}\overline{Y}_T^* - \delta_U S_{12}\overline{Y}_R^* - S_{12}\overline{Y}_T^*)^2$
$\leqslant (S_{22} + \delta_U^2 S_{11} - 2\delta_U S_{12})(S_{22}\overline{Y}_R^{*2} + S_{11}\overline{Y}_T^{*2} - 2S_{12}\overline{Y}_R^*\overline{Y}_T^* - d)$;
(3′) $(S_{22}\overline{Y}_R^* + \delta_L S_{11}\overline{Y}_T^* - \delta_L S_{12}\overline{Y}_R^* - S_{12}\overline{Y}_T^*)^2$
$\leqslant (S_{22} + \delta_L^2 S_{11} - 2\delta_L S_{12})(S_{22}\overline{Y}_R^{*2} + S_{11}\overline{Y}_T^{*2} - 2S_{12}\overline{Y}_R^*\overline{Y}_T^* - d)$.

Thus, we conclude that the two formulations are bioequivalent if and only if conditions (1′) through (3′) hold.

In the case where H_0 in (4.2.22) is rejected, $\overline{\mathbf{X}}..$ and $\mathbf{S}$ are not independent. However, in this case,

$$E[(\overline{\mathbf{X}}._1 + \overline{\mathbf{X}}._2)/2] = \boldsymbol{\mu},$$

where $(\overline{\mathbf{X}}._1 + \overline{\mathbf{X}}._2)/2 = (\overline{Y}_R, \overline{Y}_T)'$, the vector of the LS means of the reference and test formulations. It can be verified that $(\overline{\mathbf{X}}._1 + \overline{\mathbf{X}}._2)/2$ and $\mathbf{S}_1 + \mathbf{S}_2$ are independent. Therefore,

$$T_2 = \frac{2n_1n_2(n - 3)}{n} \{(\overline{\mathbf{X}}._1 + \overline{\mathbf{X}}._2)/2 - \boldsymbol{\mu}\}'$$
$$(\mathbf{S}_1 + \mathbf{S}_2)^{-1}\{(\overline{\mathbf{X}}._1 + \overline{\mathbf{X}}._2)/2 - \boldsymbol{\mu}\},$$

where $n = n_1 + n_2$ and T_2 has an F distribution with degrees of freedom 2 and $n_1 + n_2 - 3$. An exact $(1 - 2\alpha) \times 100\%$ confidence region for $\boldsymbol{\mu}$ is

$$\{(\mu_R, \mu_T)| \; T_2 < F(1 - 2\alpha, 2, n_1 + n_2 - 3)\}. \tag{4.2.24}$$

Bioequivalence is claimed if and only if the ellipse (4.2.24) is within the region bounded by two lines $\mu_T = \delta_U\mu_R$ and $\mu_T = \delta_L\mu_R$. Thus, we conclude bioequivalence if conditions (1') through (3') hold with $\overline{Y}_R^*$, $\overline{Y}_T^*$, S_{pq} and d replaced by $\overline{Y}_R$, $\overline{Y}_T$, S'_{pq} and d', where $\overline{Y}_R$ and $\overline{Y}_T$ are the LS means for the reference and test formulations, respectively, and

$$S'_{pq} = \sum_{k=1}^{2} \sum_{i=1}^{n_k} (Y_{ipk} - \overline{Y}._{pk}')(Y_{iqk} - \overline{Y}._{qk}'), \qquad p,q = 1,2;$$

and

$$d' = \frac{n(S'_{11}S'_{22} - S'_{12}S'_{12})}{2n_1n_2(n - 3)} F(1 - 2\alpha, 2, n - 3).$$

Example 4.2.4 In order to apply the above confidence region approach to the example in Section 3.6, first, the test for the presence of the period and sequence effects is given by

$$T = 3.18 < F(0.95, 2, 21) = 3.47.$$

Therefore, there is no joint presence of period and sequence effects. Furthermore, we have

$$S_{11} = 432.54, \; S_{12} = 277.65, \; S_{22} = 445.85, \quad \text{and} \quad d = 1521.40.$$

The observed ratio $\overline{Y}_T^*/\overline{Y}_R^*$ is 97.23%. For $\delta_L = 0.8$ and $\delta_U = 1.2$, it can be verified that the conditions (1') through (3') are satisfied. Hence, we conclude that the two formulations are bioequivalent based on the ± 20 rule.

4.3. THE METHODS OF INTERVAL HYPOTHESES TESTING

4.3.1. Interval Hypotheses

As mentioned earlier, the assessment of bioequivalence is based on the comparison of bioavailability profiles between formulations. However, in practice, it is recognized that no two formulations will have exactly the same bioavailability profiles. Therefore, if the profiles of the two formulations differ by less than a (clinically) meaningful limit, the profiles of the two formulations may be considered equivalent. Based on this idea, Schuirmann (1981) first introduced the use of interval hypotheses for assessing bioequivalence.

The interval hypotheses for bioequivalence can be formulated as

$$H_0: \mu_T - \mu_R \leq \theta_L \quad \text{or} \quad \mu_T - \mu_R \geq \theta_U$$
$$\text{vs} \quad H_a: \theta_L < \mu_T - \mu_R < \theta_U \tag{4.3.1}$$

where θ_L and θ_U are some clinically meaningful limits. The concept of interval hypotheses (4.3.1) is to show bioequivalence by rejecting the null hypothesis of bioinequivalence. In most bioavailability/bioequivalence studies, θ_L and θ_U are often chosen to be 20% of the reference mean (μ_R). When the natural logarithmic transformation of the data is considered, the hypotheses corresponding to (4.3.1) can be stated as

$$H_0': \mu_T/\mu_R \leq \delta_L \quad \text{or} \quad \mu_T/\mu_R \geq \delta_U$$
$$\text{vs} \quad H_a': \delta_L < \mu_T/\mu_R < \delta_U \tag{4.3.2}$$

where $\delta_L = \exp(\theta_L)$ and $\delta_U = \exp(\theta_U)$.

Note that the test for hypotheses (4.3.2) formulated on the log scale is equivalent to testing for hypotheses (4.3.1) on the raw scale. The interval hypotheses (4.3.1) can be decomposed into two sets of one-sided hypotheses

$$H_{01}: \mu_T - \mu_R \leq \theta_L$$
$$\text{vs} \quad H_{a1}: \mu_T - \mu_R > \theta_L$$

and

$$H_{02}: \mu_T - \mu_R \geq \theta_U$$
$$\text{vs} \quad H_{a2}: \mu_T - \mu_R < \theta_U. \tag{4.3.3}$$

The first set of hypotheses is to verify that the bioavailability of the test formulation is not too low, while the second set of hypotheses is to verify that the bioavailability of the test formulation is not too high. A relatively low (or high) bioavailability may refer to the concern of efficacy (or safety) of the test formulation. If one concludes that $\theta_L < \mu_T - \mu_R$ (i.e., reject H_{01}) and $\mu_T - \mu_R < \theta_U$ (i.e., reject H_{02}), then it has been concluded that

$$\theta_L < \mu_T - \mu_R < \theta_U.$$

μ_T and μ_R, thus, are equivalent. The rejection of H_{01} and H_{02}, which leads to the conclusion of bioequivalence, is equivalent to rejecting H_0 in (4.3.1).

4.3.2. Schuirmann's Two One-Sided Tests Procedure

Schuirmann (1981, 1987) first introduced the two one-sided tests procedure based on (4.3.3) for assessing bioequivalence between formulations. The proposed two one-sided procedure suggests the conclusion of equivalence of μ_T and μ_R at the α level of significance if and only if H_{01} and H_{02} in (4.3.3) are rejected at a predetermined α level of significance. Under the normality assumptions, the two sets of one-sided hypotheses can be tested with ordinary one-sided t tests. We conclude that μ_T and μ_R are equivalent if

$$T_L = \frac{(\bar{Y}_T - \bar{Y}_R) - \theta_L}{\hat{\sigma}_d \sqrt{\dfrac{1}{n_1} + \dfrac{1}{n_2}}} > t(\alpha, n_1 + n_2 - 2)$$

and

$$T_U = \frac{(\bar{Y}_T - \bar{Y}_R) - \theta_U}{\hat{\sigma}_d \sqrt{\dfrac{1}{n_1} + \dfrac{1}{n_2}}} < -t(\alpha, n_1 + n_2 - 2). \qquad (4.3.4)$$

It should be noted that the two one-sided t tests procedure is operationally equivalent to the classical (shortest) confidence interval approach, i.e., if the classical $(1 - 2\alpha) \times 100\%$ confidence interval for $\mu_T - \mu_R$ is within (θ_L, θ_U), then both H_{01} and H_{02} are also rejected at the α level by the two one-sided t tests procedure.

Example 4.3.2 For the example in Section 3.6, we have

$$\bar{Y}_T = 80.272, \qquad \bar{Y}_R = 82.559, \quad \text{and} \quad \hat{\sigma}_d^2 = 83.623.$$

Assuming that $\bar{Y}_R$ is the true reference mean, then the lower and upper bioequivalent limits are $-\theta_L = \theta_U = 16.51$. The two sets of one-sided hypotheses corresponding to the interval hypotheses (4.3.1) are

$$H_{01}: \mu_T - \mu_R \leq -16.51 \quad \text{vs} \quad H_{a1}: \mu_T - \mu_R > -16.51$$

and

$$H_{02}: \mu_T - \mu_R \geq 16.51 \quad \text{vs} \quad H_{a2}: \mu_T - \mu_R < 16.51.$$

Thus,

$$T_L = \frac{(80.272 - 82.559) + 16.51}{(9.145)\sqrt{0.167}} = 3.810,$$

and

$$T_U = \frac{(80.272 - 82.559) - 16.51}{(9.145)\sqrt{0.167}} = -5.036.$$

Since $|T_L|$ and $|T_U|$ are both greater than $t(0.05,22) = 1.717$, the null hypotheses H_{01} and H_{02} are rejected at the 5% level of significance. Hence, bioequivalence is claimed according to the ± 20 rule.

4.3.3.   Anderson and Hauck's Test

For testing the interval hypotheses in (4.3.1), unlike Schuirmann's two one-sided tests procedure, which uses statistics T_L and T_U to evaluate H_{01} and H_{02} rather than testing H_0 directly, Anderson and Hauck (1983) proposed a test statistic which can be used to evaluate H_0 directly. In other words, we would reject the null hypothesis H_0 of bioinequivalence in favor of bioequivalence for a small p-value. The test statistic is given below:

$$T_{AH} = \frac{\overline{Y}_T - \overline{Y}_R - (\theta_L + \theta_U)/2}{\hat{\sigma}_d\sqrt{\dfrac{1}{n_1} + \dfrac{1}{n_2}}}. \qquad (4.3.5)$$

Under the normality assumptions, T_{AH} follows a noncentral t distribution with noncentrality parameter

$$\delta = \frac{\mu_T - \mu_R - (\theta_L + \theta_U)/2}{\sigma_d\sqrt{\dfrac{1}{n_1} + \dfrac{1}{n_2}}}. \qquad (4.3.6)$$

where σ_d was defined in Chapter 3.

We reject H_0 in (4.3.1) in favor of bioequivalence if

$$C_1 < T_{AH} < C_2,$$

where C_1 and C_2 satisfy

$$P[C_1 < T_{AH} < C_2 \mid \mu_T - \mu_R = \theta_U, \sigma_d]$$
$$= P[C_1 < T_{AH} < C_2 \mid \mu_T - \mu_R = \theta_L, \sigma_d]$$
$$= \alpha. \qquad (4.3.7)$$

Since C_1 and C_2 can be chosen to be $C_2 = -C_1 = C$, (4.3.9) becomes

$$P\left[|T_{AH}| < C \mid \mu_T - \mu_R = \theta_U, \sigma_d\right]$$
$$= P\left[|T_{AH}| < C \mid \mu_T - \mu_R = \theta_L, \sigma_d\right]$$
$$= \alpha. \tag{4.3.8}$$

Thus, a α level rejection region can be determined by solving (4.3.8) for C. One of the difficulties of the above test, however, is that δ in (4.3.6) is usually unknown. When δ is known, an empirical significant p value can be obtained as follows:

$$p = P\left[|T_{AH}| < |t_{AH}| \mid \mu_T - \mu_R = \theta_U, \sigma_d\right], \tag{4.3.9}$$

where t_{AH} is the observed value of T_{AH}.

H_0 is rejected at the α level of significance whenever $p < \alpha$. On the other hand, when δ is unknown, three approximations to (4.3.9) based on noncentral t, central t and normal distribution are suggested. Among these approximations, as indicated by Anderson and Hauck (1983), the central t approximation appears to be the best in terms of test power. Thus, in the following, the central t approximation to (4.3.9) will be introduced. First, the noncentrality δ can be estimated by

$$\hat{\delta} = \frac{\theta_U - \theta_L}{2\hat{\sigma}_d\sqrt{\dfrac{1}{n_1} + \dfrac{1}{n_2}}}. \tag{4.3.10}$$

Then,

$$p = P\left[|T_{AH}| < |t_{AH}| \mid \mu_T - \mu_R = \theta_U\right]$$
$$= P\left[-|t_{AH}| - \hat{\delta} < T_{AH} - \hat{\delta} < |t_{AH}| - \hat{\delta}\right]$$
$$= F_t(|t_{AH}| - \hat{\delta}) - F_t(-|t_{AH}| - \hat{\delta}), \tag{4.3.11}$$

where

$$T_{AH} - \hat{\delta} = \frac{(\bar{Y}_T - \bar{Y}_R) - \theta_U}{\hat{\sigma}_d\sqrt{\dfrac{1}{n_1} + \dfrac{1}{n_2}}},$$

and F_t is the central t distribution function with degrees of freedom $n_1 + n_2 - 2$.

Note that the approximation in (4.3.11) can also be obtained by using T_L and T_U of Schuirmann's two one-sided t statistics in (4.3.4). Assuming that $t_{AH} > 0$, then

$$|t_{AH}| - \hat{\delta} = \frac{(\bar{Y}_T - \bar{Y}_R) - (\theta_U + \theta_L)/2 - (\theta_U - \theta_L)/2}{\hat{\sigma}_d\sqrt{\dfrac{1}{n_1} + \dfrac{1}{n_2}}}$$

$$= \frac{(\bar{Y}_T - \bar{Y}_R) - \theta_U}{\hat{\sigma}_d\sqrt{\dfrac{1}{n_1} + \dfrac{1}{n_2}}} = T_U.$$

Similarly, $-|t_{AH}| - \hat{\delta} = -T_L$.
Therefore,

$$p = F_t(|t_{AH}| - \hat{\delta}) - F_t(-|t_{AH}| - \hat{\delta})$$
$$= F_t(T_U) - F_t(-T_L). \tag{4.3.12}$$

If $t_{AH} < 0$, then $|t_{AH}| - \hat{\delta} = -T_L$ and $-|t_{AH}| - \hat{\delta} = T_U$. Thus,

$$p = F_t(|t_{AH}| - \hat{\delta}) - F_t (-|t_{AH}| - \hat{\delta})$$
$$= F_t(-T_L) - F_t(T_U). \tag{4.3.13}$$

From (4.3.12) and (4.3.13), Anderson and Hauck's procedure is always more powerful than Schuirmann's procedure because p may still be smaller than α even if the p-values for T_L and T_U are greater than α. It should be noted that Anderson and Hauck (1983) claim t_{AH} is the most powerful test, but as a matter of fact, under a normality assumption there exist no unconditional most powerful, uniformly most powerful, or uniformly powerful unbiased test for the interval hypotheses in (4.3.1) when μ_T, μ_R and σ_d^2 are unknown (Lehmann, 1959; Kendall and Stuart, 1961).

In a simulation study, Anderson and Hauck (1983) showed that test procedure (4.3.11) is a more (not the most) powerful test (in the sense of having higher probability of concluding equivalence), which uniformly dominates both the classical (shortest) confidence interval approach and Westlake's symmetric confidence interval method. In addition, compared to Schuirmann's two one-sided tests procedure, Anderson and Hauck's procedure is always more powerful. This is true when μ_T and μ_R are not equivalent as well as when they are equivalent. The difference in power between the two procedures, however, becomes negligible as the measure of sensitivity

$$\nabla = \frac{\theta_U - \theta_L}{\sigma_d\sqrt{\dfrac{1}{n_1} + \dfrac{1}{n_2}}}$$

becomes larger. The drawback of Anderson and Hauck's test is probably that the true level of significance may exceed the nominal level α, particularly for small degrees of freedom associated with the mean squared error. In addition, Anderson and Hauck's test may still conclude bioequivalence even when the intra-subject variability becomes very large because it has an open-ended rejection region as indicated by Schuirmann (1987). In other words, for a study with low precision, Anderson and Hauck's test may be in favor of bioequivalence regardless of the large intra-subject variability.

Example 4.3.3 For the example presented in Section 3.6, since $\overline{Y}_T = 80.272$, $\overline{Y}_R = 82.559$, $\hat{\sigma}_d^2 = 83.623$, and $-\theta_L = \theta_U = 16.51$, the observed t_{AH} can be obtained as

$$t_{AH} = \frac{(80.272 - 82.559) - (-16.51 + 16.51)/2}{(9.145)\sqrt{0.167}} = -0.613.$$

The estimated noncentrality parameter $\hat{\delta}$ is

$$\hat{\delta} = \frac{(16.51 + 16.51)/2}{(9.145)\sqrt{0.167}} = 4.423.$$

Therefore, the empirical p-value is estimated as

$$\begin{aligned}
p &= F_t(|-0.613| - 4.423) - F_t(-|-0.613| - 4.423) \\
&= F(-3.810) - F(-5.036) \\
&= 0.000478 - 0.000024 \\
&= 0.000454.
\end{aligned}$$

Since $p < 0.05$, the interval hypotheses in (4.3.1) is rejected at the 5% level of significance. Thus, we are in favor of bioequivalence between the two formulations.

4.4. BAYESIAN METHODS

In previous sections, statistical methods for the assessment of bioequivalence were derived based on the sampling distribution of the estimate of the parameter of interest such as the direct drug effect, which is assumed to be fixed but unknown. Although statistical inference (e.g., confidence interval and hypothesis testing) on the unknown direct drug effect can be drawn based on the sampling distribution of the estimate, there is little information regarding the probability of the unknown direct drug effect being within the equivalent limits, (θ_L, θ_U). As a matter of fact, from sampling theory, the probability that the true direct drug effect is within (θ_L, θ_U) is either 0 or 1 because the true direct drug effect is either within (θ_L, θ_U) or outside (θ_L, θ_U). To have a certain assurance on the probability of the direct drug effect being within (θ_L, θ_U), a Bayesian approach

(Box and Tiao, 1973), which assumes that the unknown direct drug effect is a random variable and follows a prior distribution, is useful.

In practice, before a bioavailability/bioequivalence study is conducted, investigators usually have some prior knowledge of the profile of the blood or plasma concentration-time curve. For example, according to past experiments, the investigator may have some information regarding (i) the inter-subject and the intra-subject variabilities and (ii) the ranges of AUC or C_{max} for the test and reference formulations. This information can be utilized to choose an appropriate prior distribution of the unknown direct drug effect. An appropriate prior distribution can reflect the investigator's belief about the formulations under study. After the study is completed, the observed data can be used to adjust the prior distribution of the direct drug effect which is called the posterior distribution. Based on the posterior distribution, a probability statement regarding the direct drug effect being within the bioequivalent limits can be made.

A different prior distribution can lead to a different posterior distribution which has an impact on statistical inference on the direct drug effect. Thus, an important issue in a Bayesian approach is how to choose a prior distribution. Box and Tiao (1973) introduced the use of a locally uniform distribution over a possible range of AUC or C_{max} as a (noninformative) prior distribution. A noninformative prior distribution assumes that there is an equally likely chance for any two points within the possible range being the true state of the location of the direct drug effect. In this case, the resultant posterior distribution can be used to provide the true state of the location of a direct drug effect. In practice, however, it is also desirable to provide an interval showing a range in which most of the distribution of a direct drug effect will fall. We shall refer to such an interval as a highest posterior density (HPD) interval. The HPD interval is also known as a credible interval (Edwards et al., 1963) and a Bayesian confidence interval (Lindley, 1965). An HPD interval possesses the following properties (Box and Tiao, 1973):

(i)   The density for every point inside the interval is greater than that for every point outside the interval;

(ii) For a given probability distribution, the interval is the shortest.

It can be verified that the above two properties imply each other. In the following sections, three Bayesian methods with different noninformative priors will be discussed under the raw data model.

4.4.1. The Rodda and Davis Method

Given the results of a bioavailability/bioequivalence study, Rodda and Davis (1980) proposed a Bayesian evaluation to estimate the probability of a clinically important difference, i.e., the probability that the true direct drug effect will fall within the bioequivalent limits is estimated. As indicated in Chapter 3, under the assumption of normality and equal carry-over effects, $\bar{d}_{.1}$, $\bar{d}_{.2}$ and $(n_1 + n_2 - 2)\hat{\sigma}_d^2$ are independently distributed as $N(\theta_1, \sigma_d^2/n_1)$, $N(\theta_2, \sigma_d^2/n_2)$ and $\sigma_d^2\chi^2(n_1 + n_2 - 2)$, where

$$\theta_1 = \tfrac{1}{2}[(P_2 - P_1) + (F_T - F_R)],$$
$$\theta_2 = \tfrac{1}{2}[(P_2 - P_1) + (F_R - F_T)].$$

Note that $F = \theta_1 - \theta_2 = (\mu + F_T) - (\mu + F_R)$
$$= \mu_T - \mu_R.$$

Assuming that the noninformative prior distribution for θ_1, θ_2, and $\log(\sigma_d)$ are approximately independent and locally uniformly distributed, then the joint posterior distribution of θ_1, θ_2, and σ_d^2, given data $\mathbf{Y} = \{Y_{ijk}, i = 1,2, \ldots, n_k; j,k = 1,2\}$, is

$$p(\theta_1, \theta_2, \sigma_d^2 | \mathbf{Y}) = p(\theta_1 | \sigma_d^2, \bar{d}_{.1}) p(\theta_2 | \sigma_d^2, \bar{d}_{.2}) p(\sigma_d^2 | \hat{\sigma}_d^2), \tag{4.4.1}$$

where

$$p(\theta_i | \sigma_d^2, \bar{d}_{.i}) = N(\bar{d}_{.i}, \hat{\sigma}_d^2/n_i), \qquad i = 1,2.$$
$$p(\sigma_d^2 | \hat{\sigma}_d^2) = (n_1 + n_2 - 2)\hat{\sigma}_d^2\chi^{-2}(n_1 + n_2 - 2),$$

where $\chi^{-2}(n_1 + n_2 - 2)$ is the distribution of the inverse of $\chi^2(n_1 + n_2 - 2)$. Therefore, the joint distribution of $\mu_T - \mu_R \, (= F)$ and σ_d^2 is given by

$$p(\mu_T - \mu_R, \sigma_d^2 | \mathbf{Y}) = p(\mu_T - \mu_R | \sigma_d^2, \bar{d}_{.1} - \bar{d}_{.2}) p(\sigma_d^2 | \hat{\sigma}_d^2), \tag{4.4.2}$$

where

$$p(\mu_T - \mu_R | \sigma_d^2, \bar{d}_{.1} - \bar{d}_{.2}) = N\left[\bar{d}_{.1} - \bar{d}_{.2}, \hat{\sigma}_d^2\left(\frac{1}{n_1} + \frac{1}{n_2}\right)\right]$$

$$= N\left[\bar{Y}_T - \bar{Y}_R, \hat{\sigma}_d^2\left(\frac{1}{n_1} + \frac{1}{n_2}\right)\right].$$

The marginal posterior distribution of F, given data $\mathbf{Y}$, is

$$p(\mu_T - \mu_R | \mathbf{Y}) = \frac{(\hat{\sigma}_d^2 m)^{-1/2}}{B(1/2, \nu/2)\sqrt{\nu}} \left\{1 + \frac{[(\mu_T - \mu_R) - (\bar{Y}_T - \bar{Y}_R)]^2}{\nu\hat{\sigma}_d^2 m}\right\}^{-(\nu+1)/2}, \tag{4.4.3}$$

where $m = 1/n_1 + 1/n_2$, $\nu = n_1 + n_2 - 2$ and $-\infty < \mu_T - \mu_R < \infty$. Expression (4.4.3) indicates that

$$T_{RD} = \frac{(\mu_T - \mu_R) - (\overline{Y}_T - \overline{Y}_R)}{\hat{\sigma}_d \sqrt{\dfrac{1}{n_1} + \dfrac{1}{n_2}}} \tag{4.4.4}$$

has a central student t distribution with $n_1 + n_2 - 2$ degrees of freedom. From (4.4.4), the probability of F being within the bioequivalent limits θ_L and θ_U can be estimated by

$$P_{RD} = P\{\theta_L < \mu_T - \mu_R < \theta_U\}$$
$$= F_t(t_U) - F_t(t_L), \tag{4.4.5}$$

where F_t is the cumulative distribution function of a central t variable with $n_1 + n_2 - 2$ degrees of freedom, and

$$t_U = \frac{\theta_U - (\overline{Y}_T - \overline{Y}_R)}{\hat{\sigma}_d \sqrt{\dfrac{1}{n_1} + \dfrac{1}{n_2}}}, \quad \text{and}$$

$$t_L = \frac{\theta_L - (\overline{Y}_T - \overline{Y}_R)}{\hat{\sigma}_d \sqrt{\dfrac{1}{n_1} + \dfrac{1}{n_2}}}. \tag{4.4.6}$$

The lower and upper limits of the $(1 - 2\alpha) \times 100\%$ HPD interval are given by

$$L_{RD} = (\overline{Y}_T - \overline{Y}_R) - t(\alpha, n_1 + n_2 - 2)\,\hat{\sigma}_d \sqrt{\frac{1}{n_1} + \frac{1}{n_2}},$$

$$U_{RD} = (\overline{Y}_T - \overline{Y}_R) + t(\alpha, n_1 + n_2 - 2)\,\hat{\sigma}_d \sqrt{\frac{1}{n_1} + \frac{1}{n_2}}. \tag{4.4.7}$$

Hence, it is verified that the $(1 - 2\alpha) \times 100\%$ HPD interval in (4.4.7) is numerically equivalent to the $(1 - 2\alpha) \times 100\%$ classical interval, given in (4.2.2), obtained from the sampling theory. However, the interpretation of these two intervals is totally different. For example, a 90% classical confidence interval for F indicates that, in the long run, if the study is repeatedly carried out for a large number of times, 90% of the times the interval will contain the unknown direct drug effect $\mu_T - \mu_R$. On the other hand, based on the posterior distribution of $\mu_T - \mu_R$, the chance of $\mu_T - \mu_R$ being within the lower and upper limits of a 90% HPD interval is 90%.

From (4.4.5) and (4.4.7), two formulations are bioequivalent with 90% assurance if $p_{RD} > 0.90$ or (L_{RD}, U_{RD}) is within (θ_L, θ_U). Furthermore, $p_{RD} > 0.90$ if and only if (L_{RD}, U_{RD}) is within (θ_L, θ_U). Therefore, (4.4.5) is equivalent to (4.4.7) in decision-making of bioequivalence.

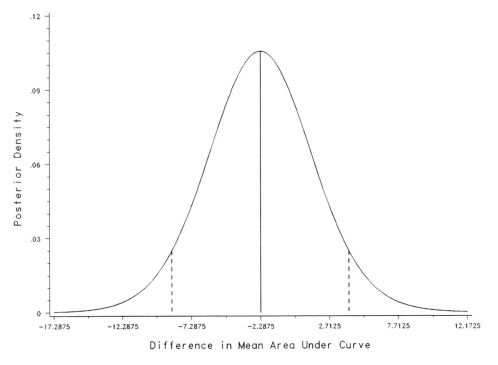

Figure 4.4.1 Posterior density of mean area under curve and highest posterior density interval by Rodda and Davis (1980).

Example 4.4.1 To illustrate Rodda and Davis's Bayesian approach, consider the example in Section 3.6. The data gives

$$\overline{Y}_T = 80.272, \qquad \overline{Y}_R = 82.559, \quad \text{and} \quad \hat{\sigma}_d^2 = 83.623.$$

From (4.4.4), we have

$$t_U = \frac{16.51 - (80.272 - 82.559)}{(9.145)\sqrt{0.167}} = 5.036, \quad \text{and}$$

$$t_L = \frac{-16.51 - (80.272 - 82.559)}{(9.145)\sqrt{0.167}} = -3.810.$$

Therefore, $P_{RD} = F_t(5.036) - F_t(-3.810) = 0.9995$ which is greater than 0.9. The 90% HPD interval is $(-8.698, 4.123)$, and we conclude that the two formulations are bioequivalent with at least 90% assurance.

The posterior density function for F with a noninformative prior, given data **Y**, is graphed in Figure 4.4.1.

4.4.2. Mandallaz and Mau's Method

Alternatively, Mandallaz and Mau (1981) proposed a Bayesian method for assessing bioequivalence for the ratio (μ_T/μ_R) (see also Fluehler et al., 1981; 1983)
under model (4.1.1) with the following assumptions:

   (i)   There are no carry-over effects;
   (ii) The effects from subjects are fixed;
   (iii) The number of subjects in each sequence are the same, i.e., $n_1 = n_2$
         $= n$.

Under these assumptions, $\overline{Y}_T$ and $\overline{Y}_R$ are independently normally distributed
with means μ_T and μ_R and variance σ_d^2/n, where

$$\mu_T = \mu + F_T \quad \text{and} \quad \mu_R = \mu + F_R.$$

Suppose that the prior distribution for μ_T, μ_R, and σ_d has the following improper
vague prior distribution of the form:

$$dP = (d\mu_T)(d\mu_R)(d\sigma_d/\sigma_d).$$

The posterior distribution of $\delta = \mu_T/\mu_R$, given data $\mathbf{Y}$, can then be approximated
by the following statistic

$$T_{MM} = \frac{\hat{\delta} - \delta}{\hat{\sigma}_d\sqrt{(1 + \delta^2)/n\overline{Y}_R^2}}, \tag{4.4.8}$$

where $\hat{\delta} = \overline{Y}_T/\overline{Y}_R$ and T_{MM} follows a central t distribution with $2(n - 1)$ degrees
of freedom.

Therefore, two formulations are considered bioequivalent with 90% assurance
if

$$P_{MM} = P\{\delta_L < \delta < \delta_U\}$$
$$= F_t(t_{\delta_L}) - F_t(t_{\delta_U}) > 0.90, \tag{4.4.9}$$

where

$$t_{\delta_U} = \frac{\hat{\delta} - \delta_U}{\hat{\sigma}_d\sqrt{(1 + \delta_U^2)/n\overline{Y}_R^2}},$$

$$t_{\delta_L} = \frac{\hat{\delta} - \delta_L}{\hat{\sigma}_d\sqrt{(1 + \delta_L^2)/n\overline{Y}_R^2}}, \quad \text{and}$$

$F_t(\cdot)$ is the cumulative distribution function of a central t variable with $2(n - 1)$ degrees of freedom.

   Note that T_{MM} in (4.4.8) can be rewritten as

$$T_{MM} = \frac{\overline{Y}_T - \delta\overline{Y}_R}{\hat{\sigma}_d\sqrt{(1 + \delta^2)/n}}, \tag{4.4.10}$$

which is the pivotal quantity used for the construction of a $(1 - 2\alpha) \times 100\%$ confidence interval of δ in the fixed Fieller's method (Schuirmann, 1989). The Mandallaz and Mau method leads to the same decision for bioequivalence as that of the fixed Fieller's method when $n_1 = n_2 = n$. To show this, it is sufficient to prove that if δ in (4.4.8) is replaced by L_4 in (4.2.19), then $T_{MM} = t(\alpha, 2(n - 1))$. Assuming that $G < 1$, the numerator of (4.4.8) with δ replaced by L_4 is

$$\hat{\delta} - L_4 = \frac{\sqrt{G_1}}{1 - G_1} [\sqrt{G_1}\hat{\delta} + \sqrt{\hat{\delta}^2 + (1 - G_1)}].$$

Furthermore, it can be easily verified that

$$1 + L_4^2 = \frac{1}{(1 - G_1)^2} [\sqrt{G_1}\hat{\delta} + \sqrt{\hat{\delta}^2 + (1 - G_1)}].$$

Thus, the denominator of (4.4.8) can be expressed as

$$\frac{\hat{\sigma}_d}{\sqrt{n}\overline{Y}_R(1 - G_1)} [\sqrt{G_1}\hat{\delta} + \sqrt{\hat{\delta}^2 + (1 - G_1)}].$$

Since $n_1 = n_2$, we have

$$G_1 = [t(\alpha, 2(n - 1))]^2 \left(\frac{MS_{Intra}/2n}{\overline{Y}_R^2}\right)$$

$$= [t(\alpha, 2(n - 1))]^2 [\hat{\sigma}_d^2/(n\overline{Y}_R^2)].$$

Therefore,

$$\frac{\hat{\delta} - L_4}{\hat{\sigma}_d\sqrt{(1 + L_4^2)/(n\overline{Y}_R^2)}} = \frac{\sqrt{G_1} [\sqrt{G_1}\hat{\delta} + \sqrt{\hat{\delta}^2 + (1 - G_1)}]}{\frac{\hat{\sigma}_d}{\sqrt{n}\overline{Y}_R} [\sqrt{G_1}\hat{\delta} + \sqrt{\hat{\delta}^2 + (1 - G_1)}]}$$

$$= t(\alpha, 2(n - 1)).$$

When $n_1 = n_2$, since the least squares means $\overline{Y}_T$ and $\overline{Y}_R$ are the same as the raw means, denoted by $\overline{Y}_T^*$ and $\overline{Y}_R^*$, the above procedure holds with $\overline{Y}_T$ and $\overline{Y}_R$ replaced by $\overline{Y}_T^*$ and $\overline{Y}_R^*$. However, when $n_1 \neq n_2$, $\overline{Y}_T$ and $\overline{Y}_R$ are not the same as $\overline{Y}_T^*$ and $\overline{Y}_R^*$. In this case, under the assumptions of no carry-over effects and fixed subject effects, we have

$$E[\overline{Y}_T^* - \delta\overline{Y}_R^*] = \frac{1}{n_1 + n_2} [(n_1 - \delta n_2)P_1 + (n_2 - \delta n_1)P_2]. \qquad (4.4.11)$$

Hence,

$$T_{MM}^* = \frac{\overline{Y}_T^* - \delta\overline{Y}_R^*}{\hat{\sigma}_d\sqrt{(1 + \delta^2)/n}}$$

is no longer distributed as a central t distribution with $n_1 + n_2 - 2$ degrees of freedom. With $n_1 \neq n_2$, T_{MM} in (4.4.10) still has a central t distribution since $E[\overline{Y}_T - \delta\overline{Y}_R] = 0$. In this case, however, whether the posterior distribution of δ can be approximated by T_{MM} under the same improper vague prior distribution remains unknown.

Example 4.4.2 For the example in Section 3.6, we have $\overline{Y}_T = 80.272$, $\overline{Y}_R = 82.559$ and $\hat{\delta} = 0.972$. Since $\hat{\sigma}_d^2 = 83.623$, and $n_1 = n_2 = 12$, we have

$$t_{\delta L} = \frac{0.972 - 0.8}{\sqrt{(83.623)(1 + 0.64)/(12)(82.559)^2}} = 4.208, \quad \text{and}$$

$$t_{\delta U} = \frac{0.972 - 1.2}{\sqrt{(83.623)(1 + 1.44)/(12)(82.559)^2}} = -4.559.$$

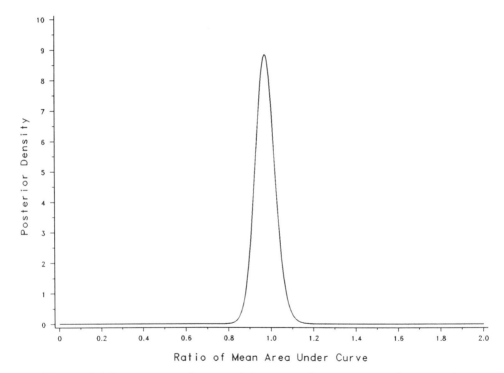

Figure 4.4.2 Posterior density of the ratio of mean area under curve by Mandallaz and Mau (1981).

Therefore,

$$P_{MM} = F_t(4.208) - F_t(-4.559) = 0.9997,$$

which is greater than 0.90. Thus, we conclude that the two formulations are bioequivalent with at least 90% assurance.

The posterior density function for δ with an improper vague prior distribution, given data Y, is graphed in Figure 4.4.2.

4.4.3. Grieve's Method

In Sections 4.4.1 and 4.4.2, the prior distribution considered in Rodda and Davis's method is a noninformative prior distribution for the population means of the period difference and intra-subject variability, while an improper vague prior distribution for the direct population means and intra-subject variability is used in Mandallaz and Mau's method. Both prior distributions only involve the intra-subject variability, which, however, do not account for the inter-subject variability. Grieve (1985) developed an alternative Bayesian method under model (3.1.1), which utilizes the information about the inter-subject variability, σ_s^2.

In model (3.1.1), it is assumed that the fixed effects such as the direct drug effect, the carry-over effect, and the period effect satisfy the following constraints:

$$F_T + F_R = 0, \quad C_T + C_R = 0, \quad \text{and} \quad P_1 + P_2 = 0.$$

Let $F' = F_T = -F_R$, $C' = C_T = -C_R$ and $P' = P_1 = -P_2$. We then have

the direct drug effect: $F = F_T - F_R = 2F'$,

the carry-over effect: $C = C_T - C_R = 2C'$,

the period effect: $P = P_1 - P_2 = 2P'$.

The parameters, μ, F', C', and P' can be estimated by

$$\hat{\mu} = \tfrac{1}{4}[\overline{Y}_{\cdot 11} + \overline{Y}_{\cdot 21} + \overline{Y}_{\cdot 12} + \overline{Y}_{\cdot 22}],$$

$$\hat{F}' = \tfrac{1}{4}[-\overline{Y}_{\cdot 11} + \overline{Y}_{\cdot 21} + \overline{Y}_{\cdot 12} - \overline{Y}_{\cdot 22}],$$

$$\hat{P}' = \tfrac{1}{4}[-\overline{Y}_{\cdot 11} + \overline{Y}_{\cdot 21} - \overline{Y}_{\cdot 12} + \overline{Y}_{\cdot 22}],$$

$$\hat{C}' = \tfrac{1}{2}[-\overline{Y}_{\cdot 11} - \overline{Y}_{\cdot 21} + \overline{Y}_{\cdot 12} + \overline{Y}_{\cdot 22}],$$

Note that $F' = (\mu_T - \mu_R)/2$ and $\hat{F}' = (\overline{Y}_T - \overline{Y}_R)/2$.

Let $\sigma_A^2 = \sigma_e^2 + 2\sigma_s^2$. Consider the joint noninformative prior distribution of μ, F', P', C', σ_e^2, and σ_A^2 as follows:

$$p(\mu, F', P', C', \sigma_e^2, \sigma_A^2) \propto \frac{1}{\sigma_e^2 \sigma_A^2}. \qquad (4.4.12)$$

Under the normality assumptions, $(\hat{\mu}, \hat{C}')'$, $(\hat{P}', \hat{F}')'$, SS_{Inter}, and SS_{Intra} are independently distributed, i.e.,

$$\begin{pmatrix} \hat{\mu} \\ \hat{C} \end{pmatrix} \sim N\left(\begin{pmatrix} \mu \\ C' \end{pmatrix}, \frac{\sigma_A^2}{2} \begin{pmatrix} m/4 & m_1/2 \\ m_1/2 & m/2 \end{pmatrix} \right),$$

$$\begin{pmatrix} \hat{P}' \\ \hat{F}' \end{pmatrix} \sim N\left(\begin{pmatrix} P' \\ F' - C'/2 \end{pmatrix}, \frac{\sigma_e^2}{8} \begin{pmatrix} m & m_1 \\ m_1 & m \end{pmatrix} \right),$$

$$SS_{Inter} \sim \sigma_A^2 \chi^2(n_1 + n_2 - 2),$$

$$SS_{Intra} \sim \sigma_e^2 \chi^2(n_1 + n_2 - 2), \tag{4.4.13}$$

where

$$m = \frac{1}{n_1} + \frac{1}{n_2} \quad \text{and} \quad m_1 = \frac{1}{n_1} - \frac{1}{n_2}.$$

Thus, the joint posterior distribution of F', C', σ_e^2, and σ_A^2, given data $\mathbf{Y}$, after integrating with respect to μ and P', is given by

$$p(F', C', \sigma_e^2, \sigma_A^2 | \mathbf{Y}) \propto (\sigma_e^2 \sigma_A^2)^{-((n_1 + n_2) - 1/2)}$$

$$\exp\left\{ -\frac{1}{2} \left(\frac{Q_1}{\sigma_e^2} + \frac{Q_2}{\sigma_A^2} \right) \right\}, \tag{4.4.14}$$

where

$$Q_1 = \frac{8}{m} \left(F' - \frac{C'}{2} - \hat{F}' \right)^2 + SS_{Intra}, \quad \text{and}$$

$$Q_2 = \frac{2}{m} (C' - \hat{C}')^2 + SS_{Inter}. \tag{4.4.15}$$

From (4.4.14), the joint posterior distribution of F', C', given $\mathbf{Y}$, can be obtained as

$$p(F', C' | \mathbf{Y}) \propto \left[\left(\frac{Q_1}{SS_{Intra}} \right) \left(\frac{Q_2}{SS_{Inter}} \right) \right]^{-((n_1 + n_2 - 1)/2)}. \tag{4.4.16}$$

Note that (4.4.16) is the product of the marginal posterior distribution of C', given data $\mathbf{Y}$, and the conditional posterior distribution of F', given C' and $\mathbf{Y}$, i.e.,

$$p(F', C' | \mathbf{Y}) = p(C' | \mathbf{Y}) \, p(F' | C', \mathbf{Y}),$$

where

$$p(C' | \mathbf{Y}) \propto \left(\frac{Q_2}{SS_{Inter}} \right)^{-((n_1 + n_2 - 1)/2)}, \tag{4.4.17}$$

$$p(F'|C',Y) \propto \left(\frac{Q_1}{SS_{Intra}}\right)^{-((n_1+n_2-1)/2)}.$$ (4.1.18)

Let $t(\alpha,\beta,\nu)$ denote a shifted and scaled t distribution with ν degrees of freedom, location parameter α, and scale parameter $\sqrt{\beta}$ (for more details of $t(\alpha,\beta,\nu)$, see Grieve, 1985; Jones and Kenward, 1989). The variance of $t(\alpha,\beta,\nu)$ is given by $\nu\beta/(\nu - 2)$. Therefore, it can be verified that the marginal posterior distribution of the carry-over effect C' is

$$p(C'|Y) = t\left(\hat{C}', \frac{mSS_{Inter}}{2(n_1 + n_2 - 2)}, n_1 + n_2 - 2\right).$$ (4.4.19)

Hence, the standardized form of C' is given by

$$T_{GC} = \frac{C' - \hat{C}'}{\sqrt{\dfrac{mSS_{Inter}}{2(n_1 + n_2 - 4)}}},$$

which has a central t distribution with $n_1 + n_2 - 2$ degrees of freedom. Since

$$\frac{SS_{Inter}}{2(n_1 + n_2 - 2)} = \frac{\hat{\sigma}_u^2}{4},$$

we have

$$T_{GC} = \sqrt{\frac{n_1 + n_2 - 4}{n_1 + n_2 - 2}}\, [-T_C],$$

where T_C and $\hat{\sigma}_u^2$ are defined in Section 3.2.1. Similarly, it can be verified that the conditional posterior distribution of F' given in (4.4.18) is

$$p(F'|C',Y) = t\left(\hat{F}' + \frac{C'}{2}, \frac{mSS_{Intra}}{8(n_1 + n_2 - 2)}, n_1 + n_2 - 2\right).$$ (4.4.20)

The standardized form of F' given $C' = 0$ is then given by

$$T_{GF1} = \frac{F' - \hat{F}'}{\sqrt{\dfrac{mSS_{Intra}}{8(n_1 + n_2 - 4)}}},$$ (4.4.21)

where T_{GF1} has a central t distribution with $n_1 + n_2 - 2$ degrees of freedom. Note that

$$\frac{mSS_{Intra}}{8(n_1 + n_2 - 4)} = \frac{m}{4}\left[\frac{n_1 + n_2 - 2}{n_1 + n_2 - 4}\right]\hat{\sigma}_d^2,$$

where

$$\hat{\sigma}_d^2 = \frac{1}{2} MS_{Intra} = \frac{1}{2} \frac{SS_{Intra}}{n_1 + n_2 - 2}.$$

Hence,

$$T_{GFI} = \sqrt{\frac{n_1 + n_2 - 4}{n_1 + n_2 - 2}} \, T_{RD}. \tag{4.4.22}$$

The above expression (4.4.22) indicates that the conditional posterior distribution of the standard form of F' by Grieve (assuming that there are no carry-over effects) and the posterior distribution of the standard form of F derived by Rodda and Davis only differ by a constant of $\sqrt{(n_1 + n_2 - 4)/(n_1 + n_2 - 2)}$. This difference is due to the choice of the prior distribution. In Rodda and Davis's method, the prior distribution is chosen to account for the intra-subject variability because the period differences, which only depend on the intra-subject variability, are irrelevant to the inter-subject variability. However, in Grieve's method, the prior distribution considered involve both the inter-subject and the intra-subject variabilities. Furthermore, in Rodda and Davis's method, it is assumed that the logarithm of σ_e is locally uniformly distributed, i.e., the prior distribution is proportional to the inverse of σ_e. In Grieve's method, the prior distribution is proportional to the inverse of σ_e^2. In practice, such a difference usually has little impact on the posterior distribution of the direct formulation effect provided that there are no carry-over effects and the sample size is not too small.

In the case where the carry-over effects are present, the marginal posterior distribution of F' is rather complicated and can only be obtained numerically by integrating the joint posterior distribution of F' and C' (4.4.16) with respect to C'. Based on an approach proposed by Patil (1965), Grieve (1985) showed that the marginal posterior distribution of F', given $\mathbf{Y}$, can be approximated by

$$p(F'|\mathbf{Y}) \approx t\left(\hat{F}' + \frac{\hat{C}'}{2}, \frac{m\beta_0}{8\nu_0}, \nu_0 \right), \tag{4.4.23}$$

where

$$\nu_0 = \frac{(SS_{Intra} + SS_{Inter})^2 (n_1 + n_2 - 6)}{(SS_{Intra})^2 + (SS_{Inter})^2} + 4,$$

and

$$\beta_0 = \frac{(\nu_0 - 2)(SS_{Intra} + SS_{Inter})}{(n_1 + n_2 - 4)}.$$

Therefore,

$$T_{GF2} = \frac{F' - (\hat{F}' + \hat{C}'/2)}{\sqrt{\dfrac{m\beta_0}{8(\nu_0 - 2)}}}, \qquad (4.4.24)$$

where T_{GF2} has an approximate central t distribution with ν_0 degrees of freedom.

The $(1 - 2\alpha) \times 100\%$ HPD intervals for F' based on the posterior distribution of F' assuming that $C' = 0$ and the marginal posterior distribution of F' are given by

$$\hat{F}' \pm t(\alpha, n_1 + n_2 - 2)\sqrt{\frac{mSS_{Intra}}{8(n_1 + n_2 - 4)}},$$

and

$$\left(\hat{F}' + \frac{\hat{C}'}{2}\right) \pm t(\alpha, \nu_0)\sqrt{\frac{m\beta_0}{8(\nu_0 - 2)}}, \qquad (4.4.25)$$

respectively.

The probability of the true direct drug effect being within (θ_L, θ_U) can be calculated based on either (4.4.21) or (4.4.24). That is, when there are no carry-over effects,

$$\begin{aligned} P_{GF1} &= P\{\theta_L < F < \theta_U\} \\ &= F_t(t_{0U}) - F_t(t_{0L}), \end{aligned} \qquad (4.4.26)$$

where

$$t_{0L} = \frac{\theta_L/2 - \hat{F}'}{\sqrt{\dfrac{mSS_{Intra}}{8(n_1 + n_2 - 4)}}},$$

$$t_{0U} = \frac{\theta_U/2 - \hat{F}'}{\sqrt{\dfrac{mSS_{Intra}}{8(n_1 + n_2 - 4)}}}.$$

When the carry-over effects are present, the probability can be approximated based on (4.4.24) which is given by

$$\begin{aligned} P_{GF2} &= P\{\theta_L < F < \theta_U\} \\ &= F_t(t_{MU}) - F_t(t_{ML}), \end{aligned} \qquad (4.4.27)$$

where

$$t_{ML} = \frac{\theta_L/2 - (\hat{F}' + \hat{C}'/2)}{\sqrt{\dfrac{m\beta_0}{8(\nu_0 - 2)}}},$$

$$t_{MU} = \frac{\theta_U/2 - (\hat{F}' + \hat{C}'/2)}{\sqrt{\dfrac{m\beta_0}{8(\nu_0 - 2)}}}.$$

Note that the degrees of freedom ν_0 associated with the marginal posterior distribution of F' might not be an integer rather than a real number. This, however, causes no problem in computing (4.4.27) because most of the standard statistical software available such as SAS procedure PROC PROBT can calculate the probability of a t distribution with non-integer degrees of freedom.

According to the ± 20 rule, the decision on bioequivalence can be made based on either (4.4.26) or (4.4.27), i.e., we conclude bioequivalence if

$$P_{GF1} > 0.90 \quad \text{or} \quad P_{GF2} > 0.90. \tag{4.4.28}$$

Example 4.5.3 Again, consider the example in Section 3.6. Table 3.6.1 leads to

$$\hat{F} = -2.288, \qquad \hat{C} = -9.592, \quad \text{and}$$
$$SS_{Inter} = 16211.49, \qquad SS_{Intra} = 3679.43.$$

Thus, $\hat{F}' = -1.144$ and $\hat{C}' = -4.796$. If we ignore the carry-over effects, i.e., assume $C' = 0$, then the conditional posterior distribution for F' is

$$p(F'|\overline{Y}, C' = 0) = t(-1.144, 3.484, 22).$$

The 90% HPD interval for F' is given by $(-4.506, 2.218)$, or equivalently, the 90% HPD interval for F is $(-9.011, 4.436)$.

When there are carry-over effects, i.e., $C' \neq 0$, the mean of the marginal posterior distribution of F' is

$$\hat{F}' + \frac{\hat{C}'}{2} = -1.144 - \frac{4.796}{2} = -3.542.$$

Furthermore,

$$\nu_0 = \frac{(16211.49 + 3679.43)^2(12 + 12 - 6)}{(16211.49)^2 + (3679.43)^2} + 4 = 29.770, \quad \text{and}$$

$$\beta_0 = \frac{(29.770 - 2)(16211.49 + 3679.43)}{(12 + 12 - 4)} = 27619.$$

Thus,

$$\frac{m\beta_0}{8\nu_0} = \frac{(1/6)(27619)}{(8)(29.770)} = 19.328.$$

Therefore,

$$p(F'|Y) \approx t(-3.542, 19.328, 29.770).$$

The 90% HPD interval for F' is then given by $(-11.269, 4.186)$. Thus, the 90% HPD interval for F is $(-22.539, 8.372)$.

The probability of the true direct formulation effect being within the bioequivalent limits under $p(F'|Y, C' = 0)$ and $p(F'|Y)$ are given by

$$P_{GF1} = F_t(4.801) - F_t(-3.633) = 0.9992,$$

and

$$P_{GF2} = F_t(2.592) - F_t(-1.036) = 0.8383,$$

respectively.

As a result, under $p(F'|Y, C' = 0)$, we conclude that the two formulations are bioequivalent since $P_{GF1} = 0.9992$ is greater than 0.90. However, there is not enough evidence for concluding bioequivalence under $p(F'|Y)$ since the probability $P_{GF2} = 0.8383$ is less than 0.90.

The posterior distribution of F' under models (4.4.20) and (4.4.23) are plotted in Figure 4.4.3. The difference between $p(F'|Y, C' = 0)$ and $p(F'|Y)$ is rather dramatic. However, these two posterior distributions represent two extreme cases of the prior belief of carry-over effects.

Thus far, all of the Bayesian approaches discussed are derived under the assumption of no carry-over effects except for the method of using P_{GF2}. As indicated in Section 3.2, a preliminary test for the presence of the carry-over effects can be assessed by testing the null hypothesis of $H_0: C = 0$ (note that $C = 0$ implies $C_T = C_R = 0$ under the constraint of $C_T + C_R = 0$). However, failing to reject H_0 at the 5% level of significance does not mean that the nonsignificant carry-over effects will have little or no impact on the statistical inference on the direct drug effect. As can be seen from the above example, P_{GF1} (without carry-over effects) and P_{GF2} (with carry-over effects) led to a totally different conclusion of bioequivalence. To account for this difference in carry-over effects, Grieve (1985) introduced the following alternative approach with use of a Bayes factor. Let π be the prior odds of no carry-over effects, i.e.,

$$\pi = \frac{P(M_0)}{P(M_1)}, \tag{4.4.29}$$

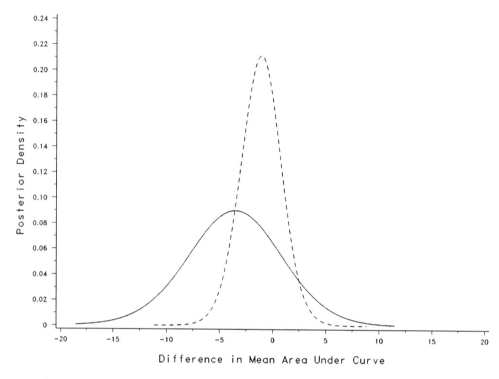

Figure 4.4.3   Posterior density of mean area under curve by Grieve's method (1985).

where M_0 and M_1 are the model without and with carry-over effects, respectively. The posterior probability for the two models are given by

$$p(M_0|\mathbf{Y}) = \frac{\pi B_{01}}{1 + \pi B_{01}},$$

and

$$p(M_1|\mathbf{Y}) = \frac{1}{1 + \pi B_{01}}, \tag{4.4.30}$$

where B_{01} is the Bayes factor (for a standard 2×2 crossover design), which is given by

$$B_{01} = \left(\frac{3}{2m}\right)^{1/2} \left(1 + \frac{F_C}{n_1 + n_2 - 2}\right)^{-((n_1+n_2)/2)}, \tag{4.4.31}$$

and F_C is previously defined in Section 3.5.

Thus, a Bayesian inference on F' can be made based on a mixture of the two posterior distributions in (4.4.30) given below:

$$p_M(F'|Y) = \frac{\pi B_{01}}{1 + \pi B_{01}} p(F'|Y,M_0) + \frac{1}{1 + \pi B_{01}} p(F'|Y,M_1), \qquad (4.4.32)$$

where $p(F'|Y,M_0)$ is given in (4.4.20) with $C' = 0$ and $p(F'|Y,M_1)$ is given in (4.4.23). As a result, it can be verified that the impact of the difference in carry-over effects on the inference of the direct drug effect is a function of the prior belief of the carry-over effects, i.e., $1/(1 + \pi)$. More details can be found in Grieve (1985).

Selwyne, Dempster, and Hall (1981) also proposed a Bayesian method by considering the following joint informative prior distribution for μ, F', C', σ_e^2, and σ_A^2:

$$p(\mu,F',C',\sigma_e^2,\sigma_A^2) \propto \sigma_e^{-2}\sigma_A^{-2} \exp\left\{ -\frac{C'}{2\sigma_{C'}^2} \right\}. \qquad (4.4.33)$$

The above joint prior distribution involves the prior information of C'. It has been criticized by many researchers (e.g., see Grieve (1985); Racine-Poon et al. (1986)) that if the prior information of C' is available, so is the prior knowledge of F' which should be incorporated in (4.4.33). This method, however, is rather complicated and hence may not be of practical interest.

In addition to the above Bayesian approaches, some other Bayesian approaches are also available in the literature. For example, Buonaccorsi and Gatsonis (1988) derived a general result of Bayesian inference for ratios of coefficients in linear models for fixed subject effects with bioequivalence as a special application. Srinivasan and Langenberg (1986) proposed an indifference zone approach based on the two-stage decision procedure of Beckhofer, Dunnett and Sobel (1954). Racine-Poon et al. (1986, 1987) considered a two-stage Bayesian approach based on the idea that information from a first-stage Bayesian approach can be used to form a predictive distribution for the outcome of a second-stage experiment.

4.5. NONPARAMETRIC METHODS

In Sections 4.2, 4.3, and 4.4, statistical methods for assessing bioequivalence between formulations were derived under the assumption that $\{S_{ik}\}$ and $\{e_{ijk}\}$ are mutually independent and normally distributed with mean 0 and variance σ_s^2 and σ_e^2. Under these normality assumptions, confidence intervals and tests for interval hypotheses were obtained based on either a two-sample t statistic or an F statistic. In practice, however, one of the difficulties commonly encountered in comparing formulations is whether the assumption of normality (for raw or untransformed data) or lognormality (for log transformed data) is valid. If the normality (or lognormality) is seriously violated, the approach based on a two-sample t statistic or an F statistic is no longer justified. In this situation, a distribution-free (or nonparametric) method is useful. In this section, a nonparametric version of the two one-sided tests procedure for testing interval hypotheses, namely, Wilcoxon-Mann-Whitney two one-sided tests, will be derived. The Hodges-Lehmann's estimator associated with the Wilcoxon rank sum test will be used to construct a $(1 - 2\alpha) \times 100\%$ confidence interval for $\mu_T - \mu_R$, the difference in average bioavailability. In addition, the possible use of a bootstrap resampling procedure will be discussed.

4.5.1. Wilcoxon-Mann-Whitney Two One-Sided Tests Procedure

Since a standard 2×2 crossover design consists of a pair of dual sequences (i.e., RT and TR), a distribution-free rank sum test can be applied directly to the two one-sided tests procedure (Hauschke et al., 1990; Cornell, 1990; Liu, 1991). We shall refer to this approach as Wilcoxon-Mann-Whitney two one-sided tests procedure. Let $\theta = \mu_T - \mu_R$. The two sets of hypotheses in (4.3.3) can then be rewritten as

$$H_{01}: \theta_L^* \leq 0$$

$$\text{vs} \quad H_{a1}: \theta_L^* > 0$$

and

$$H_{02}: \theta_U^* \geq 0$$

$$\text{vs} \quad H_{a2}: \theta_U^* < 0, \tag{4.5.1}$$

where $\theta_L^* = \theta - \theta_L$ and $\theta_U^* = \theta - \theta_U$.

As discussed in Chapter 3, the estimates of θ_L^* and θ_U^* can be obtained as a linear function of period differences d_{ik}, $i = 1,2, \ldots, n_k$, $k = 1,2$. Let

$$b_{hik} = \begin{cases} d_{ik} - \theta_h & h = L,U, \text{ for subjects in sequence 1} \\ d_{ik} & \text{for subjects in sequence 2.} \end{cases} \tag{4.5.2}$$

When there are no carry-over effects, the expected value and variance of b_{hik}, where $h = L,U$, $i = 1,2, \ldots, n_k$, and $k = 1,2$, are given by

$$E(b_{hik}) = \begin{cases} \frac{1}{2} [(P_2 - P_1) + (\theta - 2\theta_h)] & \text{for } k = 1 \\ \frac{1}{2} [(P_2 - P_1) - \theta] & \text{for } k = 2, \end{cases} \tag{4.5.3}$$

and

$$V(b_{hik}) = V(d_{ik}) = \sigma_d^2 = \sigma_e^2/2.$$

It can be seen that

$$E(b_{hi1}) - E(b_{hi2}) = (\theta - \theta_h) = \theta_h^*.$$

Thus, for a fixed h, $\{b_{hi1}\}$ and $\{b_{hi2}\}$ have the same distribution except for the difference ($= \theta_h^*$) in location of the true formulation effect. In this case, Wilcoxon-Mann-Whitney rank sum test (Wilcoxon, 1945; Mann and Whitney, 1947) for two-sample location problem can be directly applied to test each of the two sets of hypotheses given in (4.5.1).

Consider the first set of hypotheses in (4.5.1),

$$H_{01}: \theta_L^* \leq 0 \quad \text{vs} \quad H_{a1}: \theta_L^* > 0.$$

Wilcoxon-Mann-Whitney test statistic can be derived based on $\{b_{Li1}\}$, $i = 1,2, \ldots ,n_1$ and $\{b_{Li2}\}$, $i = 1,2, \ldots ,n_2$. Let $R(b_{Lik})$ be the rank of b_{Lik} in the combined sample $\{b_{Lik}\}$, $i = 1,2, \ldots ,n_k$, $k = 1,2$. Also, let R_L be the sum of the ranks of the responses for subjects in sequence 1, i.e.,

$$R_L = \sum_{i=1}^{n_1} R(b_{Li1}).$$

Thus, Wilcoxon-Mann-Whitney test statistic for H_{01} is given by

$$W_L = R_L - \frac{n_1(n_1 + 1)}{2}.$$

We then reject H_{01} if

$$W_L > w(1 - \alpha), \tag{4.5.4}$$

where $w(1 - \alpha)$ is the $(1 - \alpha)$th upper percentile of the distribution of W_L which can be obtained in Table A.5 in Appendix A. Similarly, for the second set of hypotheses in (4.5.1),

$$H_{02}: \theta_U^* \geq 0 \quad \text{vs} \quad H_{a2}: \theta_U^* < 0,$$

we reject H_{02} if

$$W_U = R_U - \frac{n_1(n_1 + 1)}{2} < w(\alpha), \tag{4.5.5}$$

where R_U is the sum of the ranks of $\{b_{Uik}\}$ for subjects in the first sequence. Hence, bioequivalence is concluded if both H_{01} and H_{02} are rejected, i.e.,

$$W_L > w(1 - \alpha) \quad \text{and} \quad W_U < w(\alpha). \tag{4.5.6}$$

The expected values and variances for W_L and W_U under the null hypotheses H_{01} and H_{02}, when there are no ties, are given by

$$E(W_L) = E(W_U) = \frac{n_1 n_2}{2},$$

$$V(W_L) = V(W_U) = \tfrac{1}{12} n_1 n_2 (n_1 + n_2 + 1). \tag{4.5.7}$$

When there are ties among observations, average ranks can be assigned to compute W_L and W_U. In this case, however, the expected values and variances of W_L and W_U become

$$E(W_L) = E(W_U) = \frac{n_1 n_2}{2},$$

$$V(W_L) = V(W_U) = \tfrac{1}{12} n_1 n_2 (n_1 + n_2 + 1 - Q), \qquad (4.5.8)$$

where

$$Q = \frac{1}{(n_1 + n_2)(n_1 + n_2 - 1)} \sum_{\nu=1}^{q} (r_\nu^3 - r_\nu),$$

where q is the number of tied groups and r_ν is the size of the tied group ν. Note that if there are no tied observations, $q = n_1 + n_2$, $r_\nu = 1$ for $\nu = 1, 2, \ldots, n$, and $Q = 0$, then (4.5.8) reduces to (4.5.7).

Since W_L and W_U are symmetric about their mean $(n_1 n_2)/2$, we have

$$w(1 - \alpha) = n_1 n_2 - w(\alpha). \qquad (4.5.9)$$

Table A.5 gives the percentiles for $\alpha = 0.001, 0.005, 0.01, 0.025, 0.05$, and 0.1 of the Wilcoxon-Mann-Whitney test statistics. Based on (4.5.9), the percentiles for $\alpha = 0.90, 0.95, 0.975, 0.99, 0.995$, and 0.999 can be easily obtained. For example, suppose there are 12 subjects in both sequences. From Table A.5, the 5th percentile is 43. Thus, the 95th percentile is given by $w(0.95) = (12)(12) - 43 = 101$.

When $n_1 + n_2$, the total number of subjects, is large (say, $n_1 + n_2 > 40$) and the ratio of n_1 and n_2 is close to 1/2, a large sample approximation using the standard normal can be used to approximate (4.5.6) for bioequivalence testing, i.e., we may conclude bioequivalence if

$$Z_L > z(\alpha) \quad \text{and} \quad Z_U < -z(\alpha),$$

where $z(\alpha)$ is the αth upper percentile of a standard normal distribution, and

$$Z_L = \frac{W_L - E(W_L)}{\sqrt{V(W_L)}} = \frac{R_L - \left[\dfrac{n_1(n_1 + n_2 + 1)}{2}\right]}{\sqrt{\tfrac{1}{12} n_1 n_2 (n_1 + n_2 + 1)}},$$

$$Z_U = \frac{W_U - E(W_U)}{\sqrt{V(W_U)}} = \frac{R_U - \left[\dfrac{n_1(n_1 + n_2 + 1)}{2}\right]}{\sqrt{\tfrac{1}{12} n_1 n_2 (n_1 + n_2 + 1)}}. \qquad (4.5.10)$$

Note that the variances in Z_L and Z_U should be replaced with that given in (4.5.8) if there are ties.

Example 4.5.1 To illustrate the above Wilcoxon-Mann-Whitney two one-sided tests procedure, consider the example in Section 3.6. Table 4.5.1 lists the ranks of b_{hik}. The observed reference mean is assumed to be the true mean μ_R. According to the ± 20 rule, the bioequivalence limits are given by

$$-\theta_L = \theta_U = (0.2)(82.559) = 16.51.$$

Thus, the interval hypotheses are

$$H_0: \theta < -16.51 \quad \text{or} \quad \theta > 16.51$$
$$\text{vs} \quad H_a: -16.51 \leq \theta \leq 16.51,$$

which leads to the following two sets of hypotheses

$$H_{01}: \theta + 16.51 \leq 0 \quad \text{vs} \quad H_{a1}: \theta + 16.51 > 0,$$

and

$$H_{02}: \theta - 16.51 \geq 0 \quad \text{vs} \quad H_{a2}: \theta - 16.51 < 0.$$

From Table 4.5.1, R_L and R_U can be found to be 207 and 91, respectively. Therefore, W_L and W_U are

$$W_L = 207 - \frac{(12)(12 + 1)}{2} = 129,$$

and

$$W_U = 91 - \frac{(12)(12 + 1)}{2} = 13.$$

From Table A.5 in Appendix A, the 5th percentile of the Wilcoxon-Mann-Whitney statistic for $n_1 = n_2 = 12$ is 43. Based on (4.5.8), the 95th percentile of the Wilcoxon-Mann-Whitney statistic can be obtained as

$$w(0.95) = (12)(12) - 43 = 101.$$

Since $W_L = 129 > w(0.95) = 101$ and $W_U = 13 < w(0.05) = 43$, both two one-sided null hypotheses (i.e., H_{01} and H_{02}) are rejected at the 5% level of significance. We then conclude bioequivalence with 90% assurance.

4.5.2. Distribution-Free Confidence Interval Based on the Hodges-Lehmann Estimator

A distribution-free $(1 - 2\alpha) \times 100\%$ confidence interval for θ can be obtained based on the Hodges-Lehmann estimator (Randles and Wolfe, 1979). Let $D_{i,i'}$,

Table 4.5.1 Ranks of b_{hik} for Data in Table 3.6.1

Seq.	Subject number	Period I	Period II	Subject total	P.D.[a]	$2 \times b_{Lik}$[b]	$R(b_{Lik})$	$2 \times b_{Uik}$	$R(b_{Uik})$
1									
RT	1	74.675	73.675	148.350	−1.000	32.024	20	−34.024	10
RT	4	96.400	93.250	189.650	−3.150	29.874	19	−36.174	8
RT	5	101.950	102.125	204.075	0.175	33.199	21	−32.849	11
RT	6	79.050	69.450	148.500	−9.600	23.424	17	−42.624	6
RT	11	79.050	69.025	148.075	−10.025	22.999	16	−43.049	5
RT	12	85.950	68.700	154.650	−17.250	15.774	9	−50.274	2
RT	15	69.725	59.425	129.150	−10.300	22.724	14	−43.329	3
RT	16	86.275	76.125	162.400	−10.150	22.874	15	−43.174	4
RT	19	112.675	114.875	227.550	2.200	35.224	22	−30.824	12
RT	20	99.525	116.250	215.775	16.725	49.749	23	−16.299	13
RT	23	89.425	65.175	153.600	−25.250	7.774	7	−58.274	1
RT	24	55.175	74.575	129.750	19.400	52.424	24	−13.624	16

114

2							
TR	2	74.825	37.350	112.175	−37.475	1	−37.475	7
TR	3	86.875	51.925	138.800	−34.950	2	−34.950	9
TR	7	81.675	72.175	153.850	−9.500	5	−9.500	17
TR	8	92.700	77.500	170.200	−15.200	3	−15.200	14
TR	9	50.450	71.875	122.325	21.425	12	21.425	22
TR	10	66.125	94.025	160.150	27.900	18	27.900	24
TR	13	122.450	124.975	247.425	2.525	6	2.525	18
TR	14	99.075	85.225	184.300	−13.850	4	−13.850	15
TR	17	86.350	95.925	182.275	9.575	8	9.575	19
TR	18	49.925	67.100	117.025	17.175	11	17.175	21
TR	21	42.700	59.425	102.125	16.725	10	16.725	20
TR	22	91.725	114.050	205.775	22.325	13	22.325	23

[a]P.D. $= 2 \times$ (Period Difference).

[b]Assume the observed reference mean is the true mean and $-\theta_L = \theta_U = (0.2)(82.5594) = 16.5119$.

$i = 1,2, \ldots ,n_1;\ i' = 1,2, \ldots ,n_2$ be all possible pairwise differences of the period differences between sequence 1 and sequence 2, i.e.,

$$D_{i,i'} = d_{i1} - d_{i'2}, \qquad i = 1,2, \ldots ,n_1;\quad i' = 1,2, \ldots ,n_2.$$

It can be verified that each $D_{i,i'}$ is an unbiased estimate of θ (i.e., $\mu_T - \mu_R$), i.e.,

$$E(D_{i,i'}) = \theta$$

Denote the ordered set of n_1n_2 differences $D_{i,i'}$ by

$$D(1) < D(2) < \cdots < D(n_1n_2).$$

The median of $\{D(i),\ i = 1,2, \ldots ,n_1n_2\}$ is then a distribution-free point estimator of $\theta = \mu_T - \mu_R$, which is also known as the Hodges-Lehmann estimator (Hodges and Lehmann, 1963), i.e.,

$$\tilde{\theta} = \begin{cases} \dfrac{1}{2}\left[D\left(\dfrac{n_1n_2}{2}\right) + D\left(\dfrac{n_1n_2}{2} + 1\right) \right], & \text{if } n_1n_2 \text{ is even} \\[4mm] D\left(\dfrac{n_1n_2 - 1}{2} + 1\right), & \text{if } n_1n_2 \text{ is odd.} \end{cases} \qquad (4.5.11)$$

The lower and upper limits for the $(1 - 2\alpha) \times 100\%$ distribution-free confidence interval for $\theta = \mu_T - \mu_R$ are then given by

$$L_w = D(w(\alpha))$$

and

$$U_w = D(w(1 - \alpha) + 1), \qquad\qquad\qquad (4.5.12)$$

where $D(w(\alpha))$ and $D(w(1 - \alpha) + 1)$ are the $[w(\alpha)]$th and $[w(1 - \alpha) + 1]$th order statistics of $D(1), D(2), \ldots ,D(n_1n_2)$.

For the assessment of bioequivalence, the distribution-free confidence interval approach is equivalent to the Wilcoxon-Mann-Whitney two one-sided tests procedure since, based on a theorem of Lehmann (1975, p. 87) and (4.5.9), it can be verified that

(i) $D(w(\alpha)) > \theta_L$ if and only if $W_L > n_1n_2 - w(\alpha) = w(1 - \alpha)$, and
(ii) $D(w(1 - \alpha) + 1) < \theta_U$ if and only if $W_U < n_1n_2 - w(1 - \alpha) = w(\alpha)$.

Thus, two approaches reach the same decision in determination of bioequivalence of two formulations.

When the total sample size $n_1 + n_2$ is large, the αth upper percentile $w(\alpha)$ can be approximated by

$$w(\alpha) = \frac{n_1 n_2}{2} + z(\alpha)\sqrt{\tfrac{1}{12} n_1 n_2 (n_1 + n_2 + 1)}. \tag{4.5.13}$$

Hauschke et al. (1990) examined the true coverage probabilities of (L_w, U_w) for $n_1, n_2 = 4, \ldots, 12$. The results indicate that the distribution-free confidence interval (L_w, U_w) is somewhat conservative since its coverage probability is always greater than $(1 - 2\alpha)$ level due to the fact that the distribution of Wilcoxon-Mann-Whitney statistic is discrete. The true coverage probability of the 90% distribution-free confidence interval for $n_1, n_2 = 4, 5, \ldots, 12$ are given in Hauschke et al. (1990).

Steinijans and Diletti (1983) proposed an alternative distribution-free procedure based on the crossover differences O_{ik} as defined in Chapter 3 using a Wilcoxon signed rank statistic (Wilcoxon, 1945) and a Walsh average (Walsh, 1949). This approach may not be of practical interest since it requires (i) equal period effects, and (ii) the distribution of the crossover difference to be symmetric, which are not true in general.

Example 4.5.2 For the example described in Section 3.6, from Table 4.5.1, the median of all possible pairwise differences can be found to be -3.263. Since $w(0.05) = 43$ and $w(0.95) + 1 = 102$, the lower and upper 90% distribution-free confidence intervals are

$$L_w = D(43) = -10.625 \quad \text{and} \quad U_w = D(102) = 4.838,$$

respectively. Thus, $(L_w, U_w) = (-10.625, 4.838)$ is within the equivalence limits $(\theta_L, \theta_U) = (-16.51, 16.51)$. Hence, we conclude that the two formulations are bioequivalent. It can be seen that this conclusion agrees with the confidence interval approaches discussed in Section 4.2. However, (L_w, U_w) is wider than the shortest confidence interval which is given by $(-8.698, 4.123)$. This may in part be explained by the fact that the actual coverage probability for (L_w, U_w) is 91.13% which is higher than the nominal level of 90%.

4.5.3. Bootstrap Confidence Interval

Chow (1990) proposed several alternative approaches for the assessment of bioequivalence with respect to various situations depending upon whether the normality assumptions for $\{S_{ik}\}$ and $\{e_{ijk}\}$ are met. When the distributions of $\{S_{ik}\}$ and $\{e_{ijk}\}$ are both unknown, a nonparametric approach for constructing a $(1 - 2\alpha) \times 100\%$ confidence interval for the ratio $\delta = \mu_T/\mu_R$ using the bootstrap resampling technique (Efron, 1982) was derived under model (4.2.20). We shall refer to such a confidence interval as a bootstrap confidence interval.

To establish this procedure, the following results are needed. The proofs of these results are straightforward and hence omitted.

Theorem 4.5.1 Assume that $n_k/n \to \lambda_k$, $k = 1,2$, where $n = n_1 + n_2$ and $0 < \lambda_k < 1$. Then

$$\sqrt{n}\left(\begin{bmatrix} \overline{\mathbf{X}}_{\cdot 1} \\ \overline{\mathbf{X}}_{\cdot 2} \end{bmatrix} - \begin{bmatrix} \boldsymbol{\mu}_1 \\ \boldsymbol{\mu}_2 \end{bmatrix}\right) \to N(0,\Sigma),$$

where

$$\Sigma = \begin{pmatrix} \lambda_1\Sigma_1 & 0 \\ 0 & \lambda_2\Sigma_2 \end{pmatrix},$$

and $\boldsymbol{\mu}_k = E(\mathbf{X}_{ik})$ and $\Sigma_k = \text{Cov}(\mathbf{X}_{ik})$, a 2×2 positive definite matrix.

Theorem 4.5.2 Let

$$\hat{\delta} = \frac{\overline{Y}_{\cdot 21} + \overline{Y}_{\cdot 12}}{\overline{Y}_{\cdot 11} + \overline{Y}_{\cdot 22}} = \frac{\overline{Y}_T}{\overline{Y}_R}.$$

Then,

$$\sqrt{n}\,(\hat{\delta} - \delta) \to N(0, \sigma^2),$$

where $\sigma^2 = (C_{22} - 2C_{12}\delta + C_{11}\delta^2)/(2\mu_R)^2$ and C_{uv} is the (u,v)th element of $\lambda_1\Sigma_1 + \lambda_2\Sigma_2$.

Theorem 4.5.3

$$\frac{n_1}{n}\,S_1 + \frac{n_2}{n}\,S_2 \to \lambda_1\Sigma_1 + \lambda_2\Sigma_2 \quad \text{almost surely,}$$

where S_k are given in Section 4.2.4.

Theorem 4.5.4 Let $\hat{C}_{uv}$ be the (u,v)th element of $(n_1/n)S_1 + (n_2/n)S_2$. Then

$$\hat{\sigma}^2 \to \sigma^2 \quad \text{almost surely,}$$

where $\hat{\sigma}^2 = (\hat{C}_{22} - 2\hat{C}_{12}\hat{\delta} + \hat{C}_{11}\hat{\delta}^2)/(2\overline{Y}_R)^2$.

Combining the above results, we have the following result:

Theorem 4.5.5

$$\sqrt{n}\,(\hat{\delta} - \delta)/\hat{\sigma} \to N(0,1).$$

Hence, an approximate $(1 - 2\alpha) \times 100\%$ confidence interval for δ is given by

$$\hat{\delta} \pm z(\alpha)\hat{\sigma}/\sqrt{n}.$$

The bootstrap procedure can then be applied on this approximate confidence interval as follows:

Step 1. For given X_{ik}, $i = 1,2, \ldots, n_k$, $k = 1,2$, draw an i.i.d. bootstrap sample with replacement $\{Z_{i1}^b, i = 1,2, \ldots, n_1\}$ from $\{X_{i1}, 1 = 1,2, \ldots, n_1\}$ and an i.i.d. sample $\{Z_{i2}^b, i = 1,2, \ldots, n_2\}$ from $\{X_{i2}, i = 1,2, \ldots, n_2\}$.

Step 2. Based on $\{Z_{ik}^b, i = 1, \ldots, n_k; k = 1,2\}$, calculate $\hat{\delta}_b$, $\hat{\sigma}_b$, and

$$T_b = \frac{\sqrt{n}(\hat{\delta}_b - \hat{\delta})}{\hat{\sigma}_b}.$$

Step 3. Repeated Step 1 and Step 2 a large number of times (say B times), i.e., $b = 1,2, \ldots, B$. The bootstrap confidence interval for δ with approximate confidence level of $1 - 2\alpha$ is given by

$$(\hat{\delta} + \omega(\alpha)\hat{\sigma}/\sqrt{n}, \hat{\delta} + \omega(1 - \alpha)\hat{\sigma}/\sqrt{n}),$$

where $\omega(\alpha)$ is the αth quantile of the histogram $\{T_b, b = 1, \ldots, B\}$.

Based on a simulation study, Chow, Cheng, and Shao (1990) showed that the coverage probability for the bootstrap confidence interval is uniformly close to the nominal ones for several combinations of parameters under study and for various distribution assumptions on $\{S_{ik}\}$ and $\{e_{ijk}\}$. However, the drawback of this method is that the bootstrap sampling is based on the asymptotic results (i.e., asymptotic variance) which does not provide the actual variability associated with $\hat{\delta}$.

Example 4.5.3 To illustrate the use of this approach, consider the example given in Section 3.6. The estimates for δ and σ and the critical values $\omega(\alpha)$ and $\omega(1 - \alpha)$ obtained from B = 1000 bootstrap samples are given below:

$$\hat{\delta} = 0.972, \quad \hat{\sigma} = 0.217, \quad \text{and}$$

$$\omega(0.05) = -1.926, \quad \omega(0.95) = 1.543.$$

Thus, the 90% bootstrap confidence interval is given by (92.8%, 104.1%), which is in favor of the conclusion of bioequivalence.

Note that the above bootstrap procedure can be applied to the confidence interval discussed in Section 4.2. when the normality assumptions of $\{S_{ik}\}$ and $\{e_{ijk}\}$ are seriously violated. In this case, however, the finite sample performance of the bootstrap confidence interval should be examined.

4.6. DISCUSSION AND OTHER ALTERNATIVES

4.6.1. Discussion

In this chapter, we have introduced several statistical methods for the assessment of bioequivalence according to the ± 20 rule. Basically, these methods (except

Table 4.6.1 Summary of Methods for Assessment of Bioequivalence in Average Bioavailability

Approach	Method	Parameter	Statistics	Decision on bioequivalence
Confidence interval	Classical	$\mu_T - \mu_R$	$CI_1 = (\bar{Y}_T - \bar{Y}_R) \pm S$	$CI_1 \in (\theta_L, \theta_U)$
	Westlake	μ_T	$CI_3 = \pm k_1 S \pm (\bar{Y}_T - \bar{Y}_R)$	$CI_3 \in (\theta_L, \theta_U)$
	Fieller	μ_T/μ_R	$CI_2 = \dfrac{1}{1-G}\left[\left(\hat{\delta} - G\dfrac{S_{TR}}{S_{RR}}G^*\right) \pm \left(t\dfrac{\sqrt{wS_{RR}^2}}{\bar{Y}_R}\right)\right]$	$CI_2 \in (0.8,\ 1.2)$
	Fixed Fieller	μ_T/μ_R	$CI_2 = \dfrac{1}{1-G}\left\{\hat{\delta} \pm t\dfrac{\sqrt{wMS_{Intra}}}{\bar{Y}_R}[\hat{\delta}^2 + (1 - G_1)]^{\frac{1}{2}}\right\}$	$CI_2 \in (0.8,\ 1.2)$
	Chow-Shao	(μ_R, μ_T)	confidence region (CR) for (μ_R, μ_T)	CR bounded by $0.8\mu_R$ and $1.2\mu_R$
	Nonparameteric	$\mu_T - \mu_R$	$CI_1 = (D_{(w(\alpha))}, D_{(w(1-\alpha)+1)})$	$CI_1 \in (\theta_L, \theta_U)$
Hypotheses testing	Schuirmann	$\mu_T - \mu_R$	$T_L = \dfrac{\bar{Y}_T - \bar{Y}_R - \theta_L}{\bar{Y}_T - \bar{Y}_R - \theta_U}$, $T_U = \dfrac{\bar{Y}_T - \bar{Y}_R - \theta_U}{}$ /S	$T_L > t$ and $T_U > t$
	Anderson-Hauck	$\mu_T - \mu_R$	$p = F_t(T_U) - F_t(-T_L)$ or $p = F_t(-T_L) - F_t(T_U)$	$p < \alpha$
	Nonparametric	$\mu_T - \mu_R$	$W_L = R_L - [n_1(n_1 + 1)/2]$, $W_U = R_U - [n_1(n_1 + 1)/2]$	$W_L > w(1 - \alpha)$, $W_U < w(\alpha)$
Bayesian	Rodda-Davis	—	$P_{RD} = F_t(t_U) - F_t(t_L)$	$P_{RD} > 0.90$
	Mandallaz-Hau	—	$P_{MM} = F_t(t_{\delta_L}) - F_t(t_{\delta_U})$	$P_{MM} > 0.90$
	Grieve (C = 0)	—	$P_{GF1} = F_t(t_{0U}) - F_t(t_{0L})$	$P_{GF1} > 0.90$

Note: $t = t(\alpha, n_1 + n_2 - 2)$, $S = \hat{\sigma}_d \sqrt{1/n_1 + 1/n_2}$, $\hat{\delta} = \bar{Y}_T/\bar{Y}_R$, $-\theta_L = \theta_U = 0.2\mu_R$, and others can be found in text.

for P_{GF2}) were derived under the assumption of equal carry-over effects which are summarized in Table 4.6.1. Some of these methods are actually operationally equivalent in the sense that they will reach the same decision on bioequivalence. For example, Schuirmann's two one-sided tests procedure based on a t-test is equivalent to the $(1 - 2\alpha) \times 100\%$ classical (shortest) confidence interval which is, in turn, equivalent to the $(1 - 2\alpha) \times 100\%$ HPD interval of Bayesian approach proposed by Rodda and Davis. In addition, the p-value of Anderson and Hauck's procedure, as shown in Section 4.3.3, can be easily obtained from T_L and T_U, the test statistics for Schuirmann's two one-sided tests procedure. Moreover, the $(1 - 2\alpha) \times 100\%$ confidence interval for μ_T/μ_R based on the fixed Fieller's theorem is equivalent to the Bayesian method proposed by Mandallaz and Mau. In Bayesian approaches, the pivotal quantity T_{GF1} proposed by Grieve only differs by a small constant $\sqrt{(n_1 + n_2 - 4)/(n_1 + n_2 - 2)}$ from T_{RD} which was proposed by Rodda and Davis. These two methods are essentially the same for decision-making of bioequivalence. For nonparametric methods, the two one-sided procedure based on Wilcoxon-Mann-Whitney test is equivalent to the distribution-free $(1 - 2\alpha) \times 100\%$ confidence interval based on Lehmann-Hodges estimator. Although some of these methods are operationally equivalent in the process of decision-making, their interpretations, however, are quite different. For example, the $(1 - 2\alpha) \times 100\%$ confidence intervals obtained based on sampling theory have a completely different interpretation from the HPD interval obtained from Bayesian methods. Although the $(1 - 2\alpha) \times 100\%$ confidence interval for $\mu_T - \mu_R$ is statistically equivalent to the hypothesis testing for $H_0: \mu_T = \mu_R$ vs $H_a: \mu_T \neq \mu_R$, the use of a confidence interval as a decision tool for bioequivalence needs to be carefully examined because the assurance in terms of coverage probability is referred to as $\mu_T - \mu_R$ within the confidence limits, not to the bioequivalence limits stated in the ± 20 rule.

To illustrate the use of these methods, the example described in Section 3.6 was used. The results from each method are summarized in Tables 4.6.2 through 4.6.5. All of the methods led to the conclusion of bioequivalence except for Grieve's Bayesian method using P_{GF2} which assumes that the carry-over effects are present. This indicates that the presence of the carry-over effects has some impact on the assessment of bioequivalence. Thus, a preliminary test for the presence of the carry-over effects is necessary in order to choose an appropriate statistical method for assessing bioequivalence. For the example in Section 3.6, test statistic T_C in (3.2.2) for the hypothesis $H_0: C_R = C_T$ indicates that there is little evidence for the presence of unequal carry-over effects $|T_C| = 0.612 <$ $t(0.025,22) = 2.074$, $p = 0.547$). Thus, we conclude that there are no carry-over effects (i.e., $C_R = C_T = 0$) under the constraint $C_R + C_T = 0$. Note that the methods introduced in this chapter except for P_{GF2} were derived under the assumption of no carry-over effects. In practice, there is a reasonable assumption

Table 4.6.2 Summary of Results from Confidence Interval Approach for Example 3.6

Method	Parameter	Decision-making[a]	90% CI	Conclusion
Shortest	$\mu_T - \mu_R$	$CI_1 \in (\theta_L, \theta_U)$[b]	$CI_1 = (-8.698, 4.123)$	BE[c]
	μ_T/μ_R	$CI_2 \in (\delta_L, \delta_U)$	$CI_2 = (89.46\%, 104.99\%)$	BE
Westlake	μ_T	$CI_3 \in (-\Delta, \Delta)$	$CI_3 = (-7.413, 7.413)$	BE
Fieller	μ_T/μ_R	$CI_2 \in (\delta_L, \delta_U)$	$CI_2 = (89.78\%, 105.19\%)$	BE
Fixed Fieller	μ_T/μ_R	$CI_2 \in (\delta_L, \delta_U)$	$CI_2 = (89.85\%, 105.20\%)$	BE
Chow and Shao	(μ_R/μ_T)	$T_1 < F(1 - 2\alpha, 2, n - 2)$ and conditions $1'$ through $3'$ in Section 4.2.4 hold	—	BE

[a]CI_1, CI_2, and CI_3 are confidence intervals for $\mu_T - \mu_R$, μ_T/μ_R, and μ_T, respectively.
[b]In most cases, $-\theta_L = \theta_U = 0.2\mu_R$; $\delta_L = 80\%$ and $\delta_U = 120\%$; and $\Delta = 20\%$.
[c]BE = Bioequivalence.

since the washout period can be chosen to be long enough to eliminate the possible carry-over effects.

Among the methods introduced, the method of the confidence interval appears to be the most commonly used method for the assessment of bioequivalence. However, it should be noted that (i) the equivalence in average bioavailability does not imply the equivalence in the variability of bioavailability, and (ii) the

Table 4.6.3 Summary of Results from the Method of Interval Hypotheses for Example 3.6

Method	Hypotheses	Decision-making	P-Value[a]	Conclusion
Schuirmann's two one-sided tests	$H_{01}: \mu_T - \mu_R \le \theta_L$ $H_{02}: \mu_T - \mu_R \ge \theta_U$	$p_1 < 0.05$ and $p_2 < 0.05$	$p_1 = 0.00002$ $p_2 = 0.00048$	BE
Anderson and Hauck	$H_{01}: \mu_T - \mu_R \le \theta_L$ or $\mu_T - \mu_R \ge \theta_U$	$p < 0.05$	$p = 0.00045$	BE

[a]p_1 and p_2 are the observed p-values under H_{01} and H_{02}, respectively, while p is the observed p-value under H_0.

Table 4.6.4 Summary of Results from the Bayesian Approach for Example 3.6

Method	Prior	Decision-making	Posterior prob.	Conclusion
Rodda and Davis	Noninformative	$P_{RM} > 0.90$	$P_{RD} = 0.9995$	BE
Mandallaz and Mau	Improper vague prior	$P_{MM} > 0.90$	$P_{MM} = 0.9997$	BE
Grieve	Joint noninformative prior	$P_{GF1} < 0.90$ or $P_{GF2} < 0.90$	$P_{GF1} = 0.9992$[a]	BE
			$P_{GF2} = 0.8383$	not BE

[a]P_{GF1} assumes that there are no carry-over effects, while P_{GF2} is obtained in the presence of carry-over effects.

probability of correctly concluding bioequivalence (or the constructed confidence interval being within the equivalent limits) may not be of the desired level (say, 90%). More discussion regarding the assessment of bioequivalence in terms of average bioavailability as well as the variability of bioavailability will be given in Chapter 7.

Appendix B.1 gives the SAS programs for computation of all procedures discussed in this chapter except for the confidence region in (4.2.4) and bootstrap resampling procedure in (4.5.3). A SAS program for Grieve's Bayesian procedure is given in Appendix B.2. Note that the SAS programs for Westlake's symmetric confidence interval, Anderson-Hauck's procedure, and Mandallaz-Mau's Bayesian method are modified based upon the programs given in Metzler (1988).

Table 4.6.5 Summary of Results from Nonparameteric Approaches for Example 3.6

Method	Parameter or hypotheses	Decision-making	P-Value or 90% CI	Conclusion
W-M-W two one-sided tests[a]	H_{01}: $\mu_T - \mu_R \leq \theta_L$ H_{02}: $\mu_T - \mu_R \geq \theta_U$	$p_1 < 0.05$ and $p_2 < 0.05$	$p_1 < 0.01$ $p_2 < 0.01$	BE
Hodge-Lehmann	$\mu_T - \mu_R$	$CI_1 \in (\theta_L, \theta_U)$	$(-10.63, 4.84)$	BE
Bootstrap	μ_T/μ_R	$CI_2 \in (\delta_L, \delta_U)$	$(92.8\%, 104.1\%)$	BE

[a]W-M-W = Wilcoxon-Mann-Whitney.

4.6.2. Other Alternatives

As indicated in Section 1.3, the extent and rate of absorption of a drug product are often used to assess bioequivalence between formulations of the drug product. The extent and rate of absorption of the drug product are measured by the pharmacokinetic parameters such as AUC, C_{max}, and t_{max} which can be obtained from the blood or plasma concentration-time curve. For the assessment of bioequivalence, separate univariate analysis of each parameter is usually performed based on the methods introduced above. However, it should be noted that AUC, C_{max}, and t_{max} are some summarized characteristics of the observed blood or plasma concentration-time curve. Thus, for a single dose bioavailability/bioequivalence study, instead of studying each parameter alone, it may be more appropriate to compare the profiles of the blood or plasma concentration-time curves for the test and reference formulations by use of a repeated measures model (Johnson and Wichern, 1982). In other words, we may consider the blood or plasma concentrations at each sampling time following a dose. For each subject, these concentrations are considered repeated measures. Analysis with a repeated measures model is a comparison of the average concentrations over time. Based on this model, the difference in the shape of the profile of the blood or plasma concentration-time curve between formulations can be examined by testing the interaction between formulation and sampling times. A significant interaction between formulation and sampling times may indicate that the rates of absorption, distribution, and elimination of the drug are different between formulations. Although a repeated measures model may be appropriate, the following are some difficulties which often occur in practice:

   (i)   this model requires restrictive assumptions about the covariance matrix;
   (ii) it is very difficult to get a reliable estimate of the covariance matrix because of the small number of subjects and large number of time points;
   (iii) unlike AUC, average concentrations over time is not a measure of the average amount of drug absorbed in the body;
   (iv) missing values at some sampling time points often occur which lead to an unbalanced situation and consequently complicate the analysis.

Therefore, for the assessment of bioequivalence using the repeated measures model based on the blood or plasma concentrations, further research is needed to address the above concerns.

Several multivariate approaches are also available for the analysis of the blood or plasma concentrations. For example, the method proposed by Grizzle and Allen (1969) can be applied to the blood or plasma concentrations which require no special assumptions on the structure of the covariance matrix. Snee (1972) proposed a method which combines the analysis of variance and principal components for the analysis of continuous curves. This method can be applied to compare the shape of the blood or plasma concentration-time curve between formulations which require less restrictive assumptions about the covariance matrix. Metzler (1974) indicated that other multivariate techniques such as a multivariate randomization test could be considered and investigated for their possible application to bioavailability studies.

Based on the observed blood or plasma concentrations, another approach, which is often considered as an adjunct to bioavailability studies, is the fitting of a pharmacokinetic model (McQuarrie, 1967; Jacques, 1972). However, the method involves the estimation of parameters from a nonlinear model (one compartment model or multi-compartment model). It is often very difficult to have some desired statistical properties (such as unbiasness and uniformly minimum variance) for estimates of the parameters of interest (e.g., the rate of absorption) due to the nature of nonlinearity. Besides, since the model may vary from subject to subject, statistical inferences such as estimate and confidence interval of the parameter obtained from a model, which may be able to adequately describe the drug kinetics, may not be meaningful due to the inter-subject and the intra-subject variabilities. Thus, how to account for the inter-subject and intra-subject variabilities in pharmacokinetic modeling is an important issue. To date, little attention has been paid to this difficulty. Further research in this area would be worthwhile.

5

Power and Sample Size Determination

5.1. INTRODUCTION

As indicated in Chapter 1, one of the major objectives of a bioavailability/
bioequivalence study comparing two formulations (e.g., a test formulation and
a reference formulation) of a drug product is to determine whether the two
formulations are bioequivalent. During the planning stage of a bioavailabil-
ity/bioequivalence study, the following questions regarding the study sample size
are of particular interest to the clinicians:

(1) how many subjects are needed in order to have a desired power (e.g.,
     80%) establishing bioequivalence between two formulations within clin-
     ically important limits (e.g., 20% of the reference mean)?
(2) what's the "trade off" if only a small number of subjects are available
     for the study due to limited budget and/or some medical considerations?

In order to provide answers to these questions, a statistical evaluation for sample
size determination is often employed. The most commonly used approach is to
perform a pre-study power calculation based on an estimate of the intra-subject
from previous studies. In other words, an appropriate sample size is chosen to
meet the desired power for assessment of bioequivalence within clinically important limits, assuming that the estimate of intra-subject variability is the true
intra-subject variability.

Determination of a sample size depends upon the power function of some test statistics for the hypotheses to be evaluated for assessment of bioequivalence. A commonly used approach is to choose a sample size based upon a power function of the test statistic for the hypothesis of equality between formulation effects (i.e., $\mu_T = \mu_R$). However, as indicated in the previous chapter, the method of hypothesis testing for equality between formulation effects (or point hypotheses) is not an appropriate statistical method for assessment of bioequivalence. The sample size obtained based upon a power function of the test for point hypotheses may not be large enough to provide sufficient power, if other appropriate statistical methods such as the classical confidence interval approach and Schuirmann's two one-sided tests procedure for interval hypotheses are used.

For assessment of bioequivalence, as indicated in the previous chapter, the classical confidence interval approach and Rodda and Davis's Bayesian method are (operationally) equivalent to Schuirmann's two one-sided t tests procedure for interval hypotheses in the sense that they lead to the same conclusion of bioequivalence. Therefore, in this chapter, without loss of generality, a sample size determination will be evaluated based upon the power of Schuirmann's two one-sided t tests procedure for interval hypotheses using the ± 20 rule for assessment of bioequivalence.

In the next section, the concept of type I and type II errors which occur in interval hypotheses testing will be introduced. In Section 5.3, the powers and sizes of Schuirmann's two one-sided t tests procedure and Anderson and Hauck's procedure will be examined. Statistical properties for a commonly used power approach, which is based on the 80/20 rule, for assessment of bioequivalence will be examined. Also included will be the comparison of relative efficiencies among methods introduced in the previous chapter. In Section 5.4, a formula for sample size determination will be provided based upon the power function of Schuirmann's two one-sided t tests procedure. Since the calculation of power for Schuirmann's two one-sided t tests procedure at $\theta = \theta_0 \neq 0$ requires a complicated numerical integration, an empirical power, which can easily be obtained through a simulation study will be introduced in order to approximate the true power. The power of Schuirmann's two one-sided t tests procedure is also compared with that of Anderson and Hauck's procedure in this section.

5.2. HYPOTHESES AND TYPE I AND TYPE II ERRORS

5.2.1. Hypotheses Testing

In bioavailability studies, a hypothesis is a postulation, assumption, or statement that is made about the population regarding a drug product under study such as a test formulation and a reference formulation of the drug product. The statement that there is a carry-over effect (i.e., $C_R \neq C_T$), for example, is a hypothesis regarding the existence of unequal carry-over effects of the drug under study. Another example is the statement that there is a direct formulation effect (i.e., $F \neq 0$). This is a hypothesis regarding the treatment effect. A random sample is usually drawn through a bioavailability study to evaluate hypotheses about the drug product. To perform a hypothesis testing, the following steps are essential:

(1) Choose the hypothesis that is to be questioned, denoted by H_0, where H_0 is usually referred to as the null hypothesis;

(2) Choose an alternative hypothesis, denoted by H_a, where H_a is usually the hypothesis of particular interest to the investigators;

(3) Select a test statistic and define the rejection region (or a rule) for decision making about when to reject the null hypothesis and when to fail to reject it.

(4) Draw a random sample by conducting a bioavailability study;

(5) Calculate test statistic(s);

(6) Make conclusion(s) according to the pre-determined rule specified in (3).

5.2.2. Type I and Type II Errors

Basically, two kinds of errors occur when testing hypotheses. If the null hypothesis is rejected when it is true, then a type I error has occurred. If the null hypothesis is not rejected when it is false, then a type II error has been made. The probabilities of making type I and type II errors are given below:

α = P(type I error)

= P(reject H_0 when H_0 is true),

β = P(type II error)

= P(fail to reject H_0 when H_0 is false).

The probability of making a type I error, α, is called the level of significance. In practice, α is also known as the producer's risk, while β is sometimes referred to as the consumer's risk. Table 5.2.1 summarizes the relationship between type I and type II errors when testing hypotheses.

Power of the test is defined as the probability of correctly rejecting H_0 when H_0 is false, i.e.,

Power = $1 - \beta$

= P(reject H_0 when H_0 is false).

To illustrate the relationship between α and β (or power), a plot based upon the hypothesis of equality is presented in Figure 5.2.1, which shows various β's

Table 5.2.1 Relationships Between Type I and Type II Errors

I. General case

		If H_0 is	
		True	False
When	Fail to reject	No error	Type II error
	Reject	Type I error	No error

II. Bioequivalence trial

	True state H_0	
Decision	Bioinequivalent	Bioequivalent
Bioinequivalent (Fail to reject H_0)	Right decision	Type II error
Bioequivalent (Reject H_0)	Type I error	Right decision

under H_0 (i.e., AUC $=$ 100) for various alternatives (i.e., AUC $=$ 102, 104, and 106) at $\alpha = 5\%$ and 10%. It can be seen that α decreases as β increases and α increases as β decreases. The only way of decreasing both α and β is to increase the sample size. In practice, since a type I error is usually considered to be a more important and/or serious error which one would like to avoid, a typical approach in hypothesis testing is to control α at an acceptable level and try to minimize β by choosing an appropriate sample size. In other words, the null hypothesis can be tested at a pre-determined level (or nominal level) of significance with a desired power. From Figure 5.2.1, it can be seen that for a fixed α, β increases when H_a moves toward H_0. This means that we will not have sufficient power to detect a small difference between H_0 and H_a. On the other hand, β decreases when H_a moves away from H_0, increasing the test power.

5.2.3. Hypotheses Setting

In practice, the null hypothesis H_0 and the alternative hypothesis H_a are sometimes reversed and evaluated for different interests. However, it should be noted that a test for H_0 vs H_a is not equivalent to a test for $H_0' = H_a$ vs $H_a' = H_0$. Two tests under different null hypotheses may lead to a totally different conclusion. For example, a test for H_0 vs H_a may lead to the rejection of H_0 in favor of H_a. However, a test for $H_0' = H_a$ vs $H_a' = H_0$ may reject the null hypothesis. Thus, the choice of the null hypothesis and the alternative hypothesis may have some

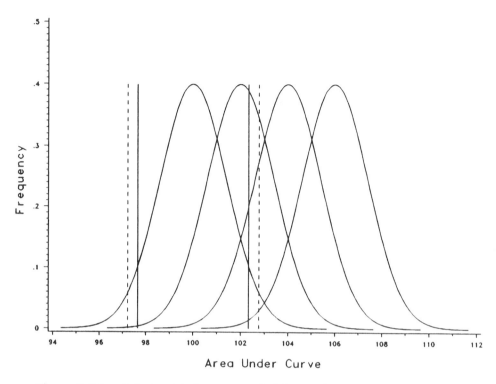

Figure 5.2.1   Relationship between probabilities of type I and II errors. (Hypothesis of equality. The null hypothesis is mean AUC = 100. Solid line corresponds to the test at the 10% nominal level. Dashed line corresponds to the test at the 5% nominal level.)

impact on the parameter to be tested. The following criteria are commonly used as rule of thumb for choosing the null hypothesis.

Rule 1: Choose H_0 based upon the importance of a type I error. Under this rule, we believe that a type I error is more important and serious than that of a type II error. We would like to control the chance of making a type I error at a tolerable limit (i.e., α). Thus, H_0 is chosen so that the maximum probability of making a type I error, i.e., P(reject H_0 when H_0 is true), will not exceed α level.

Rule 2: Choose the hypothesis we wish to reject as H_0 (Colton, 1974; Ott, 1984; Ware et al. 1986). The purpose of this rule is to establish H_a by rejecting H_0. Note that we will never be able to prove that H_0 is true even though the data fail to reject it.

In some cases, for a given set of hypotheses, it may be easy to determine whether a type I error is more important and/or serious than a type II error. If a type II error appears to be more important and/or serious than a type I error, Rule 1 suggests that the null hypothesis and the alternative hypothesis be reversed. In many cases, however, the relative importance of the type I error and the type II error is usually very subjective. In this case, Rule 2 is useful in choosing H_0 and H_a. To illustrate the use of these two criteria, consider the following three examples.

Example 5.2.1 Example of Doctor-Patient Suppose there is a very sick patient who is about to die and his life is dependent upon life-support equipment. The situation that exists here is that when no sign of life (e.g., no pulse or heart beat) is detectable on the equipment, the physician will have to make a decision whether the patient is still alive. The consequence of this decision is that if the patient is pronounced dead, the life-support equipment will be removed; otherwise, the life-support equipment will remain and a rescue action will be taken. When a decision is made as to whether the patient is dead or still alive, the following errors may occur:

(1) The doctor declares the patient to be dead when in fact the patient is still alive. As a consequence of this wrong decision, the life-support equipment will be removed and the patient will then be dead. The cost of this error is a human life.

(2) The doctor concludes that the patient is still alive when in fact the patient is dead. In this case, the life-supporting equipment will remain. The consequence of this wrong decision is that the life-support equipment will not be available for other patients who are in critical condition and need the support of this equipment.

A wrong decision on the death of a patient when he or she is still alive is clearly more serious than the mistake made by concluding that the patient is alive when in fact he or she is dead. A doctor will want to avoid or at least minimize the chance of making this error of wrongly concluding the death of a patient. In this case, Rule 1 may be used to choose H_0 as follows:

H_0: the patient is alive.

vs H_a: the patient is dead.

Therefore, the probability of making a type I error will not exceed a tolerable limit, e.g., 1%. In this case, in order to reject the null hypothesis that the patient is alive in favor of the alternative hypothesis that the patient is dead, strong evidence or more information (e.g., brain death) is needed before a decision can be made.

However, it should be noted that we cannot prove that the patient is still alive even if we fail to reject H_0. Based on Rule 2, we are, in fact, more interested in proving that the patient is dead. On the other hand, if we are interested in proving that the patient is alive, we may want to consider the following hypotheses based on Rule 2:

H_0: the patient is dead.

vs H_a: the patient is alive.

Example 5.5.2   Example of Jury-Criminal   Suppose there is a suspect who is charged with first degree murder and is currently on trial. The situation that exists here is that if the suspect is found guilty, he or she will then be sentenced to death (or a life sentence). On the other hand, if he or she is found innocent, he or she will then be released. The jury will have to determine whether he or she is guilty. Again, there are two kinds of mistakes that the jury can make:

(1)   The jury can conclude that the suspect is guilty when, in fact, he or she is innocent. The consequence of this mistake is that the person may lose his or her life or spend the rest of his or her life in jail. Human rights would be seriously violated.

(2)   The suspect is set free when, in fact, he or she is guilty of first degree murder. The consequence of this mistake is that the suspect may again kill or threaten other people's lives. In this situation, the safety of the society is of great concern.

Unlike the doctor-patient example, it is not clear which mistake is more important and serious. Different people may have different viewpoint(s) on this issue. However, if you believe that human rights are more important than anything else, you probably want to minimize the chance of making the first kind of error. In other words, based on Rule 1, you are interested in testing the following hypotheses:

H_0: the suspect is innocent.

vs H_a: the suspect is guilty.

To prove the suspect is guilty by rejecting H_0, more evidence will be needed so that the chance of making a type I error is controlled at a tolerable limit, e.g., 1%. Based on the above hypotheses, we won't be able to prove that the suspect is innocent even when we fail to reject H_0. However, in the current judicial system, a suspect is assumed innocent until he or she is proven guilty. Note that, based on Rule 2, the purpose of the above hypotheses is to prove that the suspect is guilty.

On the other hand, if the safety of the community is of greatest concern, we may want to minimize the probability of making the second kind of mistake. Therefore, the hypotheses are set up as follows:

H_0: the suspect is guilty.

vs H_a: the suspect is innocent.

To avoid making the mistake of wrongly concluding the suspect is innocent when in fact he or she is guilty, a careful evaluation is necessary before the suspect is concluded to be innocent. The purpose of the above hypotheses is, therefore, to prove that the suspect is innocent.

Example 5.2.3 Example of Bioequivalence-Bioinequivalence Similarly, the following two errors occur in the assessment of bioequivalence when comparing two formulations in average bioavailabilities:

(1) We conclude bioequivalence when in fact the test formulation is not bioequivalent to the reference formulation;
(2) We conclude bioinequivalence when in fact the test formulation is bioequivalent to the reference formulation.

In the interest of controlling the chance of making a type I error, the FDA may consider (1) is more important than (2) and consequently prefer the following hypotheses:

H_0: Bioinequivalence

vs H_a: Bioequivalence.   (5.2.1)

On the other hand, pharmaceutical companies may want to eliminate the probability of wrongly rejecting the null hypothesis of bioequivalence. Thus, the following hypotheses are used:

H_0: Bioequivalence

vs H_a: Bioinequivalence.   (5.2.2)

It is very subjective whether (1) is more important than (2) or (2) is more important than (1) when comparing two formulations of the same drug product.

In bioavailability studies, Rule 2 is usually applied to choose H_0. For example, when a new formulation is developed by the innovator, the innovator will want to show bioequivalence between the new formulation and the reference formulation by disproving the hypothesis of bioinequivalence. In this case, hypotheses (5.2.1) may be considered. On the other hand, if the formulation is prepared by generic companies, the innovator will be in favor of the hypothesis of bioinequivalence. In this latter case, hypothesis (5.2.2) is preferred.

5.3. POWER AND RELATIVE EFFICIENCY

5.3.1. Power and Size of Tests

Since, based on Rule 2 for hypotheses setting, the purpose of a bioavailability study is to establish bioequivalence between formulations, the hypotheses in (5.2.1) are considered, which can be stated in terms of interval hypotheses in (4.3.1).

Without loss of generality, the equivalence limits (θ_L, θ_U) can be chosen such that $-\theta_L = \theta_U = \Delta > 0$. Thus, the hypotheses (5.2.1) become:

$$H_0: \theta \leq -\Delta \quad \text{or} \quad \theta \geq \Delta \text{ (i.e., bioinequivalence)}$$

$$\text{vs} \quad H_a: -\Delta < \theta < \Delta \text{ (i.e., bioequivalence)}, \qquad (5.3.1)$$

where $\theta = \mu_T - \mu_R$.

From hypotheses (5.3.1), it can be seen that the true difference in formulation means, θ, could be any real number between $-\infty$ and ∞. Under H_0, θ is either in $(-\infty, -\Delta]$ or $[\Delta, \infty)$, while θ could be any number between $-\Delta$ and Δ under H_a. Let

$$\Omega_0 = \{(-\infty, -\Delta] \cup [\Delta, \infty)\}, \quad \text{and}$$
$$\Omega_a = \{(-\Delta, \Delta)\}, \qquad (5.3.2)$$

where Ω_0 and Ω_a are usually referred to as the parameter space corresponding to H_0 and H_a, respectively.

The hypotheses in (5.3.1) are then equivalent to

$$H_0: \theta \in \Omega_0$$
$$\text{vs} \quad H_a: \theta \in \Omega_a. \qquad (5.3.3)$$

A test is said to have a significance level of α if the probability of committing a type I error is less than or equal to α, $0 < \alpha < 1$, for $\theta \in \Omega_0$, i.e.,

P{reject H_0 given bioinequivalence}

= P{reject H_0 given $\theta \in \Omega_0$}

$\leq \alpha$.

A test is said to have a size of α if

$$\max_{\theta \in \Omega_0} P\{\text{reject } H_0 \text{ given } \theta \in \Omega_0\} = \alpha.$$

Similarly, the probability of correctly concluding bioequivalence or power of the test for hypotheses (5.3.1) or (5.3.3) is given by

$$\phi(\theta) = 1 - \beta$$
$$= P\{\text{reject } H_0 \text{ given } \theta \in \Omega_a\}$$
$$= P\{\text{reject } H_0 \text{ given bioequivalence}\},$$

where β is the probability of wrongly concluding bioequivalence, i.e.,

β = P{fail to reject H_0 given $\theta \in \Omega_a$}

= P{fail to reject H_0 given bioequivalence}.

5.3.2. Power of Schuirmann's Two One-Sided t Tests Procedure

Let

$$Y = \overline{Y}_T - \overline{Y}_R, \quad m = \sqrt{\frac{1}{n_1} + \frac{1}{n_2}},$$

$$t = t(\alpha, n_1 + n_2 - 2), \quad \text{and} \quad s = \hat{\sigma}_d.$$

Schuirmann's two one-sided t tests procedure for hypotheses (5.3.1) or (5.3.3) then leads to the rejection of H_0 at the α level of significance if

$$t_1 = \frac{Y + \Delta}{ms} > t \quad \text{and} \quad t_2 = \frac{Y - \Delta}{ms} < -t.$$

In other words, the null hypothesis, H_0, of bioinequivalence is rejected at the α level of significance if the observed value of Y is within the following rejection region:

$$\mathcal{R} = \{(Y,s)| -\Delta + tms < Y < \Delta - tms\}. \tag{5.3.4}$$

Note that the acceptance region (i.e., the region that leads to the conclusion of bioinequivalence) is the complement of $\mathcal{R}$, i.e.,

$$\mathcal{R}^c = \{(Y,s)| Y \leq -\Delta + tms \quad \text{or} \quad Y \geq \Delta - tms\}.$$

Schuirmann (1987) examined the rejection region $\mathcal{R}$ in terms of the relationship between Y (observed difference in LS means between formulations) and ms (standard error). For example, Figure 5.3.1 gives the rejection region (the triangular area) at the α level of significance for the case where $\Delta = 20$ and $n_1 = n_2 = 6$. The results indicate that we will never reject H_0 (or conclude bioequivalence) if the standard error of Y (i.e., ms) is greater than Δ/t, even when $Y = 0$. On the other hand, if the standard error of Y is small, a large value of Y would possibly lead to the rejection of H_0 as long as Y is within $(-\Delta, \Delta)$.

Based on hypotheses (5.3.1) or (5.3.3), the power for Schuirmann's two one-sided tests procedure (or the probability of correctly concluding bioequivalence) is given below:

$$\phi_s(\theta) = P\{\text{reject } H_0 \text{ given bioequivalence}\}$$

$$= P\{(Y,s) \in \mathcal{R} \text{ given } \theta \in \Omega_a\}. \tag{5.3.5}$$

The power can be evaluated at $\theta = \theta_0$ as follows:

$$\phi_s(\theta_0) = P\left\{ \frac{-\theta_0}{m\sigma_d} + \frac{-\Delta + tms}{m\sigma_d} < \frac{Y - \theta_0}{m\sigma_d} < \frac{-\theta_0}{m\sigma_d} + \frac{\Delta - tms}{m\sigma_d} \right\}$$

$$= E\left[P\left\{ \frac{-\theta_0}{m\sigma_d} - \frac{(\Delta - tms)}{m\sigma_d} < \frac{Y - \theta_0}{m\sigma_d} \right. \right.$$

$$\left. \left. < \frac{-\theta_0}{m\sigma_d} + \frac{\Delta - tms}{m\sigma_d} \,\middle|\, s \right\} \right],$$

(5.3.6)

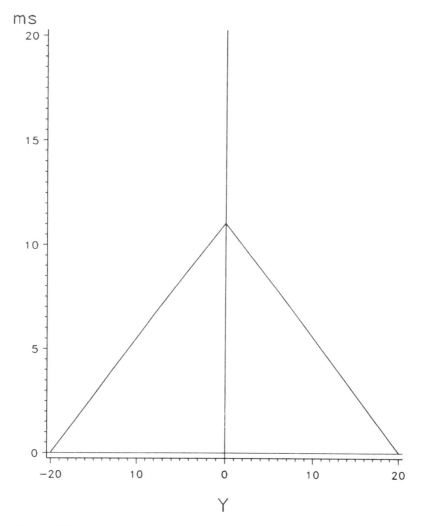

Figure 5.3.1 Rejection region of Schuirmann's two one-sided tests procedure for $\Delta = 20$, $n_1 = n_2 = 6$, and the 0.05 nominal level. (From Schuirmann (1987).)

where the expectation is taken over the distribution of s^2/σ_d^2 which is a $\chi^2(n_1 + n_2 - 2)/(n_1 + n_2 - 2)$.

Denote

$$a = \frac{\theta_0}{m\sigma_d}, \quad b = \frac{\Delta - tms}{m\sigma_d}, \quad \text{and} \quad Z = \frac{Y - \theta_0}{m\sigma_d}.$$

Then, (5.3.6) becomes

$$\begin{aligned}
\phi_s(\theta_0) &= E\ [P\{-a - b < Z < -a + b \mid s\}] \\
&= E\ [\Phi(-a + b) - \Phi(-a - b)], \quad\quad (5.3.7)
\end{aligned}$$

where Φ is the cumulative distribution function of the standard normal distribution.

If $\theta = -\theta_0$, the power is given by

$$\phi_s(-\theta_0) = E\ [\Phi(a + b) - \Phi(a - b)].$$

Since $\Phi(a + b) - \Phi(a - b) = \Phi(-a + b) - \Phi(-a - b)$, the power function $\phi_s(\theta)$ is symmetric about $(\theta_L + \theta_U)/2$. In particular, $\phi_s(\theta)$ is symmetric about 0 when $-\theta_L = \theta_U > 0$. It should be noted that the maximum power occurs at $\theta = 0$. It can be seen that the power decreases as θ moves away from 0, i.e., $\phi_s(\theta') < \phi_s(\theta'')$ if either $\theta' > \theta'' > 0$ or $\theta' < \theta'' < 0$. Therefore, the size of Schuirmann's two one-sided t tests procedure can be evaluated at $\theta = \Delta$ or $\theta = -\Delta$, i.e.,

$$\max_{\theta \in \phi_0} \phi_s(\theta) = \phi_s(\Delta) = \phi_s(-\Delta).$$

Note that

$$\begin{aligned}
\phi_s(\Delta) &= P\left\{\frac{-2\Delta}{ms} + t < \frac{Y - \Delta}{ms} < -t\right\} \\
&\leq P\left\{\frac{Y - \Delta}{ms} < -t\right\} \\
&= \alpha. \quad\quad (5.3.8)
\end{aligned}$$

As a result, Schuirmann's two one-sided t tests procedure is a test of significance level α. However, in general, it does not provide a size of α because

$$\max_{\theta \in \phi_0} \phi_s(\theta) = \phi_s(\Delta) \leq \alpha.$$

For a given nominal level α, if the probability of concluding bioequivalence (or power of the test) is always greater than or equal to α when the true unknown mean difference between formulations is within the pre-specified equivalence limits but less or equal to α if it is outside the limits, then the test is said to be an unbiased test, i.e.,

$$\phi(\theta) \leq \alpha \quad \text{if} \quad \theta \leq -\Delta \quad \text{or} \quad \theta \geq \Delta, \text{ and}$$

$$\phi(\theta) \geq \alpha \quad \text{if} \quad -\Delta < \theta < \Delta, \tag{5.3.9}$$

where $\phi(\theta)$ is the power function of the test. The power function $\phi_s(\theta)$ of Schuirmann's two one-sided t tests is an unbiased test for the interval hypotheses (5.3.1) or (5.3.3) because $\phi_s(\theta)$ decreases as θ moves away from 0 in either direction and $\max_{\theta \in \Omega_0} \phi_s(\theta) \leq \alpha$.

Schuirmann (1987) also examined $\phi_s(\theta)$ in terms of the so-called index of sensitivity, which is defined as the ratio of width of the bioequivalence interval (2Δ) to standard error of the difference in LS means, i.e.,

$$\nabla = \frac{2\Delta}{\sigma_d \sqrt{\dfrac{1}{n_1} + \dfrac{1}{n_2}}}. \tag{5.3.10}$$

According to the ± 20 rule, $\Delta/\mu_R = 20\%$. Therefore, ∇ can be expressed in terms of $CV = \sigma_e/\mu_R$ as follows:

$$\nabla = \frac{2\Delta/\mu_R}{(\sigma_e/\mu_R)\sqrt{\dfrac{1}{2}\left(\dfrac{1}{n_1} + \dfrac{1}{n_2}\right)}}$$

$$= \frac{40}{CV\sqrt{\dfrac{1}{2}\left(\dfrac{1}{n_1} + \dfrac{1}{n_2}\right)}}. \tag{5.3.11}$$

Table 5.3.1 gives values of ∇ for different combinations of CVs n_1, and n_2 which are encountered in most bioequivalence studies. The results indicate that

Table 5.3.1 Values of the Index of Sensitivity

Sample size	CV						
	10%	15%	20%	25%	30%	35%	40%
8	8.00	5.33	4.00	3.20	2.67	2.29	2.00
10	8.94	5.96	4.47	3.58	2.98	2.56	2.34
12	9.80	6.53	4.90	3.92	3.27	2.80	2.45
14	10.58	7.06	5.29	4.23	3.53	3.02	2.65
16	11.31	7.54	5.66	4.53	3.77	3.23	2.83
18	12.00	8.00	6.00	4.80	4.00	3.43	3.00
20	12.65	8.43	6.32	5.06	4.22	3.61	3.16
22	13.27	8.84	6.63	5.31	4.42	3.79	3.32
24	13.86	9.24	6.93	5.54	4.62	3.96	3.46

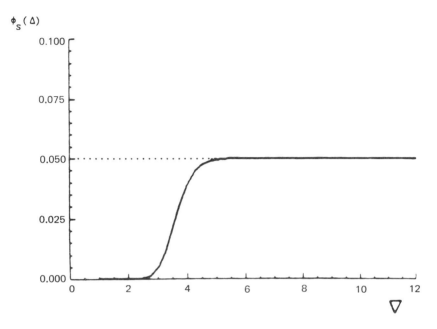

Figure 5.3.2 $\phi_s(\Delta)$ versus ∇ for Schuirmann's two one-sided tests procedure for $n_1 = n_2 = 22$, and the nominal level $= 0.05$. (From Schuirmann (1987).)

∇ increases as sample size increases and as CV decreases. Schuirmann (1987) examined the size of two one-sided test at $\theta = \Delta$ as a function of ∇. For example, Figure 5.3.2 plots $\phi_s(\Delta)$ against ∇ for $n_1 = n_2 = 22$ (adapted from Schuirmann (1987)). It appears that the size is close to the nominal level α when ∇ approaches 5 or 6. In view of Table 5.3.1, 38% of ∇'s (24 out of 63) are less than 5. In other words, the size of Schuirmann's two one-sided t tests procedure for about 38% of the combinations is lower than the nominal level α. Furthermore, Figure 5.3.3 presents the power curve $\phi_s(\theta)$ for the case $\nabla = 4$ and $n_1 = n_2 = 22$. It can be seen that the maximum power which occurs at $\theta = 0$ is about 25%. This is due to a tremendous intra-subject variability (about 47%). When the intra-subject variability is large, the size of Schuirmann's two one-sided t tests procedure is less than the nominal level and the procedure may lose power for correctly concluding bioequivalence.

Note that the evaluation of $\phi_s(\theta)$ at $\theta = \theta_0$ (5.3.6) is rather complicated, requiring a numerical integration from s^2/σ_d^2 to $(\Delta/m\sigma_d)^2/t^2$ as indicated by Schuirmann (1987) and Müller-Cohrs (1990). Phillips (1990) suggested a method using a bivariate noncentral t distribution, in which the correlation between the two variables is 1, to calculate the power of Schuirmann's two one-sided t tests procedure. The idea is to show that T_L and T_U of (4.3.4) follows a bivariate

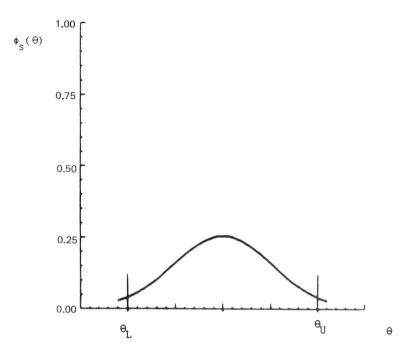

Figure 5.3.3   Power function $\phi_s(\theta)$ of Schuirmann's two one-sided tests procedure for $\nabla = 4$, $\Delta = 20$, $n_1 = n_2 = 22$, and the nominal level $= 0.05$. (From Schuirmann (1987).)

follows a bivariate noncentral t distribution with $n_1 + n_2 - 2$ degrees of freedom and noncentrality parameters

$$NC_L = \frac{\theta + \Delta}{m\sigma_d}, \quad \text{and}$$

$$NC_U = \frac{\theta - \Delta}{m\sigma_d}. \tag{5.3.12}$$

Let X be a normal random variable with mean 0 and variance σ_d^2. $Y + \Delta$ and $Y - \Delta$ can then be expressed in terms of X as follows:

$$Y + \Delta = X + \theta + \Delta \sim N(\theta + \Delta, m\sigma_d^2), \quad \text{and}$$

$$Y - \Delta = X + \theta - \Delta \sim N(\theta - \Delta, m\sigma_d^2). \tag{5.3.13}$$

It can be seen that the correlation between $Y + \Delta$ and $Y - \Delta$ is equal to 1. Now, $(n_1 + n_2 - 2)s^2/\sigma_d^2$ is distributed as $\chi^2(n_1 + n_2 - 2)$ and is independent of Y. Therefore, the joint distribution of $T_L = (Y + \Delta)/ms$ and $T_U = (Y - \Delta)/ms$ follows a bivariate noncentral t distribution with $n_1 + n_2 - 2$ degrees of freedom, correlation $= 1$, and noncentrality parameters NC_L and NC_U. Thus, the power $\phi_s(\theta)$ at $\theta = \theta_0$ and $\sigma_d^2 = \sigma_0^2$ can be evaluated by

$$P\{T_L > t \quad \text{and} \quad T_U < -t \mid \theta_0, \sigma_0^2\}. \tag{5.3.14}$$

Owen (1965) showed that the integral of a bivariate noncentral t distribution can be expressed as the difference of the integrals between two univariate noncentral distributions when the correlation is 1. Although Owen's formulae are rather complicated and involve a definite integral, these formulae can be programmed using statistical softwares such as SAS (Phillips, 1990).

From the above discussion, the computation of power for Schuirmann's two one-sided t tests procedure requires special programs for numerical integration based on either (5.3.6) or (5.3.14). To examine the performance of power, instead, we may consider an empirical power, which can easily be obtained by conducting a simulation study with fixed θ, σ_d, and sample size, to estimate the true power. This simulation study can be carried out using any standard statistical package such as SAS. The procedure for obtaining an empirical power is outlined below:

Step 1. Generate a random sample of size $n_1 + n_2$ according to model (4.1.1) with pre-specified μ_T, μ_R, σ_d^2, and σ_s^2.

Step 2. For a given equivalence limit Δ (or θ_L and θ_U), calculate T_L and T_U.

Step 3. Repeat steps 1 and 2 a large number of times, say B times.

Step 4. The empirical power is the proportion of B random samples such that $T_L > t$ and $T_U < -t$.

5.3.3. Power of Anderson and Hauck's Test Procedure

Let $p_1 = P(T_L > t)$ and $p_2 = P(T_U < -t)$. Anderson-Hauck's test procedure for interval hypotheses (5.3.1) rejects H_0 and concludes bioequivalence at the α level of significance if

$$|p_1 - p_2| < \alpha. \tag{5.3.15}$$

Anderson-Hauck's test procedure is always more powerful than Schuirmann's two one-sided tests procedure at the same nominal level α because Schuirmann's procedure rejects H_0 if

$$p_1 < \alpha \quad \text{and} \quad p_2 < \alpha. \tag{5.3.16}$$

However, Anderson-Hauck's procedure has two major drawbacks which are criticized by the FDA (see Attachment No. 5 of the Report by the Bioequivalence Task Force, January, 1988), which are described below.

Frick (1987) pointed out that the real size of the Anderson Hauck's procedure does not depend upon the pre-specified equivalence limit Δ but only upon the

Table 5.3.2 The Actual Size of Anderson-Hauck's Procedure at the Nominal Levels of α = 0.05 and α = 0.01

DF	$\alpha = 0.05$	$\alpha = 0.01$
10	0.0613	0.0140
20	0.0560	0.0121
50	0.0525	0.0109
100	0.0513	0.0105

Source: Frick (1987) and Müller-Cohrs (1990).

sample size (or degrees of freedom) and the nominal level α. Table 5.3.2 lists the actual sizes of Anderson-Hauck's procedure at nominal levels α = 0.05 and α = 0.01 for various degrees of freedom. The results indicate that the actual size of Anderson-Hauck's procedure is always larger than the nominal level (also see Müller-Cohrs, 1990). Therefore, Anderson-Hauck's procedure can lead to

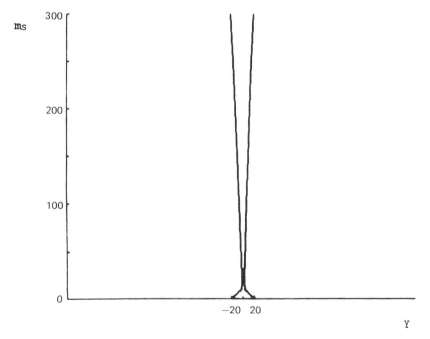

Figure 5.3.4 Rejection region of Anderson-Hauck's procedure for Δ = 20, $n_1 = n_2 = 6$, and the 0.05 nominal level. (From Schuirmann (1987).)

a high probability of committing a type I error (wrongly concluding bioequivalence).

Furthermore, Figure 5.3.4 gives the rejection region of the Anderson-Hauck's procedure at $\alpha = 0.05$ nominal level with $\Delta = 20$, $n_1 = n_2 = 6$. The nonconvex shape of the rejection region indicates that any observed LS mean difference $\overline{Y}_T - \overline{Y}_R$ might reject the hypothesis of bioinequivalence and conclude bioequivalence if the observed standard error of $\overline{Y}_T - \overline{Y}_R$ is sufficiently large (also see Rocke (1984); Schuirmann (1987); Müller-Cohrs (1990)).

Like Schuirmann's two one-sided t tests procedure, calculation of the power of the Anderson-Hauck's procedure also requires a complicated numerical integration. Müller-Cohrs (1990) compared the power of Schuirmann's two one-sided tests procedure with that of Anderson-Hauck's procedure at the same real size. The results indicate that Schuirmann's procedure is slightly more powerful than Anderson-Hauck's procedure if the index of sensitivity is greater than 6 (or CV is less than 20%) and the sample size is moderate (e.g., 18–24). This suggests that Schuirmann's procedure is preferred over Anderson-Hauck's procedure for assessment of bioequivalence in average bioavailability in the case where the sample size is between 18–24 and a prior CV is less than 20%.

5.3.4. Power Approach For Assessing Bioequivalence

The power approach, according to the 80/20 rule, consists of two steps. The first step is to test the hypothesis of equality between two formulations at the α level, i.e.,

$$H_0': \theta = 0 \quad \text{vs} \quad H_a': \theta \neq 0, \tag{5.3.17}$$

where $\theta = \mu_T - \mu_R$.

If H_0' is rejected at the α level of significance, we can then not conclude that the two formulations are bioequivalent. However if H_0' is not rejected, we proceed to examine whether the power for detection of a difference of $\Delta = 0.2\mu_R$ is greater than 80%. We conclude bioequivalence if the power is greater than 80%. Let

$$T_p = \frac{Y}{ms} = \frac{\overline{Y}_T - \overline{Y}_R}{\hat{\sigma}_d \sqrt{\dfrac{1}{n_1} + \dfrac{1}{n_2}}}. \tag{5.3.18}$$

Also, let $t(\alpha/2)$ be the upper $\alpha/2$ quantile of a central t distribution with $n_1 + n_2 - 2$ degrees of freedom. Then, the power approach concludes that two formulations are bioequivalent if

(i) $|T_p| < t(\alpha/2)$, and

(ii) $P\{|T_p| > t(\alpha/2) \text{ given } |\theta| = \Delta\} > 0.80.$ (5.3.19)

Note that the power in (ii) of (5.3.19) can be approximated by a central t distribution, i.e.,

$$P\{|T_p| > t(\alpha/2) \text{ given } |\theta| = \Delta\}$$
$$\cong P\{T > t(\alpha/2) - (\Delta/ms)\},$$

where T has a central t distribution with $n_1 + n_2 - 2$ degrees of freedom.

Therefore, the power approach leads to the rejection of bioinequivalence between the two formulations if

$$t(\alpha/2) - \frac{\Delta}{ms} < -t(0.2).$$

As a result, no value of Y (or $\overline{Y}_T - \overline{Y}_R$) will lead to the conclusion of bioequivalence if

$$ms \geq \frac{\Delta}{t(\alpha/2) + t(0.2)}. \tag{5.3.20}$$

Note that (i) of (5.3.19) is the acceptance region of the null hypothesis of equality which is bounded by $Y = mst(\alpha/2)$ and $Y = -mst(\alpha/2)$. Hence, the region for conclusion of bioequivalence using the power approach based upon the 80/20 rule is the intersection of

$$\{(Y,s) \mid -mst(\alpha/2) < Y < mst(\alpha/2)\} \quad \text{and} \quad \left\{s \mid s < \frac{\Delta/m}{t(\alpha/2) + t(0.2)}\right\}$$

Figure 5.3.5 provides the rejection region of the power approach for $\Delta = 20$, $\alpha = 0.05$ and $n_1 + n_2 = 12$. The shape of the rejection region is an upside-down triangle which indicates a major drawback of the power approach. As ms (or the standard error of $\overline{Y}_T - \overline{Y}_R$) decreases, the rejection region of the power approach becomes smaller and smaller. When ms = 0, Y has to be 0 in order to conclude bioequivalence. This suggests that if the same LS mean difference is observed in two bioequivalence studies, the probability of correctly concluding bioequivalence is smaller for the study with smaller intra-subject variability than the study with larger intra-subject variability. This undesirable property is a direct consequence of the incorrect use of T_p for testing the hypothesis of equality for the assessment of bioequivalence. The test statistic T_p is the uniformly most powerful unbiased test for the hypothesis of equality. It has the maximum power for detection of a difference among all unbiased tests when there is difference between two formulations. The power of T_p for detecting a difference increases as ms decreases. This implies that the rejection region for the hypothesis of equality

$$\{(Y,s) \mid Y > mst(\alpha/2) \quad \text{and} \quad Y < -mst(\alpha/2)\}$$

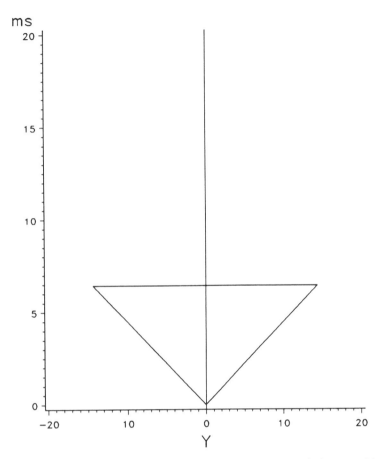

Figure 5.3.5 Rejection region of the power approach for $\Delta = 20$, $n_1 = n_2 = 6$, and the nominal level $= 0.05$. (From Schuirmann (1987).)

increases as ms decreases. However, (i) of (5.3.19) is just the complement of the rejection region for the hypothesis of equality which becomes smaller as ms decreases.

Schuirmann (1987) pointed out that the power approach for assessment of bioequivalence cannot control the size at the nominal level. The size of the power approach is an increasing function of degrees of freedom. As an example, Table 5.3.3 lists the size of the power approach at $\theta = \Delta$ for various degrees of freedom. The results indicate that it requires a large sample size (degrees of freedom $= 50$) to reach a size of 0.1016. The size of the power approach converges to 0.2 when sample size approaches infinity.

Table 5.3.3   The Size of
Power Approach for the
Interval Hypotheses at
$\theta = \Delta$

DF	Size
10	0.0605
16	0.0722
20	0.0779
26	0.0847
30	0.0884
40	0.0958
50	0.1016
100	0.1188

Source: Schuirmann (1987).

5.3.5. Relative Efficiency

In previous sections, test powers for Schuirmann's two one-sided t tests procedure, Anderson-Hauck's test procedure and the power approach were discussed. It should be noted that Schuirmann's two one-sided t tests procedure and Anderson-Hauck's test procedure are based upon interval hypotheses, while the power approach is based upon the hypothesis of equality. One of the major drawbacks of Anderson-Hauck's test procedure is that it has an open-ended rejection region as ms increases. For the power approach, the region for conclusion of bioequivalence becomes smaller as ms decreases. Among these test procedures, Schuirmann's two one-sided t tests procedure appears to be a reasonable approach for interval hypotheses despite the fact that it is somewhat conservative when the index of sensitivity is small. It should be noted that Schuirmann's two one-sided t tests procedure is equivalent to the classical confidence approach as well as Rodda and Davis's Bayesian method in the sense that they all reach the same conclusion on bioequivalence. In other words, the probability of correctly concluding bioequivalence is the same for these methods.

For interval hypotheses testing, little has been done in the comparison of Schuirmann's two one-sided tests procedure based upon a t test with that based upon a Wilcoxon-Mann-Whitney rank sum test. For the hypothesis of equality, Wilcoxon-Mann-Whitney rank sum test is a locally most powerful test for detection of a location shift under logistic distribution (Randles and Wolfe, 1979). Its asymptotic Pitman's relative efficiency against a two sample t test is $3/\pi = 95.5\%$ under normal distribution. However, it is not known whether the same property remains for the nonparametric two one-sided procedure for interval hypotheses discussed in 4.5.1. To compare the performance of a nonparametric two one-sided procedure based upon the Wilcoxon-Mann-Whitney rank sum test with parametric two one-sided procedure based upon a t test, a small simulation study was conducted to examine the size and power of the two procedures at the $\alpha = 0.05$ nominal level. The sizes were evaluated at either $\mu_T = 80$ and $\mu_R = 100$ or $\mu_T = 120$ and $\mu_R = 100$, i.e., $\theta = \Delta = \pm 20$. The powers were obtained at $\theta = 0$ or $\mu_T = \mu_R = 100$. A total of 1000 normal random samples were generated for each combinations of μ_T, μ_R, CV (10%, 20%, 30%, and 40%), and sample size $n_1 = n_2 = 9, 12$. We computed the proportion of 1000 random samples which leads to a conclusion of bioequivalence for each combination. Table 5.3.4 summarizes the results of the simulation study. From Table 5.3.4, it seems that the size of the nonparametric procedure is larger than the parametric, while the power of the parametric procedure is greater than that of

Table 5.3.4 The Empirical Size and Power of Parametric and Nonparametric Two One-Sided Procedures Under Normal Distribution

CV	μ_T	μ_R	$n_1 = n_2 = 9$ Parametric[a]	Non-parametric[b]	$n_1 = n_2 = 12$ Parametric	Non-parametric
10%	80	100	6.2%	6.4%	3.8%	4.8%
	120	100	5.2%	5.6%	4.2%	4.6%
	100	100	100.0%	100.0%	100.0%	100.0%
15%	80	100	5.8%	6.3%	4.6%	5.1%
	120	100	5.9%	6.3%	4.4%	4.7%
	100	100	98.0%	96.8%	99.1%	98.6%
20%	80	100	4.4%	4.8%	5.7%	4.9%
	120	100	4.9%	5.3%	5.3%	5.7%
	100	100	77.1%	75.9%	89.9%	88.5%
30%	80	100	4.0%	4.5%	3.8%	4.3%
	120	100	3.4%	4.1%	4.6%	4.3%
	100	100	25.9%	27.2%	43.2%	42.0%
40%	80	100	0.6%	1.5%	1.5%	2.1%
	120	100	1.2%	1.7%	1.9%	2.0%
	100	100	3.0%	4.8%	10.7%	9.2%

Note: Size and Power are evaluated at $\mu_T - \mu_R = \pm 20$ and $\mu_T = \mu_R = 100$, respectively. Calculation was done based upon 1,000 normal random samples.
[a]Calculated based upon 90% classical confidence interval.
[b]Calculated based upon 90% nonparametric confidence interval.

the nonparametric procedure. However, the differences in size and power between the two procedures are very small (less than 2%) for all combinations considered. Therefore, even for the normal distribution, the Wilcoxon-Mann-Whitney two one-sided tests procedure is very competitive. However, further investigation is worthwhile for this comparison with other distributions besides a normal distribution.

Schuirmann (1989) compared the methods of confidence approach for three intervals (L_2, U_2), (L_3, U_3), and (L_4, U_4) given in Section 4.2, which are for the interval hypotheses (4.3.2) expressed by a ratio, in terms of their actual sizes and powers. He referred to (L_2, U_2), (L_3, U_3), and (L_4, U_4) as the approximate method, the exact method, and the fixed Fieller's method, respectively. Note that the fixed Fieller's method is equivalent to the Bayesian method proposed by Mandallaz and Mau (1981) as shown in the previous chapter. Some of these results are summarized in Figures 5.3.6 through 5.3.10, which will be discussed separately below.

Figure 5.3.6 plots the actual size of these three intervals against Δ for $r = 1$ and $\rho = 0.9$, where $\Delta = \overline{Y}_R / \sqrt{w\sigma_R^{*2}}$, r is the ratio of the standard deviation of $\overline{Y}_T$ to $\overline{Y}_R$, ρ is the correlation between $\overline{Y}_T$ and $\overline{Y}_R$, and σ_R^{*2} is the variance of $\overline{Y}_R$. The results indicate that these three intervals are very conservative when Δ is small. However, the exact method can always control the actual size under the nominal level of $\alpha = 0.05$. The approximate method produces a larger actual size compared to $\alpha = 0.05$ when $\delta_U = 1.2$ (f $= 0.2$ in the figures), while it controls the actual size under the nominal level of 0.05 when $\delta_L = 0.8$ (f $= -0.2$ in the figures). The fixed Fieller's method has an actual size greater than $\alpha = 0.05$ as Δ increases.

Figure 5.3.7 plots the actual size of the three methods against Δ for $\rho = 0.9$ and $r = 0.5$ which is the case where the intra-subject variability of the test formulation is smaller than that of the reference formulation. It appears that the exact method controls actual size under the nominal level, while the approximate and fixed Fieller's methods are somewhat conservative (size < 0.05) when $\delta_L = 0.8$ and a little bit liberal (size > 0.05) when $\delta_U = 1.2$. Schuirmann (1989) also showed that when $\rho = 0.9$ and $r = 2$, which is the case where the intra-subject variability of the test formulation is greater than that of reference formulation, both the approximate and fixed Fieller's methods are liberal (size > 0.05) when $\delta_L = 0.8$ and conservative (size < 0.05) when $\delta_U = 1.2$.

Figure 5.3.8 plots the actual size of the three procedures against ρ for $r = 1$, $\Delta = 40$, $\delta_L = 0.8$, and $\delta_U = 1.2$. The actual size of the exact Fieller's method under these conditions is independent of ρ and exactly at the nominal level of 0.05. As ρ approaches 1, the actual size of the approximate method and the fixed Fieller's method converges to 0.5.

Figure 5.3.9 gives the same graphs as those in Figure 5.3.6 except that it provides the actual size for the case of unknown variance for fixed Fieller's

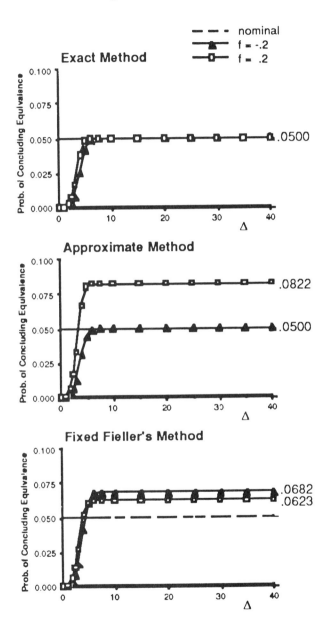

Figure 5.3.6   Actual size of exact Fieller's, approximate, and fixed Fieller's methods as a function of Δ, for $\rho = 0.9$, $r = 1$. (From Schuirmann (1989).)

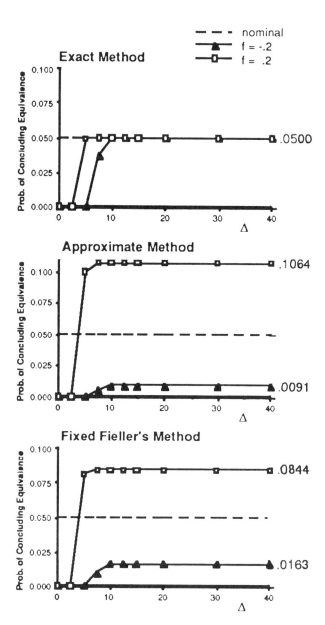

Figure 5.3.7 Actual size of exact Fieller's, approximate, and fixed Fieller's methods as a function of Δ, for $\rho = 0.9$, $r = 0.5$. (From Schuirmann (1989).)

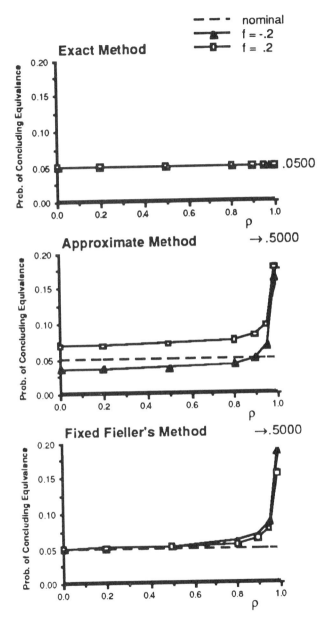

Figure 5.3.8   Actual size of exact Fieller's, approximate, and fixed Fieller's methods as a function of ρ for Δ = 40, r = 1. (From Schuirmann (1989).)

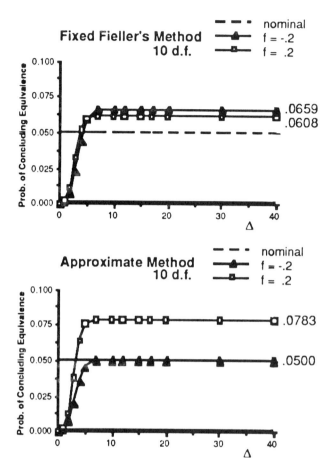

Figure 5.3.9   Actual size of fixed Fieller's and approximate methods as a function of Δ, for $\rho = 0.9$, $r = 1$ and $n_1 = n_2 = 6$. (From Schuirmann (1989).)

method and the approximate method when $n_1 = n_2 = 6$. The results from Figure 5.3.9 are similar to those from Figure 5.3.6 discussed earlier.

Finally, Figure 5.3.10 presents the power curves of the three methods over the range from $\delta_L = 0.8$ to $\delta_U = 1.2$ for $r = 1$, $\rho = 0.9$, $\Delta = 10, 20$, and 40. It is deceptive to conclude that the power curves of the three methods are symmetric about 0 because they are not, due to the results from Figures 5.3.6, 5.3.8, and 5.3.9. In addition, these power curves are those compared at the same nominal level of 0.05, not at the same actual size. Therefore, it is not fair to say that these three methods have the same power as those shown in Figures

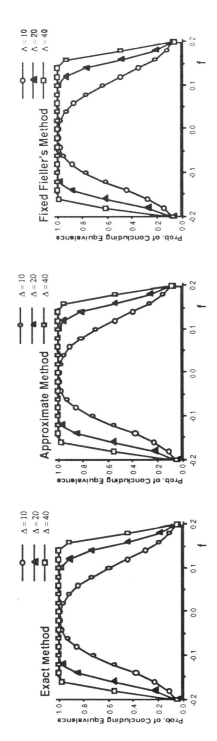

Figure 5.3.10 Full power curves of exact Fieller's, approximate, and fixed Fieller's methods as a function of f, for ρ = 0.9, r = 1. (From Schuirmann (1989).)

5.3.6, 5.3.8, and 5.3.9. The actual size of the approximate method and the fixed Fieller's method is much larger than 0.05 when $\rho = 0.9$.

Based on the above arguments, among these three methods, the exact Fieller's method is the recommended method for interval hypotheses expressed by a ratio in the assessment of bioequivalence for average bioavailability under a raw data model.

5.4. SAMPLE SIZE DETERMINATION

In bioavailability studies, How large a sample size is required in order to have a desired power to establish bioequivalence within meaningful limits? is a question of particular interest to the investigator. To provide an answer to this question, we will first discuss the calculation of sample size for the hypothesis of equality under a standard 2×2 crossover model (4.1.1).

5.4.1. Sample Size for Point Hypotheses

Let α be the nominal level of significance, i.e., the probability of committing a type I error which one is willing to tolerate; and $\phi = 1 - \beta$, the power one wishes to have in order to detect a difference of at least Δ magnitude, where β is the probability of making a type II error. In the interest of balance, we assume $n_1 = n_2 = n_e$. In other words, each sequence will be allocated the same number of subjects at random. The sample size per sequence n_e for the hypothesis of equality (5.3.17) can be determined by the formulation given below.

$$n_e \geq 2[t(\alpha/2,2n - 2) + t(\beta,2n - 2)]^2[\hat{\sigma}_d/\Delta]^2, \tag{5.4.1}$$

where $\hat{\sigma}_d$ can usually be obtained from previous studies. According to the power approach based on the 80/20 rule, the sample size should be large enough to provide a power of 80% for detection of a difference of the magnitude at least 20% of the unknown reference mean. Thus, formula (5.4.1) can be simplified as

$$n_e \geq [t(\alpha/2,2n - 2) + t(\beta,2n - 2)]^2[CV/20]^2, \tag{5.4.2}$$

where

$$CV = 100 \times \frac{\sqrt{2\hat{\sigma}_d^2}}{\mu_R} = 100 \times \frac{\sqrt{MSE}}{\mu_R}.$$

The total number of subjects required for a standard 2×2 crossover design is $N = 2n_e$. Since the degrees of freedom $(2n - 2)$ in both (5.4.1) and (5.4.2) are unknown, a numerical iterative procedure is required to solve for n_e. To illustrate this, let us consider the following example.

Example 5.4.1 Suppose we would like to conduct a bioequivalence study to compare bioavailability of a new formulation with a reference formulation as discussed in Example 3.6.1. The design was chosen to be a standard 2×2 crossover design and the 80/20 rule will be used to determine bioequivalence in average bioavailability between two formulations. The next question then is how many subjects are needed in order to have 80% power to detect a 20% difference. Based on data from Example 3.6.1, we have

$$CV = 100 \times \frac{\sqrt{167.246}}{82.559} = 15.66.$$

Let us first guess $n_e = 9$. This gives degrees of freedom $2n - 2 = 18 - 2 = 16$, $t(0.025,16) = 2.12$ and $t(0.2,16) = 0.865$. By (5.4.2),

$$n_e = (2.12 + 0.865)^2 (15.66/20)^2 = 5.5 \cong 6.$$

We then start with $n_e = 6$ and repeat the same calculation which gives

$$\text{degrees of freedom} = 12 - 2 = 10,$$
$$t(0.025,10) = 2.228,$$
$$t(0.2,10) = 0.879.$$

Again, by (5.4.2), we have

$$n_e = (2.228 + 0.879)^2 (15.66/20)^2 = 5.9 \cong 6,$$

which is very close to the previous solution.

Therefore, a total of $N = (2)(6) = 12$ subjects are needed based on the 80/20 rule.

As we have pointed out earlier, the power approach based on the 80/20 rule is an ad hoc method for assessment of bioequivalence which may not be statistically valid. The sample size determined by (5.4.1) or (5.4.2) may not be large enough to provide sufficient power if other methods for interval hypotheses such as Schuirmann's two one-sided tests procedure are used.

5.4.2. Sample Size for Interval Hypotheses

As discussed in Chapter 4, the classical (or shortest) confidence interval, Schuirmann's two one-sided tests procedure, as well as Rodda and Davis's Bayesian method can lead to the same conclusion for determination of bioequivalence in average bioavailability. Therefore, in this section, we will focus on sample size determination based upon Schuirmann's two one-sided tests procedure for interval hypotheses.

As indicated in Section 5.3, the power function $\phi_s(\theta)$ for Schuirmann's two one-sided t tests procedure is symmetric about 0 when the ± 20 rule is used for

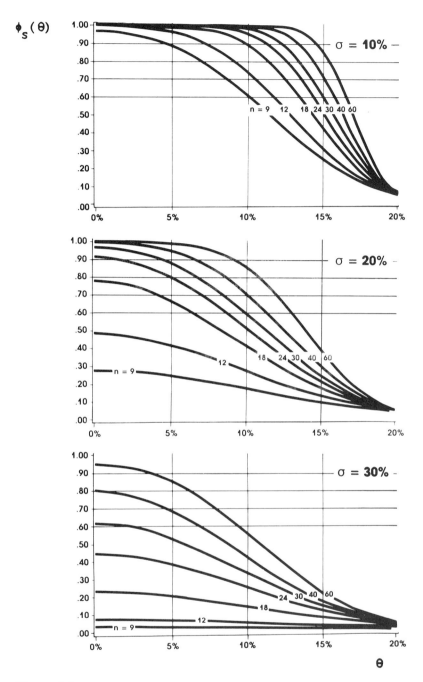

Figure 5.3.11 Power curves of the Schuirmann's two one-sided tests procedure for total sample size of 9, 12, 18, 24, 30, 40, and 60 at the 0.05 nominal level and $\Delta = 0.2\mu_R$. (From Phillips (1990).)

the assessment of bioequivalence. Furthermore, the intra-subject variability has an impact on the power function. To illustrate this, Phillips (1990) provided several graphs for the power of Schuirmann's two one-sided t tests procedure for various sample sizes and CV's. Some graphs are presented in Figure 5.3.11. Since calculation for the exact power for Schuirmann's two one-sided t tests procedure requires complicated numerical integration as discussed in Section 5.3, the sample size determination based on the power function is complicated and difficult to obtain. However, an approximate sample size based on the power function can be obtained using some familiar traditional methods (Liu and Chow, 1992a).

We first consider the case where $\theta = \mu_T - \mu_R = 0$ and $n_1 = n_2 = n$. In this case,

$$\frac{Y}{\sqrt{\frac{2}{n} \hat{\sigma}_d^2}}$$

has a central t distribution with $2n - 2$ degrees of freedom. The power at $\theta = 0$ is then given by

$$
\begin{aligned}
\phi_s(0) &= P\left\{ -\Delta + t(\alpha, 2n - 2)\sqrt{\frac{2}{n} \hat{\sigma}_d^2} < Y < \Delta - t(\alpha, 2n - 2)\sqrt{\frac{2}{n} \hat{\sigma}_d^2} \right\} \\
&= P\left\{ \frac{-\Delta}{\sqrt{\frac{2}{n} \hat{\sigma}_d^2}} + t(\alpha, 2n - 2) < \frac{Y}{\sqrt{\frac{2}{n} \hat{\sigma}_d^2}} < \frac{\Delta}{\sqrt{\frac{2}{n} \hat{\sigma}_d^2}} - t(\alpha, 2n - 2) \right\}
\end{aligned}
$$

$$(5.4.3)$$

Since a central t distribution is symmetric about 0, the lower and upper endpoints of (5.4.3) are also symmetric about 0, i.e.,

$$\frac{\Delta}{\sqrt{\frac{2}{n} \hat{\sigma}_d^2}} - t(\alpha, 2n - 2) = -\left\{ \frac{-\Delta}{\sqrt{\frac{2}{n} \hat{\sigma}_d^2}} + t(\alpha, 2n - 2) \right\}.$$

Therefore, $\phi_s(0) \geq 1 - \beta$ implies that

$$\left| \frac{\Delta}{\sqrt{\frac{2}{n} \hat{\sigma}_d^2}} - t(\alpha, 2n - 2) \right| \geq t(\beta/2, 2n - 2), \tag{5.4.4}$$

or

$$n_I \geq 2[t(\alpha, 2n - 2) + t(\beta/2, 2n - 2)]^2 \left(\frac{\hat{\sigma}_d}{\Delta} \right)^2, \tag{5.4.5}$$

where n_I is the sample size required to achieve $1 - \beta$ power at the α level of significance for interval hypotheses (4.3.1). If the ± 20 rule is used with $\Delta = 0.2\mu_R$, (5.4.5) becomes

$$n_I \geq [t(\alpha,2n - 2) + t(\beta/2,2n - 2)]^2\left(\frac{CV}{20}\right)^2. \qquad (5.4.6)$$

The sample size is exact for $\theta = 0$ either by (5.4.5) or (5.4.6). Note that Westlake (1986) also derived (5.4.5) using the $(1 - 2\alpha) \times 100\%$ classical confidence interval.

We will now consider the case where $\theta \neq 0$. Since the power of Schuirmann's two one-sided t tests procedure is symmetric about 0, without loss of generality, we will consider a sample size determination for the case where $\theta > 0$ only. When $0 < \theta = \theta_0 < \Delta$,

$$\frac{Y - \theta_0}{\sqrt{\frac{2}{n}\hat{\sigma}_d^2}}$$

has a central distribution with $2n - 2$ degrees of freedom. The power for Schuirmann's two one-sided t tests procedure at θ_0 is

$$\phi_s(\theta_0) = P\left\{\frac{-\Delta - \theta_0}{\sqrt{\frac{2}{n}\hat{\sigma}_d^2}} + t(\alpha,2n - 2) < \frac{Y - \theta_0}{\sqrt{\frac{2}{2}\hat{\sigma}_d^2}}\right.$$

$$\left. < \frac{\Delta - \theta_0}{\sqrt{\frac{2}{n}\hat{\sigma}_d^2}} - t(\alpha,2n - 2)\right\}. \qquad (5.4.7)$$

Note that, unlike the case where $\theta = 0$, the lower and upper endpoints of (5.4.7) are not symmetric about 0 because

$$-\left[\frac{\Delta - \theta_0}{\sqrt{\frac{2}{n}\hat{\sigma}_d^2}} - t(\alpha,2n - 2)\right] = \frac{-\Delta + \theta_0}{\sqrt{\frac{2}{n}\hat{\sigma}_d^2}} + t(\alpha,2n - 2)$$

$$> \frac{-\Delta - \theta_0}{\sqrt{\frac{2}{n}\hat{\sigma}_d^2}} + t(\alpha,2n - 2).$$

Therefore, if we choose

$$\frac{\Delta - \theta_0}{\sqrt{\frac{2}{n}\hat{\sigma}_d^2}} - t(\alpha,2n - 2) = t(\beta/2,2n - 2),$$

then the resultant sample size might be too large to be of practical interest. The power may be more than we require. A compromise method, which is less conservative, for sample size determination can be obtained using the following inequality:

$$\phi_s(\theta_0) \leq P\left\{\frac{Y - \theta_0}{\sqrt{\frac{2}{n}\hat{\sigma}_d^2}} < \frac{\Delta - \theta_0}{\sqrt{\frac{2}{n}\hat{\sigma}_d^2}} - t(\alpha, 2n - 2)\right\}. \tag{5.4.8}$$

As a result, $\phi_s(\theta_0) \geq 1 - \beta$ gives

$$\frac{\Delta - \theta_0}{\sqrt{\frac{2}{n}\hat{\sigma}_d^2}} - t(\alpha, 2n - 2) = t(\beta, 2n - 2),$$

or

$$n_I(\theta_0) \geq 2[t(\alpha, 2n - 2) + t(\beta, 2n - 2)]^2\left(\frac{\hat{\sigma}_d}{\Delta - \theta_0}\right)^2. \tag{5.4.9}$$

If the ± 20 rule is used with $\Delta = 0.2\mu_R$, then (5.4.9) becomes

$$n_I(\theta_0) \geq [t(\alpha, 2n - 2) + t(\beta, 2n - 2)]^2\left(\frac{CV}{20 - \theta_0'}\right)^2, \tag{5.4.10}$$

where $\theta_0' = 100 \times \theta_0/\mu_R$.

Based on (5.4.10), Table 5.4.1 gives the total sample sizes which are needed to achieve a desired power for a standard 2×2 crossover design for various combinations of θ and CV's. The values in Table 5.4.1 agree with those numbers given in Phillips (1990). An approximate formula for sample size calculations can be found in Liu and Chow (1992a). The results indicate that more subjects are needed in order to achieve the same power, if intra-subject variability and θ increase. In the following, the data presented in Section 3.6 will be used to illustrate the calculations for sample sizes in order to achieve 80% power for $\theta = 0$ and 5% of μ_R.

Example 5.4.2 First of all, let us consider the case where $\theta = 0$. We start with an initial guess of 9, i.e., $n_I(0) = 9$. This gives a degrees of freedom of $2n - 2 = 16$ and

$$t(0.05, 16) = 1.746 \quad \text{and} \quad t(0.1, 16) = 1.337.$$

By (5.4.6), we have

$$n_I(0) = (1.746 + 1.337)^2(15.66/20)^2 = 5.8 \cong 6.$$

Table 5.4.1 Sample Sizes for Schuirmann's Two One-Sided t Tests Procedure at Δ $= 0.2\mu_R$ and $\alpha = 0.05$ Nominal Level

Power	CV(%)[a]	$100 \times (\mu_T - \mu_R)/\mu_R$			
		0%	5%	10%	15%
80%	10	8	8	16	52
	12	8	10	20	74
	14	10	14	26	100
	16	14	16	34	126
	18	16	20	42	162
	20	20	24	52	200
	22	24	28	62	242
	24	28	34	74	288
	26	32	40	86	336
	28	36	46	100	390
	30	40	52	114	448
	32	46	58	128	508
	34	52	66	146	574
	36	58	74	162	644
	38	64	82	180	716
	40	70	90	200	794
90%	10	10	10	20	70
	12	10	14	28	100
	14	14	18	36	136
	16	16	22	46	178
	18	20	28	58	224
	20	24	32	70	276
	22	28	40	86	334
	24	34	46	100	396
	26	40	54	118	466
	28	44	62	136	540
	30	52	70	156	618
	32	58	80	178	704
	34	66	90	200	794
	36	72	100	224	890
	38	80	112	250	992
	40	90	124	276	1098

[a]$CV = 100 \times (\sqrt{MSE}/\mu_R)$.

Source: Liu and Chow (1992a).

Table 5.4.2 Empirical Power by Simulation for Example
5.4.2 at $\alpha = 0.05$ Nominal Level

Procedure	$n_1(0) = 6$	$n_1(5\%) = 8$
Schuirmann's two one-sided	81.8%	82.0%
Nonparametric two one-sided	82.9%	80.2%
Anderson-Hauck	93.1%	91.4%

We then use $n_1(0) = 6$ as a starting value for the next iteration. This gives

$$\text{degrees of freedom} = 10;$$
$$t(0.05,10) = 1.812;$$
$$t(0.1,10) = 1.372.$$

Therefore, by (5.4.6),

$$n_1(0) = (1.812 + 1.372)^2(15.66/20)^2 = 6.2 \cong 6.$$

Since these two iterations gives a similar result of 6 subjects per sequence, a total of 12 subjects would be needed in order to have an 80% power to detect a 20% difference for the case where $\theta = 0$.

On the other hand, when $\theta = 0.05\mu_R$, we may apply a similar iterative procedure. Let us start with $n_1(5\%) = 6$. Since $t(0.2,10) = 0.879$, by (5.4.10), we have

$$n_1(5\%) \geqslant (1.812 + 0.879)^2[15.66/(20 - 5)]^2 = 7.9 \cong 8.$$

We then use $n_1(5\%) = 8$ as an initial value for the next iteration. With $n = 8$, the degrees of freedom equals 14. Thus, $t(0.05,14) = 1.761$ and $t(0.2,14) = 0.868$. By (5.4.10),

$$n_1(5\%) \geqslant (1.761 + 0.868)^2[15.66/(20 - 5)]^2 = 7.5 \cong 8.$$

Therefore, a total of 16 subjects would be needed in order to achieve an 80% power, if θ is 5% of the reference mean.

Sample sizes $n_1(0)$ and $n_1(5\%)$ were also verified by means of empirical powers at $\alpha = 0.05$ level through a simulation study with 1000 random samples using the procedure described in Section 5.3. The results are summarized in Table 5.4.2. The results indicate that both sample sizes calculated using (5.4.4) for $\theta = 0$ and (5.4.10) for $\theta = 5\%$ are large enough to provide at least 80% power. This result is also true for Wilcoxon-Mann-Whitney two one-sided tests procedure and Anderson-Hauck's procedure. As expected, Anderson-Hauck's procedure has a better power since its actual size is greater than the nominal level.

6

Transformation and Analysis of Individual Subject Ratios

6.1. INTRODUCTION

In Chapter 4, based on the ± 20 rule, we have introduced several statistical methods for assessment of bioequivalence. Most of these methods were derived under a raw data model (i.e., model (4.1.1) or (4.1.2)) for a standard 2×2 crossover design, with normality assumptions on the between subject and within subject random variables. The intra-subject variability is assumed to be the same from subject to subject and from formulation to formulation. As a result, the responses (e.g., AUC or C_{max}) are assumed to be normally distributed. One of the difficulties commonly encountered in bioavailability studies, however, is whether the assumption of normality is valid. In many cases, distributions of the responses are positively skewed and exhibit a lack of homogeneity of variances (e.g., variance being dependent on the mean). In this situation, a log-transformation on the responses is often considered in order to reduce the skewness and to achieve an additive model with relatively homogeneous variances. This leads to a multiplicative (or a log-transformed) model. Based on the transformed data, the methods introduced in Chapter 4 can then be applied directly and followed by an antilog-transformation to assess bioequivalence.

Under a multiplicative model, the ratio of means, which is usually considered as a measure of bioequivalence, may be confounded with the period effect and/ or intra-subject variabilities. Therefore, in this chapter, we will consider several estimators for the ratio of average bioavailabilities which reflect the effect caused only by differences in the formulations. These estimators include the maximum likelihood estimator and the minimum variance unbiased estimator (Liu and Weng, 1992). These estimators will be derived under the multiplicative model with normality assumptions on the transformed data. Based on these estimators, a $(1 - 2\alpha) \times 100\%$ confidence interval for the ratio of average bioavailabilities can be obtained to assess bioequivalence.

If the normality assumptions are seriously violated and there is no period effect, Peace (1986) suggested studying individual subject ratios to remove the heterogeneity of intra-subject variability from the comparison between formulations. Under his model, however, distribution of the ratios is unknown and, therefore, the statistical procedures are not exact. Under the assumption of no period effects, Tse (1990) examined several approaches, which are derived assuming that the ratios follow a lognormal distribution, for constructing the confidence interval for mean and median of individual subject ratios. Anderson and Hauck (1990) also examined individual bioequivalence using individual subject ratios. Peace (1990) recommended the use of individual subject ratios as a preliminary test for assessment of bioequivalence. In this chapter, the ratio of least squares means (RM) and the least squares mean of individual subject ratios (MIR), which are often considered as alternative estimators for the average bioavailabilities ratio, will be examined.

This chapter is organized as follows. In the section that follows, a brief description of a multiplicative model for a standard 2×2 crossover design will be given. The maximum likelihood estimator, the minimum variance unbiased estimator, the mean of individual subject ratios and the ratio of LS means for estimation of the ratio of average bioavailabilities will be described in Sections 6.4 through 6.7. Section 6.8 will provide statistical evaluation for the performances of these methods using a simulation study. An example concerning the comparison of two erythromycin formulations is presented in Section 6.9. Finally, a brief discussion is given in Section 6.10.

6.2. MULTIPLICATIVE (OR LOG-TRANSFORMED) MODEL

As indicated earlier, the distributions of responses such as AUC and C_{max} are often positively skewed and exhibit a lack of homogeneity of variances. Therefore, the normality assumption on AUC and C_{max} may not be appropriate. In this situation, the assessment of bioequivalence based upon a raw data model and normality assumptions may not be appropriate. Therefore, to reduce the skewness and achieve an additive model with relatively homogeneous variances, a log-transformation on AUC or C_{max} is usually considered. This leads to the following multiplicative model (or log-transformed model):

$$X_{ijk} = \tilde{\mu} \, \tilde{S}_{ik} \, \tilde{P}_j \, \tilde{F}_{(j,k)} \, \tilde{C}_{(j-1,k)} \, \tilde{e}_{ijk}, \tag{6.2.1}$$

or equivalently

$$Y_{ijk} = \log(X_{ijk}) = \mu + S_{ik} + P_j + F_{(j,k)} + C_{(j-1,k)} + e_{ijk},$$

where $i = 1, 2, \ldots, n_k$ and $j, k = 1, 2$ for a standard 2×2 crossover design and μ, S_{ik}, P_j, $F_{(j,k)}$, $C_{(j-1,k)}$, and e_{ijk} are as defined in (2.5.1). From the above multiplicative model, it can be seen that $\tilde{\mu} = \exp(\mu)$, $\tilde{S}_{ik} = \exp(S_{ik})$, $\tilde{P}_j = \exp(P_j)$, $\tilde{F}_{(j,k)} = \exp(F_{(j,k)})$, $\tilde{C}_{(j-1,k)} = \exp(C_{(j-1,k)})$, and $\tilde{e}_{ijk} = \exp(e_{ijk})$. If $\{S_{ik}\}$ and $\{e_{ijk}\}$ are independent and normally distributed with covariance structure as specified in (2.5.5), then X_{ijk} is said to follow a lognormal linear model (Bradu and Mundlak, 1970; Shimizu, 1988). Model (6.2.1) can also be expressed as

$$X_{ijk} = \exp\{\mu + S_{ik} + P_j + F_{(j,k)} + C_{(j-1,k)} + e_{ijk}\}. \tag{6.2.2}$$

Note that, under the normality assumptions of $\{S_{ik}\}$ and $\{e_{ijk}\}$, the vector, $(X_{i1k}, X_{i2k})'$, of pair responses observed on the ith subject in the kth sequence follows a bivariate lognormal distribution (Crow and Shimizu, 1988).

6.3. BIOEQUIVALENCE MEASURES

In practice, the ratio of means of X_{ijk} between the test and reference formulations is usually considered as a measure of bioequivalence. Since the distribution of X_{ijk} is often positively skewed, some researchers suggest that the ratio of medians, instead of the ratio of means, be used as an alternative measure of bioequivalence because the median is a better representative of the central location of the corresponding distribution (Metzler and Huang, 1983; Chinchilli and Durham, 1989). However, the ratio of means and the ratio of medians may be confounded with the period effect and/or the intra-subject variabilities under model (6.2.1). Table 6.3.1 gives the means and medians of X_{ijk} by sequence and period assuming that there are no carry-over effects. The results indicate that means of X_{ijk} involve the inter-subject variability, σ_s^2, and the intra-subject variabilities σ_T^2 and σ_R^2 of

Table 6.3.1 Sequence-by-Period Means and Medians for Model (6.2.1)

Seq.		Period I	Period II
1	Mean	$\exp\left[(\mu + P_1 + F_R) + \frac{1}{2}(\sigma_R^2 + \sigma_s^2)\right]$	$\exp\left[(\mu + P_2 + F_T) + \frac{1}{2}(\sigma_T^2 + \sigma_s^2)\right]$
	Median	$\exp(\mu + P_1 + F_R)$	$\exp(\mu + P_2 + F_T)$
2	Mean	$\exp\left[(\mu + P_1 + F_T) + \frac{1}{2}(\sigma_T^2 + \sigma_s^2)\right]$	$\exp\left[(\mu + P_2 + F_R) + \frac{1}{2}(\sigma_R^2 + \sigma_s^2)\right]$
	Median	$\exp(\mu + P_1 + F_T)$	$\exp(\mu + P_2 + F_R)$

the log-transformed data Y_{ijk} while medians of X_{ijk} consist only of the fixed effects $\tilde{\mu}$, $\tilde{P}_j$ and $\tilde{F}_{(j,k)}$.

Therefore, for assessment of bioequivalence, we may consider the following ratio of average bioavailabilities on the original scale, which only reflects the effect caused by the difference between the two formulations:

$$\begin{aligned}
\delta &= \tilde{F} \\
&= \exp(F) \\
&= \exp(F_T - F_R) \\
&= \tilde{F}_T/\tilde{F}_R \\
&= \exp(\mu_T + F_T)/\exp(\mu_R + F_R) \\
&= \tilde{\mu}\tilde{F}_T/\tilde{\mu}\tilde{F}_R \\
&= \tilde{\mu}_T/\tilde{\mu}_R,
\end{aligned} \qquad (6.3.1)$$

where $\tilde{\mu}_T = \tilde{\mu}\tilde{F}_T$ and $\tilde{\mu}_R = \tilde{\mu}\tilde{F}_R$.

Note that there are several measures of relative average bioavailability, e.g., the ratio of the marginal test formulation mean to the reference mean which is given by

$$\delta_M = \begin{cases} \exp[P + F + \frac{1}{2}(\sigma_T^2 - \sigma_R^2)], & k = 1 \\ \exp[-P + F + \frac{1}{2}(\sigma_T^2 - \sigma_R^2)], & k = 2. \end{cases}$$

Under the assumption of no period effect, δ_M reduces to

$$\delta_M = \exp[F + \frac{1}{2}(\sigma_T^2 - \sigma_R^2)].$$

It can be seen that δ_M is a measure of average bioavailability which assesses bioequivalence by combining the information of the fixed direct formulation effect and difference in intra-subject variabilities on the logarithmic scale. When $\sigma_T^2 = \sigma_R^2$, δ_M is the ratio of the marginal median of the test formulation to the reference formulation. Another appealing measure of relative bioavailability is the mean of individual subject ratios, which is given below:

$$\delta_I = \begin{cases} \exp[P + F + 2\sigma_d^2], & k = 1 \\ \exp[-P + F + 2\sigma_d^2], & k = 2 \end{cases}$$

where $\sigma_d^2 = (\frac{1}{4})(\sigma_T^2 + \sigma_R^2)$.

If there is no period effect, δ_I reduces to

$$\delta_I = \exp(F + 2\sigma_d^2).$$

Hence, assessment of bioequivalence using δ_I involves not only the fixed direct formulation effect but also the average of intra-subject variabilities. Consequently, even if the direct fixed formulation effect (τ) on the logarithmic scale is zero, two formulations may still be concluded bioequivalent because of intra-subject variabilities. On the other hand, the median of individual subject ratio reduces to δ in the absence of period effect. Since δ involves only the fixed direct formulation effect on the logarithmic scale, we will focus on point estimation of δ in this section. In the following sections, we will consider several estimators for estimation of δ. These estimators include the maximum likelihood (ML) estimator, the minimum variance unbiased estimator (MVUE), mean of individual subject ratios (MIR), and ratio of least squares means (RM). Based upon these estimators, a corresponding $(1 - 2\alpha) \times 100\%$ confidence interval for δ may be obtained to assess bioequivalence.

6.4. MAXIMUM LIKELIHOOD ESTIMATOR

If the skewness of X_{ijk} can be removed by means of a log-transformation and the log-transformed data are approximately normally distributed, then bioequivalence can be assessed using statistical methods described in Chapters 3 and 4 based on the transformed data. In other words, a point estimate and a $(1 - 2\alpha) \times 100\%$ confidence interval for the mean formulation difference on the logarithmic scale (i.e., $\mu_T - \mu_R$) can be obtained based on the log-transformed data, which are given by $\hat{F} = \overline{Y}_T - \overline{Y}_R$ and (L_1, U_1), respectively, where L_1 and U_1 are defined as in (4.2.2). Thus, a point estimate and a $(1 - 2\alpha) \times 100\%$ confidence interval for $\delta = \hat{\mu}_T/\hat{\mu}_R$ on the original scale can be obtained by inverse transformation (i.e., anti-log) of the corresponding estimate and interval on the logarithmic scale.

From (3.3.6), we have

$$\hat{F} = \overline{Y}_T - \overline{Y}_R = \tfrac{1}{2}[(\overline{Y}_{\cdot 21} + \overline{Y}_{\cdot 12}) - (\overline{Y}_{\cdot 11} + \overline{Y}_{\cdot 22})]$$
$$= \tfrac{1}{2}[(\overline{Y}_{\cdot 21} - \overline{Y}_{\cdot 11}) - (\overline{Y}_{\cdot 22} - \overline{Y}_{\cdot 12})]$$
$$= \overline{d}_{\cdot 1} - \overline{d}_{\cdot 2},$$

where $\overline{d}_{\cdot k}$ are the sample means of period differences d_{ik} defined in (3.3.1) for the transformed data. Note that

$$d_{ik} = \tfrac{1}{2}(Y_{i2k} - Y_{i1k})$$
$$= \tfrac{1}{2}(\log X_{i2k} - \log X_{i1k})$$
$$= \tfrac{1}{2}\log(X_{i2k}/X_{i1k})$$
$$= \tfrac{1}{2}\log(r_{ik}), \tag{6.4.1}$$

where $r_{ik} = X_{i2k}/X_{i1k}$ are the period ratios, $i = 1, 2, \ldots, n_k$, $k = 1, 2$. Under the assumption of no carry-over effects, r_{ik} can be expressed as

$$r_{ik} = \frac{X_{i2k}}{X_{i1k}} = \begin{cases} \dfrac{\tilde{P}_2 \, \tilde{F}_T \, \tilde{e}_{i2k}}{\tilde{P}_1 \, \tilde{F}_R \, \tilde{e}_{i1k}} & \text{if } k = 1 \\[3ex] \dfrac{\tilde{P}_2 \, \tilde{F}_R \, \tilde{e}_{i2k}}{\tilde{P}_1 \, \tilde{F}_T \, \tilde{e}_{i1k}} & \text{if } k = 2. \end{cases} \qquad (6.4.2)$$

As indicated in Chapter 4, for a standard 2×2 crossover design, the inter-subject variability can be eliminated with the use of the period differences under an additive model (4.1.1). Similarly, the period ratios r_{ik} can remove the inter-subject variability from the comparison of bioavailability between formulations when a multiplicative model is used.

Since $\hat{F}$ is the maximum likelihood estimator of $\mu_T - \mu_R$, the estimator obtained from the inverse transformation (i.e., exponential) is also the maximum likelihood estimator for $\delta = \tilde{\mu}_T/\tilde{\mu}_R$, which is given below:

$$\hat{\delta}_{ML} = \exp(\hat{F}) = \exp(\overline{Y}_T - \overline{Y}_R). \qquad (6.4.3)$$

Note that the relationship between the difference of least squares means on the logarithmic scale and the period ratios on the original scale can be examined as follows:

Let R_k be the geometric mean of period ratios r_{ik} obtained from sequence k which is given by

$$R_k = \left(\prod_{i=1}^{n_k} r_{ik} \right)^{1/n_k}, \qquad k = 1,2.$$

It can be verified that

$$\hat{\delta}_{ML} = (R_1/R_2)^{1/2}.$$

Therefore, the best linear unbiased estimator of $\hat{F}$ on the logarithmic scale is the log transformation of the square root of the ratio of geometric means of the period ratios of sequence 1 to sequence 2, i.e., $\hat{F} = \log(\hat{\delta}_{ML})$. As a result, $\hat{\delta}_{ML}$ follows a univariate lognormal distribution with mean $\delta \exp(m\sigma_d^2/2)$, which indicates that $\hat{\delta}_{ML}$ is not an unbiased estimator of $\delta = \tilde{\mu}_T/\tilde{\mu}_R$.

The corresponding $(1 - 2\alpha) \times 100\%$ confidence interval for δ can be obtained as follows:

$$(\exp(L_1), \exp(U_1)). \qquad (6.4.4)$$

Since, under the normality assumption, the exact $(1 - 2\alpha) \times 100\%$ confidence interval, (L_1, U_1), is the shortest confidence interval for $\mu_T - \mu_R$, the exact $(1 - 2\alpha) \times 100\%$ confidence interval, $(\exp(L_1), \exp(U_1))$, is also the shortest confidence interval for $\bar{\mu}_T/\bar{\mu}_R$ (Land, 1988). According to the ± 20 rule, we conclude that the two formulations are bioequivalent if

$$\exp(L_1) > 80\% \quad \text{and} \quad \exp(U_1) < 120\%.$$

6.5. MINIMUM VARIANCE UNBIASED ESTIMATOR

As it can be seen from the previous section, the maximum likelihood estimator $\hat{\delta}_{ML}$ overestimates δ when the intra-subject variability is large and the sample size is small. The bias could be substantial. Consequently, the mean squared error of the estimate could be too large to draw a meaningful statistical inference on δ. In the interest of unbiasedness, we may consider the minimum variance unbiased estimator (MVUE) of $\delta = \bar{\mu}_T/\bar{\mu}_R$, which is given by

$$\hat{\delta}_{MVUE} = \hat{\delta}_{ML} \, \Phi_f(-mSSD), \tag{6.5.1}$$

where $m = 1/n_1 + 1/n_2$, SSD is the pooled sum of squares of the period differences of the transformed data (i.e., SSD $= (n_1 + n_2 - 2)\hat{\sigma}_d^2$), f is the degrees of freedom $(n_1 + n_2 - 2)$, and

$$\Phi_f(-mSSD) = \sum_{j=0}^{\infty} \frac{\Gamma(f/2)}{\Gamma[(f/2) + j]j!} [(-m/4)SSD]^j, \tag{6.5.2}$$

where $\Gamma(\cdot)$ is the gamma function.

From (6.5.2), although $\Phi_f(-mSSD)$ involves an infinite series, it, in fact, converges at a much faster rate than $\exp(-mSSD)$ because $\Gamma(f/2)/\Gamma[(f/2) + j!]$ < 1. To obtain $\hat{\delta}_{MVUE}$ for δ as given in (6.5.1), the calculation of $\Phi_f(-mSSD)$ is necessary. Our experience indicates that, in most cases, the first five terms of $\Phi_f(-mSSD)$ are sufficient to provide a reasonable approximation at accuracy up to 10^{-8} for a small sample size (e.g., ≤ 24). A SAS program for the calculation of $\hat{\delta}_{MVUE}$ is included in Appendix B.3.

The function of Φ_f was first introduced by Neyman and Scott (1960). The properties of Φ_f, which have been examined by several researchers (Mehran, 1973; Holye, 1968; Smith, 1988), are summarized below:

Lemma 6.5.1 If SSD is distributed as $\sigma_d^2 \chi^2(f)$, then for any number $c \neq 0$, we have

(a) $E[\Phi_f(cSSD)] = \exp[(c/2)\sigma_d^2]$;

(b) $E\{[\Phi_f(cSSD)]^2\} = \exp(c\sigma_d^2)\Phi_f(c\sigma_d^4).$ \qquad (6.5.3)

Proof A proof can be found in Neyman and Scott (1960) and Mehran (1973).

Let σ_d^2 be the variance of period differences in log scale, i.e., $\sigma_d^2 = (\sigma_T^2 + \sigma_R^2)/4$. When $\sigma_T^2 = \sigma_R^2 = \sigma_e^2$, $\sigma_d^2 = \sigma_e^2/2$. The following results can be obtained using the above lemma.

Theorem 6.5.2 Under model (6.2.1) with the assumption of no carry-over effects, the bias, mean squared error (MSE) of $\hat{\delta}_{ML}$, the variance of $\hat{\delta}_{MVUE}$, the relative efficiency of MVUE to MLE, and an unbiased estimator of the variance of MVUE are given below:

(a) $\text{Bias}(\hat{\delta}_{ML}) = \delta\{\exp[(m/2)\sigma_d^2] - 1\};$ (6.5.4)

(b) $\text{MSE}(\hat{\delta}_{ML}) = \delta^2\{[\exp(m\sigma_d^2) - 1]^2$

$\qquad + 2\exp(m\sigma_d^2/2)[\exp(m\sigma_d^2/2) - 1]\};$ (6.5.5)

(c) $\text{Var}(\hat{\delta}_{MVUE}) = \delta^2\{\exp(m\sigma_d^2)\Phi_f[(m\sigma_d^2)^2] - 1\};$ (6.5.6)

(d) The relative efficiency of MVUE to MLE is

$\text{eff}(\hat{\delta}_{MVUE}, \hat{\delta}_{ML}) = \text{MSE}(\hat{\delta}_{ML})/\text{Var}(\hat{\delta}_{MVUE});$ (6.5.7)

(e) An unbiased estimator of $\text{Var}(\hat{\delta}_{MVUE})$ is given by

$\hat{\text{Var}}(\hat{\delta}_{MVUE}) = \exp[2(\overline{Y}_T - \overline{Y}_R)]\{[\Phi_f(-mSSD)]^2 - \Phi_f(-4mSSD)\}.$
(6.5.8)

Proof

(a) The result follows from the fact that

$E(\hat{\delta}_{ML}) = \delta \exp(m\sigma_d^2/2).$

(b) Since $\text{Var}(\hat{\delta}_{ML}) = \delta^2 \exp(m\sigma_d^2)[\exp(m\sigma_d^2) - 1]$, the result follows from $\text{MSE}(\hat{\delta}_{ML}) = \text{Var}(\hat{\delta}_{ML}) + [\text{Bias}(\hat{\delta}_{ML})]^2.$

(c) $\text{Var}(\hat{\delta}_{MVUE}) = E(\hat{\delta}_{MVUE}^2) - \delta^2$

$\qquad = E\{\exp[2(\overline{Y}_T - \overline{Y}_R)]\}E\{[\Phi_f(-mSSD)]^2\} - \delta^2$

$\qquad = \{\delta^2\exp(2m\sigma_d^2)\}\{\exp(-m\sigma_d^2)\Phi_f(m^2\sigma_d^4)\} - \delta^2$

$\qquad = \delta^2\{\exp(m\sigma_d^2)\Phi_f[(m\sigma_d^2)^2] - 1\}.$

(d) and (e) can be easily verified. ∎

Since $\exp[(m/2)\sigma_d^2]$ is greater than 1, $\hat{\delta}_{ML}$ always overestimates δ. It should be noted that $\hat{\delta}_{ML}$ is asymptotically unbiased for a large sample size (i.e., n_1 and n_2 are large). In most bioequivalence studies, however, the sample size ranges from small (e.g., 6) to moderate (e.g., 24). Therefore, if the intra-subject variability σ_d^2 in log scale is rather large, then the bias of $\hat{\delta}_{ML}$ could be substantial if the ± 20 rule is used for assessment of bioequivalence.

6.6. MEAN OF INDIVIDUAL SUBJECT RATIOS

When there are no period effects, Peace (1986) and Anderson and Hauck (1990) suggested use of individual subject ratios for assessment of bioequivalence. Peace

(1990) proposed a method using the Tchebycheff inequality based upon individual subject ratios as a preliminary test for assessment of bioequivalence. The individual subject ratios for a standard 2×2 crossover design are defined as follows:

$$\tilde{r}_{ik} = \begin{cases} r_{ik} & \text{if } k = 1 \\ 1/r_{ik} & \text{if } k = 2, \end{cases} \tag{6.6.1}$$

where r_{ik} are defined in (6.4.2).

As indicated earlier, the individual subject ratios can remove inter-subject variability. However, it cannot remove the period effect. Therefore, if we use the mean of individual subject ratios to estimate δ, the bias could be substantial. Note that $\tilde{r}_{ik}$ is independently lognormally distributed, i.e., $\log(\tilde{r}_{ik})$ is independently normally distributed with mean $\mu + (F_T - F_R) + (P_2 - P_1)$ for $k = 1$ and $\mu + (F_T - F_R) + (P_1 - P_2)$ for $k = 2$ and variance $\sigma_T^2 + \sigma_R^2$. Tse (1990) compared several methods for constructing a confidence interval for δ based on mean of individual subject ratios ($\tilde{r}_{ik}$) under a lognormal distribution assumption. The results indicate that the resultant confidence intervals for δ lead to a poor coverage probability.

For estimation of δ, the least squares mean of individual subject ratios, denoted by $\hat{\delta}_{MIR}$, is usually considered, i.e.,

$$\hat{\delta}_{MIR} = \tfrac{1}{2} \sum_{k=1}^{2} \frac{1}{n_k} \sum_{i=1}^{n_k} \tilde{r}_{ik}. \tag{6.6.2}$$

Under mode (6.2.1) with the assumption of no carry-over effects, the expected value and variance of $\hat{\delta}_{MIR}$ are given by

$$E(\hat{\delta}_{MIR}) = \tfrac{1}{2} \delta \exp\left(\frac{\sigma_T^2 + \sigma_R^2}{2}\right) \cdot \{\exp(P_1 - P_2) + \exp(P_2 - P_1)\}, \tag{6.6.3}$$

$$Var(\hat{\delta}_{MIR}) = \tfrac{1}{4} V_{MIR}^* \left\{\frac{1}{n_1} \exp[2(P_2 - P_1)] + \frac{1}{n_2} \exp[2(P_1 - P_2)]\right\}, \tag{6.6.4}$$

where

$$V_{MIR}^* = \exp[2(F_T - F_R) + (\sigma_T^2 + \sigma_R^2)]\{\exp(\sigma_T^2 + \sigma_R^2) - 1\}.$$

Therefore, the bias of $\hat{\delta}_{MIR}$ is given by

$$Bias(\hat{\delta}_{MIR}) = \tfrac{1}{2} \delta \left\{\left[\exp\left(\frac{\sigma_T^2 + \sigma_R^2}{2}\right)\right] \right.$$
$$\left. [\exp(P_1 - P_2) + \exp(P_2 - P_1)] - 2\right\}. \tag{6.6.5}$$

It can be seen from (6.6.5) that (i) the bias of $\hat{\delta}_{MIR}$ is independent of the sample size and (ii) the bias involves intra-subject variabilities as well as the period effect. Hence, unlike $\hat{\delta}_{ML}$, we cannot reduce the bias by increasing the sample size. In addition, $\hat{\delta}_{MIR}$ is not an unbiased estimator of δ even when there is no period effect. These undesirable statistical properties certainly argue against its use in the assessment of bioequivalence.

To compare the relative performances in terms of bias between $\hat{\delta}_{ML}$ and $\hat{\delta}_{MIR}$, Figures 6.6.1 and 6.6.2 plot their relative biases against $\sigma_T^2 + \sigma_R^2$ for $n_1 = n_2 = 10, 15$ under the assumption of no period and carry-over effects. The results indicate that the biases of $\hat{\delta}_{ML}$ and $\hat{\delta}_{MIR}$ are an increasing function of the intra-subject variability. Therefore, one way to reduce the bias is to reduce the intra-subject variability. In addition, the bias of $\hat{\delta}_{MIR}$ is always larger than that of $\hat{\delta}_{ML}$. For example, when the total intra-subject variability (in log scale) is 1.0 and $n_1 = n_2 = 10$, the bias of $\hat{\delta}_{MIR}$ is about 60%, while the bias of $\hat{\delta}_{ML}$ is about 8%. The bias of $\hat{\delta}_{ML}$ reduces to about 3% as the sample size increases to

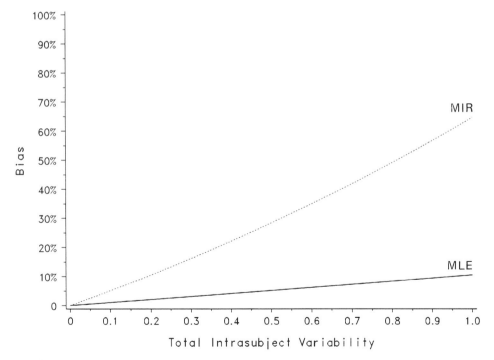

Figure 6.6.1   Bias of MLE and LS mean of individual subject ratios. (Sample size per sequence = 10; direct formulation effect in log scale = 0.)

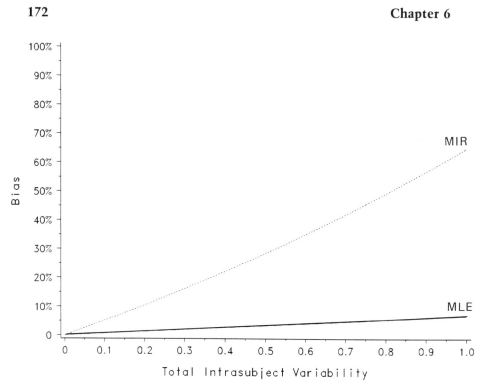

Figure 6.6.2   Bias of MLE and LS mean of individual subject ratios. (Sample size per sequence = 15; direct formulation effect in log scale = 0.)

$n_1 = n_2 = 15$, while the bias of $\hat{\delta}_{MIR}$ remains the same as the sample size increases because the bias is not a function of sample size. As a result, the bias of $\hat{\delta}_{ML}$ can be reduced by either increasing the sample size or decreasing the intra-subject variability, and the bias of $\hat{\delta}_{MIR}$ can only be reduced by decreasing the intra-subject variability.

6.7.   RATIO OF FORMULATION MEANS

From model (6.2.1), the least squares means for test and reference formulations are

$$\overline{X}_T = \tfrac{1}{2}(\overline{X}_{.21} + \overline{X}_{.12}),$$
$$\overline{X}_R = \tfrac{1}{2}(\overline{X}_{.11} + \overline{X}_{.22}).$$

Therefore, the ratio of least squares means is given by

$$\hat{\delta}_{RM} = \overline{X}_T/\overline{X}_R,$$

which is usually considered for assessment of bioequivalence. For example, in Section 4.5.3, we introduced a bootstrap method for constructing a confidence interval for δ based on $\hat{\delta}_{RM}$. It, however, should be noted that the exact distribution of $\overline{X}_T/\overline{X}_R$ is rather complicated and is, thus, not trackable. In this section, the results in Section 4.5.3 will be used to examine the asymptotic bias of $\hat{\delta}_{RM}$ under model (6.2.1).

Let $\overline{X} = (\overline{X}_{.11}, \overline{X}_{.21}, \overline{X}_{.12}, \overline{X}_{.22})'$ and $\lambda_k = n_k/n$, where $n = n_1 + n_2$; $k = 1,2$. Under model (6.2.1) and by Theorem 4.5.1, we then have

$$\overline{X} \xrightarrow{\mathscr{D}} (\mu, \Sigma/n),$$

where $\mu = (\mu_{11}, \mu_{21}, \mu_{12}, \mu_{22})'$, and

$$\Sigma = \begin{bmatrix} \lambda_1\Sigma_1 & 0 \\ 0 & \lambda_2\Sigma_2 \end{bmatrix} = \begin{bmatrix} V_{11} & V_{12} & & \\ V_{12} & V_{22} & & 0 \\ & & V_{33} & V_{34} \\ & 0 & V_{34} & V_{44} \end{bmatrix},$$

where

$$\mu_{11} = \exp[(\mu + P_1 + F_R) + (\sigma_R^2 + \sigma_s^2)/2];$$

$$\mu_{21} = \exp[(\mu + P_2 + F_T) + (\sigma_T^2 + \sigma_s^2)/2];$$

$$\mu_{12} = \exp[(\mu + P_1 + F_T) + (\sigma_T^2 + \sigma_s^2)/2];$$

$$\mu_{22} = \exp[(\mu + P_2 + F_R) + (\sigma_R^2 + \sigma_s^2)/2];$$

$$V_{11} = \exp[2(\mu + P_1 + F_R) + (\sigma_R^2 + \sigma_s^2)][\exp(\sigma_R^2 + \sigma_s^2) - 1];$$

$$V_{22} = \exp[2(\mu + P_2 + F_T) + (\sigma_T^2 + \sigma_s^2)][\exp(\sigma_T^2 + \sigma_s^2) - 1];$$

$$V_{33} = \exp[2(\mu + P_1 + F_T) + (\sigma_T^2 + \sigma_s^2)][\exp(\sigma_T^2 + \sigma_s^2) - 1];$$

$$V_{44} = \exp[2(\mu + P_2 + F_R) + (\sigma_R^2 + \sigma_s^2)][\exp(\sigma_R^2 + \sigma_s^2) - 1];$$

$$V_{12} = V_{34} = \exp[2\mu + (\sigma_R^2 + \sigma_T^2 + 2\sigma_s^2)/2][\exp(\sigma_s^2) - 1]. \qquad (6.7.1)$$

Thus, the random vector $(\overline{X}_T, \overline{X}_R)'$ is also asymptotically normal with mean vector $\mu^* = (\mu_T^*, \mu_R^*)'$ and covariance matrix

$$\frac{1}{4n}\Sigma^* = \begin{bmatrix} V_{TT} & V_{TR} \\ V_{TR} & V_{RR} \end{bmatrix},$$

where

$$\mu_T^* = \tfrac{1}{2}\{\exp[(\mu + F_T) + (\sigma_T^2 + \sigma_s^2)/2]\}\{\exp(P_1) + \exp(P_2)\};$$

$$\mu_R^* = \tfrac{1}{2}\{\exp[(\mu + F_R) + (\sigma_R^2 + \sigma_s^2)/2]\}\{\exp(P_1) + \exp(P_2)\};$$

$$V_{TT} = \exp[2(\mu + F_T) + (\sigma_T^2 + \sigma_s^2)][\exp(\sigma_T^2 + \sigma_s^2) - 1]$$
$$[\lambda_1 \exp(2P_2) + \lambda_2 \exp(2P_1)];$$
$$V_{RR} = \exp[2(\mu + F_R) + (\sigma_R^2 + \sigma_s^2)][\exp(\sigma_R^2 + \sigma_s^2) - 1]$$
$$[\lambda_1 \exp(2P_1) + \lambda_2 \exp(2P_2)];$$
$$V_{TR} = \exp[2\mu + (\sigma_T^2 + \sigma_R^2 + 2\sigma_s^2)/2][\exp(\sigma_s^2) - 1]. \tag{6.7.2}$$

The asymptotic bias of $\hat{\delta}_{RM}$ is given in the following theorem.

Theorem 6.7.1 Suppose that there are no carry-over effects. Under model (6.2.1), the asymptotic bias of $\hat{\delta}_{RM}$, denoted by $ABias(\hat{\delta}_{RM})$, is then given by

$$ABias(\hat{\delta}_{RM}) = E(\hat{\delta}_{RM}) - \delta = \delta \left\{ \exp\left[\frac{\sigma_T^2 - \sigma_R^2}{2}\right] \right.$$
$$\left. \left[1 + \frac{1}{n\pi_1} \{\pi_2[\exp(\sigma_R^2 + \sigma_s^2) - 1] - [\exp(\sigma_s^2) - 1]\}\right]\right\} - \delta, \tag{6.7.3}$$

where

$$\pi_1 = [\exp(P_1) + \exp(P_2)]^2, \quad \text{and}$$
$$\pi_2 = [\lambda_1\exp(2P_1) + \lambda_2 \exp(2P_2)].$$

Proof Consider Taylor expansion of $\hat{\delta}_{RM}$ around μ_T^*/μ_R^* up to the second order term. We then have

$$\hat{\delta}_{RM} = \frac{\overline{X}_T}{\overline{X}_R}$$
$$= \frac{\mu_T^*}{\mu_R^*} + \frac{1}{\mu_R^*}(\overline{X}_T - \mu_T^*) - \frac{\mu_T^*}{\mu_R^{*2}}(\overline{X}_R - \mu_R^*) + \frac{\mu_T^*}{\mu_R^{*3}}(\overline{X} - \mu_R^*)^2$$
$$- \frac{1}{\mu_R^{*2}}(\overline{X}_T - \mu_T^*)(\overline{X}_R - \mu_R^*) + O(n^{-2}).$$

Therefore, the asymptotic bias is

$$Bias(\hat{\delta}_{RM}) = E(\hat{\delta}_{RM}) - \delta$$
$$= \frac{\mu_T^*}{\mu_R^*} + \frac{\mu_T^*}{\mu_R^{*3}} V_{RR} - \frac{V_{TR}}{\mu_R^{*2}} + E[O(n^{-2})] - \delta.$$

Since $E[R(n^{-2})]$ is negligible as n tends to infinity, the result follows. ∎

Corollary 6.7.2 When there is no period effect and $\lambda_1 = \lambda_2 = 1/2$, then the asymptotic bias becomes

$$\text{ABias}(\hat{\delta}_{RM}) = \delta \left\{ \exp\left[\frac{\sigma_T^2 - \sigma_R^2}{2} \right] \right.$$
$$\left. \left[1 + \frac{1}{4n} \exp(\sigma_s^2)[\exp(\sigma_R^2) - 1] \right] - 1 \right\}. \quad (6.7.4)$$

Note that the asymptotic bias of $\hat{\delta}_{RM}$ contains inter-subject variability.

6.8. COMPARISON OF MLE, MVUE, MIR, AND RM

Thus far, we have introduced four estimators for δ. They are the maximum likelihood estimator ($\hat{\delta}_{ML}$), minimum variance unbiased estimator ($\hat{\delta}_{MVUE}$), mean of individual subject ratios ($\hat{\delta}_{MIR}$), and ratio of formulation means ($\hat{\delta}_{RM}$). Under model (6.2.1), the results for biases, variances and mean squared errors for $\hat{\delta}_{ML}$, $\hat{\delta}_{MVUE}$ and $\hat{\delta}_{MIR}$ are exact. For the ratio of means $\hat{\delta}_{RM}$, however, only asymptotic results are obtained. To evaluate their relative performances in small samples, a simulation study (Liu and Weng, 1992) was conducted to compare these estimators in terms of relative bias (i.e., $100 \times (\text{bias}/\delta)$), variance and mean squared error under model (6.2.1). In this simulation study, we consider combinations of three sample sizes ($n_1 = n_2 = 7, 10$, and 15), six covariance structures (in log scale), two formulation effects ($F_T - F_R = 0$ and 0.5), and two period effects ($P_1 - P_2 = 0$ and 2). For each combination, a total of 1000 random samples were generated to compute average relative bias and average variance/mean squared error for the four estimators. The random samples were obtained by generating normal random samples according to model (4.1.1) and following an inverse transformation. The results are summarized in Tables 6.8.1 (for relative bias) and 6.8.2 (for variance and mean squared error), respectively.

From Table 6.8.1, the results indicate that the biases of $\hat{\delta}_{ML}$ and $\hat{\delta}_{RM}$ decrease as either sample size increases or the intra-subject variability decreases when $\sigma_T^2 = \sigma_R^2 = \sigma_e^2$. In addition, as expected, the biases of $\hat{\delta}_{ML}$ are not affected by period effects. However, the biases of $\hat{\delta}_{RM}$ and $\hat{\delta}_{MIR}$ increase substantially when there is a period effect. For the case where $\sigma_T^2 < \sigma_R^2$, the bias of $\hat{\delta}_{ML}$ does not change much. This is because it depends only upon the total intra-subject variabilities, $\sigma_T^2 + \sigma_R^2$. On the other hand, the bias of $\hat{\delta}_{RM}$ becomes negative and does not change much regardless of the presence of a period effect. This is because the leading term of the asymptotic bias of $\hat{\delta}_{RM}$ is $\exp(\sigma_T^2 - \sigma_R^2)$ and is not a function of sample size, while the first order term which contains the period effect is a decreasing function of sample size. In general, the absolute magnitude of the bias of $\hat{\delta}_{MIR}$ is much larger than that of $\hat{\delta}_{RM}$ for the combinations considered.

The empirical relative bias of $\hat{\delta}_{MVUE}$ is always less than 2% except for one combination. The absolute magnitudes of the maximum relative biases for $\hat{\delta}_{ML}$, $\hat{\delta}_{RM}$ and $\hat{\delta}_{MIR}$ are 11.6%, -74%, and 1700%, respectively.

From Table 6.8.2, the results indicate that the empirical variances and mean squared errors of the four estimators decrease as the sample size increases and increase in the presence of formulation effect. As shown in Table 6.8.2, as expected, empirical variances and mean squared errors for $\hat{\delta}_{MVUE}$ and $\hat{\delta}_{ML}$ do not change much regardless of the presence of period effect. However, the mean squared errors of $\hat{\delta}_{MIR}$ and $\hat{\delta}_{RM}$ increase dramatically when there is period effect.

Table 6.8.1 Relative Average Bias (%) of Estimators

Covariance structure		Esimator	$F_T - F_R = 0,$ $P_1 - P_2 = 0$			$F_T - F_R = 0,$ $P_1 - P_2 = 2$		
			7	10	15	7	10	15
0.5	0.25	MVUE	0.41	0.11	0.39	-1.0	<0.1	0.37
	0.5	MLE	1.4	1.2	1.2	0.79	1.2	0.45
		RM	1.7	1.3	1.8	6.2	5.1	2.2
		MIR	27.8	28.9	29.0	375.9	382.4	380.0
1	0.5	MVUE	0.85	-0.98	-0.25	0.12	0.84	0.29
	1	MLE	4.6	1.5	1.4	3.8	3.8	2.0
		RM	7.7	3.9	2.4	17.0	15.0	8.9
		MIR	66.5	64.6	63.2	510.6	535.7	533.9
2	1	MVUE	1.1	-1.1	-0.41	-0.83	-0.29	0.36
	2	MLE	8.8	4.0	3.0	6.6	4.8	3.7
		RM	23.1	12.5	9.7	45.6	29.2	20.1
		MIR	173.5	186.6	171.0	942.2	886.3	911.9
0.5	$\sqrt{0.125}$	MVUE	0.69	0.39	-0.31	0.77	-0.69	<0.1
	1	MLE	3.5	2.4	1.0	3.6	1.3	1.3
		RM	-14.5	-18.0	-19.4	-9.7	-12.9	-16.0
		MIR	48.2	48.3	48.4	469.0	452.7	455.6
0.5	0.5	MVUE	0.39	-0.18	0.18	0.56	0.44	1.3
	2	MLE	5.7	3.7	2.7	6.1	4.3	3.9
		RM	-39.1	-43.7	-46.1	-32.9	-38.4	-41.6
		MIR	104.5	112.0	110.6	723.1	717.2	712.0
0.5	$\sqrt{0.5}$	MVUE	-0.58	0.94	-0.79	-0.42	0.22	-0.48
	4	MLE	11.3	9.1	4.5	11.3	8.1	4.7
		RM	-66.2	-69.8	-74.0	-58.8	-64.2	-70.5
		MIR	361.2	380.5	367.6	1674.8	1643.3	1611.1

Table 6.8.1 Continued

Covariance structure		Esimator	$F_T - F_R = 0.5,$ $P_1 - P_2 = 0$			$F_T - F_R = 0.5,$ $P_1 - P_2 = 2$		
			7	10	15	7	10	15
0.5	0.25	MVUE	0.13	0.72	0.40	0.33	−0.24	−0.12
	0.5	MLE	1.9	2.0	1.2	2.1	1.0	0.72
		RM	3.0	2.3	1.7	6.1	3.9	3.5
		MIR	28.5	29.3	29.0	381.1	381.5	384.0
1	0.5	MVUE	0.20	0.93	−0.18	0.61	1.23	0.53
	1	MLE	3.9	3.5	1.5	4.2	3.7	1.1
		RM	6.0	5.98	3.0	14.5	13.0	7.2
		MIR	66.3	68.1	63.1	515.1	524.8	507.9
2	1	MVUE	<0.1	−1.8	−1.1	−0.88	0.55	0.22
	2	MLE	7.4	3.3	2.3	6.3	5.7	3.6
		RM	17.4	12.6	9.0	33.7	26.5	16.3
		MIR	173.7	171.7	168.7	862.8	895.8	930.1
0.5	$\sqrt{0.125}$	MVUE	−0.42	−0.47	0.49	−0.39	0.20	<0.1
	1	MLE	2.4	1.6	1.8	2.5	2.2	1.4
		RM	−16.6	−18.6	−18.6	−10.4	−12.2	−16.2
		MIR	47.6	49.1	49.5	457.7	453.9	457.6
0.5	0.5	MVUE	−0.21	−0.35	0.12	−0.40	1.3	<0.1
	2	MLE	5.3	3.5	2.4	5.1	5.1	2.4
		RM	−41.2	−44.1	−46.6	−32.3	−38.2	−42.2
		MIR	109.0	113.3	112.8	679.1	720.4	692.4
0.5	$\sqrt{0.5}$	MVUE	−0.87	2.4	−0.48	<0.1	1.1	−0.15
	4	MLE	10.7	10.6	4.7	11.6	9.2	5.1
		RM	−65.4	−69.8	−73.2	−57.5	−64.5	−70.5
		MIR	376.8	380.2	364.5	1619.6	1698.7	1612.3

Relative average bias (%) = $100[\text{average bias}/(\exp(F_T - F_R))]$. MVUE = minimum variance unbiased estimator, MLE = maximum likelihood estimator, RM = ratio of means, MIR = mean of individual ratios.

Source: Liu and Weng (1992).

In general, $\hat{\delta}_{MIR}$ has the largest mean squared error among the four estimators. The mean squared errors for $\hat{\delta}_{ML}$ are quite close to the variance of $\hat{\delta}_{MVUE}$.

In summary, for estimation of δ (or for assessment of bioequivalence), the minimum variance unbiased estimator should be used while the maximum likelihood estimator is competitive when the total intra-subject variabilities (in log scale) is small (e.g., < 0.5) and sample size is moderate (e.g., $n_1 = n_2 > 10$).

Table 6.8.2 Average Mean Square Error and Variance of Estimators

Covariance structure	Estimator	$F_T - F_R = 0,$ $P_1 - P_2 = 0$			$F_T - F_R = 0,$ $P_1 - P_2 = 2$		
		7	10	15	7	10	15
0.5 0.25	MVUE	0.036	0.027	0.018	0.033	0.024	0.016
0.5	MLE	0.037	0.028	0.018	0.034	0.025	0.016
	RM	0.050	0.036	0.026	0.141	0.095	0.057
	MIR	0.151	0.140	0.122	15.878	16.082	15.380
1 0.5	MVUE	0.068	0.045	0.034	0.066	0.054	0.034
1	MLE	0.076	0.048	0.035	0.072	0.057	0.036
	RM	0.146	0.098	0.069	0.511	0.344	0.180
	MIR	0.756	0.649	0.540	33.440	35.580	33.420
2 1	MVUE	0.153	0.098	0.066	0.171	0.104	0.075
2	MLE	0.182	0.110	0.071	0.200	0.117	0.082
	RM	0.706	0.434	0.269	6.771	1.325	0.840
	MIR	5.633	4.839	4.409	204.709	122.727	119.150
0.5 $\sqrt{0.125}$	MVUE	0.060	0.045	0.026	0.054	0.036	0.026
1	MLE	0.065	0.043	0.027	0.058	0.038	0.027
	RM	0.088	0.075	0.067	0.156	0.129	0.096
	MIR	0.418	0.361	0.320	26.504	23.840	23.280
0.5 0.5	MVUE	0.107	0.083	0.051	0.115	0.077	0.061
2	MLE	0.121	0.091	0.054	0.132	0.084	0.065
	RM	0.230	0.245	0.246	0.279	0.252	0.244
	MIR	1.911	2.000	1.691	97.100	90.890	64.377
0.5 $\sqrt{0.5}$	MVUE	0.250	0.158	0.113	0.261	0.186	0.103
4	MLE	0.323	0.193	0.127	0.336	0.223	0.116
	RM	0.506	0.526	0.570	0.477	0.486	0.537
	MIR	32.762	32.940	25.590	1200.760	698.380	529.244

Table 6.8.2 Continued

Covariance structure		Estimator	$F_T - F_R = 0.5,$ $P_1 - P_2 = 0$			$F_T - F_R = 0.5,$ $P_1 - P_2 = 2$		
			7	10	15	7	10	15
0.5	0.25	MVUE	0.101	0.072	0.044	0.097	0.071	0.041
	0.5	MLE	0.105	0.074	0.045	0.101	0.073	0.045
		RM	0.151	0.111	0.070	0.424	0.269	0.177
		MIR	0.433	0.383	0.324	44.840	43.940	42.761
1	0.5	MVUE	0.217	0.144	0.091	0.196	0.134	0.082
	1	MLE	0.236	0.155	0.094	0.214	0.144	0.085
		RM	0.475	0.350	0.184	1.021	0.789	0.498
		MIR	2.169	1.951	1.466	96.199	91.890	79.790
2	1	MVUE	0.444	0.293	0.175	0.394	0.281	0.197
	2	MLE	0.517	0.325	0.188	0.458	0.317	0.213
		RM	2.442	1.290	0.743	5.362	2.600	2.235
		MIR	17.171	15.319	12.184	371.444	346.882	340.516
0.5	$\sqrt{0.125}$	MVUE	0.155	0.102	0.078	0.159	0.108	0.078
	1	MLE	0.164	0.106	0.081	0.171	0.114	0.080
		RM	0.249	0.213	0.177	0.434	0.353	0.279
		MIR	1.096	1.003	0.921	71.582	65.301	63.510
0.5	0.5	MVUE	0.303	0.227	0.140	0.302	0.213	0.147
	2	MLE	0.343	0.247	0.148	0.344	0.236	0.156
		RM	0.662	0.668	0.684	0.738	0.687	0.675
		MIR	5.650	6.309	4.845	203.713	207.342	171.998
0.5	$\sqrt{0.5}$	MVUE	0.612	0.500	0.303	0.650	0.480	0.284
	4	MLE	0.795	0.604	0.338	0.831	0.576	0.320
		RM	1.325	1.455	1.525	1.281	1.304	1.467
		MIR	111.466	139.567	81.574	3285.000	2790.540	1428.850

MVUE = minimum variance unbiased estimator, MLE = maximum likelihood estimator, RM = ratio of means, MIR = mean of individual ratios
Source: Liu and Weng (1992).

However, it should be noted that bias of the maximum likelihood estimator involves the intra-subject variability, which may have some impact on the assessment of bioequivalence in average bioavailability when the ± 20 rule is applied.

6.9. AN EXAMPLE

To illustrate the use of estimators of δ discussed in the previous sections, we consider the AUC data from two erythromycin formulations in a bioavailability study published by Clayton and Leslie (1981). In this study, a standard 2 × 2 crossover experiment was conducted with 18 subjects to compare a new erythromycin formulation (i.e., erythromycin stearate) with a reference formulation (i.e., erythromycin base). The new formulation and the reference formulation are denoted by formulations C and D, respectively. Since no sequence identification of each subject was provided in Clayton and Leslie (1981), for the purpose of this illustration, we will adapt the order of periods given in Weiner (1989) and assign subjects 1 through 9 to sequence 1 and the remaining subjects to sequence 2. Table 6.9.1 gives the original AUC's and log-transformed AUC's as well as the individual subject ratios and their log-transformation.

Note that this data set has been analyzed by many researchers because of its

Table 6.9.1 AUCs for Two Erythromycin Formulations

Subject	Sequence	C (Stearate)		D (Base)		Ratio C/D	
		Raw	Log (raw)	Raw	Log (raw)	Raw	Log (raw)
1	CD	2.52	0.9243	5.47	1.6993	0.4607	−0.7750
2	CD	8.87	2.1827	4.84	1.5769	1.8326	0.6058
3	CD	0.79	−0.2357	2.25	0.8109	0.3511	−1.0467
4	CD	1.68	0.5188	1.82	0.5988	0.9231	−0.0800
5	CD	6.95	1.9387	7.87	2.0631	0.8831	−0.1243
6	CD	1.05	0.0488	3.25	1.1787	0.3231	−1.1299
7	CD	0.99	−0.0101	12.39	2.5169	0.0800	−2.5269
8	CD	5.60	1.7228	4.77	1.5624	1.1740	0.1604
9	CD	3.16	1.1506	1.88	0.6313	1.6809	0.5193
10	DC	3.19	1.1600	4.98	1.6054	0.6406	−0.4454
11	DC	9.83	2.2854	7.14	1.9657	1.3768	0.3197
12	DC	2.91	1.0682	1.81	0.5933	1.6077	0.4748
13	DC	4.58	1.5217	7.34	1.9933	0.6240	−0.4716
14	DC	7.05	1.9530	4.25	1.4469	1.6588	0.5061
15	DC	3.41	1.2267	6.66	1.8961	0.5120	−0.6694
16	DC	2.49	0.9123	4.76	1.5603	0.5231	−0.6480
17	DC	6.18	1.8213	7.16	1.9685	0.8631	−0.1472
18	DC	2.85	1.0473	5.52	1.7084	0.5163	−0.6611

Source: Clayton and Leslie (1981).

Table 6.9.2 Analysis of Variance for Log-Transformed Data in Table 6.9.1

Source of variation	df	SS	MS	F	P-Value
Inter-subject					
Carry-over	1	1.3053	1.3053	2.50	0.1332
Residuals	16	8.3434	0.5215	1.69	
Intra-subject					
Formulation	1	1.0470	1.0470	3.39	0.0843
Period	1	0.1959	0.1959	0.63	0.4376
Residuals	16	4.9440	0.3090		

possible violation of the normality assumption in raw data and the existence of potential outliers (see, e.g., Metzler and Huang, 1983; Hauck and Anderson, 1984; Anderson and Hauck, 1990). Furthermore, the conclusion drawn by these authors does not support Clayton and Leslie's claim on bioequivalence between the two formulations. This data set, however, will also be used for checking the normality assumption and for detection of outliers in later chapters.

Table 6.9.2 presents a table of analysis of variance for the log-transformed data. The results show no evidence of unequal carry-over effects (p-value > 0.10) and period effect (p-value > 0.4). However, the test for inter-subject variability by F_v in (3.5.13) is not significant at the 10% level. This indicates that the intra-subject variability might be larger than the inter-subject variability. A close examination of the data reveals that possible outlying observations may occur in subject 7 since the AUC's for the two formulations are quite different (0.99 vs 12.39 in the original scale or -0.01 vs 2.52 in log scale). The estimated CV is about 36.55% which is high but not unusual in bioavailability/bioequivalence studies. The index of sensitivity defined in (5.3.11) is estimated to be about 3.28. Therefore, as discussed in Section 5.3.2, this study has little power to conclude bioequivalence if Schuirmann's two one-sided procedure and the ± 20 rule are applied.

In the following, for estimation of δ, the estimators and their corresponding biases, variances, and mean squared errors will be obtained based upon the AUC data in Table 6.9.1.

I. Maximum Likelihood Estimator $\hat{\delta}_{ML}$

From Table 6.9.1, the LS means, $\hat{F}$, SSD, $\hat{\sigma}_d^2$, and the classical 90% confidence interval for $\hat{F}$, based upon the log-transformed data, can be obtained as follows:

$$\overline{Y}_T = 1.1798, \qquad \overline{Y}_R = 1.5209, \qquad \hat{F} = -0.3411,$$

$$SSD = 2.4720, \qquad \hat{\sigma}_d^2 = 0.1545, \quad \text{and}$$

$$(L_1, U_1) = (-0.6646, -0.0176).$$

Therefore, the maximum likelihood estimate of δ is given by

$$\hat{\delta}_{ML} = \exp(-0.3411) = 0.7110.$$

The 90% confidence interval for $\tilde{F}$ according to (6.4.2) is then given by

$$(\exp(-0.6646), \exp(-0.0176)) = (0.5145, 0.9826).$$

Since the confidence interval for $\tilde{F}$ is not within the bioequivalence limits (80%, 120%), we conclude that the two erythromycin formulations are not bioequivalent according to the ± 20 rule.

From Theorem 6.5.2, the estimates for bias and variance of $\hat{\delta}_{ML}$ can be obtained as follows:

$$\begin{aligned}
\text{Bias}(\hat{\delta}_{ML}) &= \hat{\delta}_{ML}[\exp(m\hat{\sigma}_d^2/2) - 1] \\
&= (0.7110)[\exp(0.1545/9) - 1] \\
&= 0.0123; \\
\hat{\text{Var}}(\hat{\delta}_{ML}) &= \hat{\delta}_{ML}^2 \exp(m\hat{\sigma}_d^2)[\exp(m\hat{\sigma}_d^2) - 1] \\
&= 0.0183.
\end{aligned}$$

Hence, the estimate of mean squared error of $\hat{\delta}_{ML}$ is given by

$$\begin{aligned}
\hat{\text{MSE}}(\hat{\delta}_{ML}) &= 0.0183 + (0.0123)^2 \\
&= 0.0184.
\end{aligned}$$

II. Minimum Variance Unbiased Estimator $\hat{\delta}_{MVUE}$

For the minimum variance unbiased estimate of δ, since $\Phi_f(-mSSD) = 0.9831$, we have

$$\begin{aligned}
\hat{\delta}_{MVUE} &= \hat{\delta}_{ML}\Phi_f(-mSSD) \\
&= (0.7110)(0.9831) \\
&= 0.6990.
\end{aligned}$$

From (6.5.7), the estimate for the variance of $\hat{\delta}_{MVUE}$ can be obtained as follows:

$$\begin{aligned}
\hat{\text{Var}}(\hat{\delta}_{MVUE}) &= \hat{\text{MSE}}(\hat{\delta}_{MVUE}) \\
&= \hat{\delta}_{ML}^2\{[\Phi_f(-mSSD)]^2 - \Phi_f(-4mSSD)\}
\end{aligned}$$

$$= (0.7110)^2[(0.9831)^2 - 0.9354]$$
$$= 0.0157.$$

As a result, the relative efficiency of $\hat{\delta}_{MVUE}$ to $\hat{\delta}_{ML}$ is estimated as

$$\text{eff}(\hat{\delta}_{MVUE}, \hat{\delta}_{ML}) = (0.0184)/(0.0157)$$
$$= 117.39\%.$$

Therefore, for this data set, for the estimation of δ, the minimum variance unbiased estimator is not only an unbiased estimator but also is about 17% more efficient than the maximum likelihood estimator.

III. Mean of Individual Subject Ratios $\hat{\delta}_{MIR}$

The mean of individual subject ratios based on the raw data is given by $\hat{\delta}_{MIR} = 0.8906$ which is much higher than $\hat{\delta}_{ML}$ and $\hat{\delta}_{MVUE}$. Since $\hat{P} = P_2 - P_1 = 0.1475$, the bias of $\hat{\delta}_{MIR}$ can be estimated using (6.6.5) which is given below:

$$\text{Bias}(\hat{\delta}_{MIR}) = \tfrac{1}{2}\hat{\delta}_{ML}\{[\exp(2\hat{\sigma}_d^2)]\,[\exp(-\hat{P}) + \exp(\hat{P})] - 2\}$$
$$= 0.2680.$$

Similarly, an estimate of the variance of $\hat{\delta}_{MIR}$ can be obtained using (6.6.4) which is given by 0.0465. Hence, the mean squared error for $\hat{\delta}_{MIR}$ can be estimated by 0.1183.

IV. Ratio of Formulation Means $\hat{\delta}_{RM}$

Since the least squares means based upon the raw data are

$$\overline{X}_T = 4.1167 \quad \text{and} \quad \overline{X}_R = 5.2311,$$

the ratio of least squares means is given by

$$\hat{\delta}_{RM} = \overline{X}_T/\overline{X}_R = 4.1167/5.2311 = 0.7870.$$

Under the assumption of no carry-over and period effects, the sample covariance matrix, based upon 17 degrees of freedom, can be obtained as follows:

$$\begin{bmatrix} 0.3093 & 0.1326 \\ 0.1326 & 0.5607 \end{bmatrix}.$$

Consequently, the estimates for $\sigma_T^2 - \sigma_R^2$, σ_R^2, and σ_s^2, by the method of moments, are 0.2514, 0.1767, and 0.1326, respectively. Therefore, by (6.7.4) in Corollary 6.7.2, the asymptotic bias is estimated by

$$\text{Bias}(\hat{\delta}_{RM}) = \hat{\delta}_{ML}\{[\exp(0.2514/2)]\{1 + \tfrac{1}{72}[\exp(0.1326)][\exp(0.1767) - 1]\}$$
$$- 1\} = 0.0977.$$

Table 6.9.3   Estimates for Biases, Variances, and Mean Squared Errors of MLE, MVUE, MIR, and RM

Estimator	Estimate	Estimated bias	Estimated variance/MSE
MVUE ($\hat{\delta}_{MVUE}$)	0.6990	—	0.01570
MLE ($\hat{\delta}_{ML}$)	0.7110	0.0123	0.01840
MIR ($\hat{\delta}_{MIR}$)	0.8906	0.2680	0.11830
RM ($\hat{\delta}_{RM}$)	0.7870	0.0977[a]	—

[a]Asymptotic bias when there is no period effect.

Source: Lin and Weng (1992).

Table 6.9.3 summarizes the results of estimates of the biases, variances, and mean squared errors for Clayton and Leslie's data. Although the estimated bias of the maximum likelihood estimator is very small, the estimated variance is about 17% larger than that of the minimum variance unbiased estimator. The estimated biases for the least squares mean of individual subject ratios and the ratio of the least squares means are both greater than 10%. The estimated mean squared error of the least squares mean of individual subject ratios is about seven times as large as those of the maximum likelihood estimator and the minimum variance unbiased estimator. These results strongly suggest that, under model. (6.2.1), the least squares mean of individual subject ratios and the ratio of the least squares means should not be used for estimation of δ.

6.10.   DISCUSSION

In this chapter, we considered several estimators for estimation of δ, the ratio of average bioavailabilities which is used as a bioequivalence measure under model (6.2.1). Based upon these estimators, a $(1 - 2\alpha) \times 100\%$ confidence intervals for δ can be obtained to assess bioequivalence. For example, based on the maximum likelihood estimator $\hat{\delta}_{ML}$, an exact $(1 - 2\alpha) \times 100\%$ confidence interval for δ is given in (6.4.4). For other estimators, although no exact confidence intervals for δ are available, approximate confidence intervals may be obtained using a nonparametric bootstrap resampling procedure. These confidence intervals can then be used to determine bioequivalence. In other words, we conclude bioequivalence if the confidence interval is within bioequivalent limits (e.g., 80% and 120%) according to the ± 20 rule.

For the analysis of individual subject ratios, it should be noted that although the least squares mean of individual subject ratios is commonly used for assessment of bioequivalence, the expected value of individual subject ratios is not equal to 1 (i.e., $E(r_{ik}) \neq 1$) even when X_{i1k} and X_{i2k} have exactly the same distribution. Since the distribution of individual subject ratios is often positively skewed, it may be more appropriate to use the median of individual subject ratios as an alternative bioequivalence measure. Tse (1990) examined the relative performances of confidence intervals for the true mean and median of individual subject ratios, which were obtained under the assumption that the ratios follow a lognormal distribution, in terms of their coverage probabilities. The results indicate that the confidence interval for the median of individual subject ratios provides a better coverage probability. As a result, it may suggest that the ratio of medians of X_{ijk} under model (6.2.1) or the true median of individual subject ratios be used as alternative measures of bioequivalence. However, statistical inference on the ratio of medians or the median of individual subject ratios is more complicated to obtain.

In practice, although a log-transformation may be able to reduce skewness and achieve an additive model with relatively homogeneous variances, the log-transformed data may not follow a normal distribution due to unknown distributions of the transformed random subject effects. Distribution of the transformed data for a different formulation may be of a different type due to different distributions of transformed random subject effects. To remove the unknown random subject effects, assuming that there are no period effects, Chow, Peace, and Shao (1991) proposed a method using log-transformed individual subject ratios followed by an inverse transformed under a semi-parametric multiplicative model. However, it should be noted that the method proposed by Chow, Peace, and Shao is similar to a method proposed by Steinijans and Diletti (1985) which is derived to construct a confidence interval for a bioavailability ratio using a Wilcoxon signed rank test.

The Assessment of Inter- and Intra-Subject Variabilities

7.1. INTRODUCTION

In previous chapters, we considered the assessment of equivalence only in average bioavailabilities between formulations. However, as indicated in Section 1.5, bioequivalence between formulations, in fact, depends upon whether the marginal distributions of the pharmacokinetic parameters of interest such as AUC or C_{max} for the two formulations are equivalent. This equivalence is usually referred to as population bioequivalence (Anderson and Hauck, 1990). Under normality assumptions, the equivalence between distributions can be determined by the equivalence between their first moment (average) and second moment (variability). For assessment of bioequivalence, however, the FDA only requires evidence of equivalence in average bioavailabilities between formulations be provided. The conclusion of bioequivalence based only upon average bioavailability may be somewhat misleading because the safety and exchangeability of the test formulations are questionable.

In practice, since individual subjects may differ widely in their response to the drug, in addition to equivalence in average bioavailability, it is important to compare the variability of bioavailability. If the variability of the test formulation is much larger than that of the reference formulation, then the safety of the test formulation may be of concern and the exchangeability between two formulations is questionable even when the two formulations are equivalent in average bioavailability. This is because the equivalence of average bioavailability does not take into account the difference in variabilities (Liu, 1991). Suppose a patient switches from the reference formulation (with a smaller intra-subject variability) to the test formulation (with a much larger intra-subject variability). It is very likely that the AUC of the patient will be outside the therapeutic range. If the AUC is below the therapeutic range, the test formulation may not be effective. On the other hand, if the AUC is above the therapeutic range, the safety of the test formulation is of great concern because it may cause severe adverse experiences. As a result, equivalence in average bioavailability does not guarantee that the two formulations are therapeutically equivalent and exchangeable. For example, for comparison of an intranasal formulation and an intravenous formulation of a drug product (e.g., insulin for diabetes or butorphanol for migraine headaches), bioequivalence (in average bioavailability) may be concluded. However, they may not be therapeutically equivalent because the intra-subject variability of the intranasal formulation is usually much higher than that of the intravenous formulation.

The objective of this chapter is then to investigate the assessment of equivalence in variabilities of bioavailability between formulations. In the next section, possible decision rules for assessing equivalence in intra-subject variabilities between formulations are discussed. Point and interval estimates for the intersubject and intra-subject variabilities are provided in Section 7.3. The commonly used Pitman-Morgan's adjusted F test for equality of variabilities is outlined in Section 7.4. A distribution-free test procedure based upon Spearman's rank correlation coefficient is also included in this section. In Section 7.5, two test procedures for interval hypotheses of equivalence in variability of bioavailability are presented. In section 7.6, a conditional random effects model is considered to assess the intra-subject variability in terms of the common CV when the intrasubject variabilities vary from subject to subject. A brief discussion is presented in Section 7.7.

7.2. VARIABILITY AND DECISION-MAKING

As indicated earlier, for assessment of bioequivalence, it is important to compare variability (the inter-subject variability and the intra-subject variability) of bioavailability between formulations since individual subjects may differ widely in their responses to the drug. Levy (1986) pointed out that knowledge of the intersubject and intra-subject variabilities may provide valuable information in the assessment of bioequivalence. For assessment of bioequivalence between formulations, however, the FDA only requires evidence of equivalence in average bioavailability. In Chapter 4, several methods were introduced for assessment of equivalence in average bioavailability. One of the primary assumptions regarding variability is that the intra-subject variabilities between formulations are assumed to be the same, i.e., $\sigma^2_T = \sigma^2_R$. It is then of interest to examine whether these methods are still valid when there is a significant difference in intra-subject variabilities. Since these methods are operationally equivalent as indicated in Chapter 4, in the following, without loss of generality, we will only examine whether the two one-sided tests procedure is valid when there is a difference in intra-subject variabilities.

For a bioavailability/bioequivalence study with a 2×2 crossover design, as indicated in Section 4.2, the two one-sided tests procedure (parametric or non-parametric) is based upon the period differences d_{ik} as defined in (3.3.1). If there is no unequal carry-over effects, under model (4.1.2), the expected value and variance are given by

$$E(d_{ik}) = \begin{cases} \frac{1}{2}\,[(P_2 - P_1) + (F_T - F_R)] & \text{for } k = 1 \\ \frac{1}{2}\,[(P_2 - P_1) + (F_R - F_T)] & \text{for } k = 2, \end{cases}$$

and

$$\mathrm{Var}(d_{ik}) = \sigma^2_d = \tfrac{1}{4}\,(\sigma^2_R + \sigma^2_T).$$

Therefore, $\{d_{i1}, i = 1,2, \ldots, n_1\}$ and $\{d_{i2}, i = 1,2, \ldots, n_2\}$ are two independent samples with equal variance σ^2_d. Note that σ^2_d is half of the average of the intra-subject variability of the test and reference formulations. As a result, the two one-sided tests procedure for assessing equivalence in average bioavailability is still valid even when the intra-subject variabilities between formulations are different, i.e., $\sigma^2_T \neq \sigma^2_R$. In other words, for assessment of equivalence in average bioavailability, the methods introduced in Chapter 4, which depend upon the period differences, completely ignore the difference of intra-subject variabilities between formulations. The difference in intra-subject variabilities, however, may have an important impact on the safety and exchangeability of the test formulation. Therefore, in addition to equivalence in average bioavailability, it is imperative to demonstrate equivalence in variability of bioavailability. However, in the current FDA guidelines there is little or no discussion regarding the assessment of equivalence in variability of bioavailability. The 75/75 rule was probably the only method to account for both average bioavailability and variability of bioavailability, but its use has been discouraged due to its undesirable statistical properties. Therefore, a decision rule for assessment of equivalence in variability of bioavailability is necessary.

Similar to the ± 20 rule, we may conclude the two formulations are equivalent in variability of bioavailability if σ^2_T is within $\pm \Delta\%$ of σ^2_R where Δ is a clinically important difference. Based upon this decision rule, we may test the following interval hypotheses to establish equivalence in variability of bioavailability:

$$\begin{aligned} &H_0: \ \sigma^2_T - \sigma^2_R \leq -\Delta\sigma^2_R \quad \text{or} \quad \sigma^2_T - \sigma^2_R \geq \Delta\sigma^2_R \\ \text{vs} \quad &H_a: \ -\Delta\sigma^2_R < \sigma^2_T - \sigma^2_R < \Delta\sigma^2_R. \end{aligned} \qquad (7.2.1)$$

The selection of Δ, of course, depends upon the characteristics of the drug and/or routes of administration. A suggestion for selection of Δ based upon the collected data is given in Section 7.7. Alternatively, we may consider interval hypotheses for the difference in variances or the ratio of variances as follows:

$$H_0: \sigma_T^2 - \sigma_R^2 \leq \theta_1 \quad \text{or} \quad \sigma_T^2 - \sigma_R^2 \geq \theta_2 \tag{7.2.2}$$
$$\text{vs} \quad H_a: \theta_1 < \sigma_T^2 - \sigma_R^2 < \theta_2,$$

and

$$H_0: \sigma_T^2/\sigma_R^2 \leq \delta_1 \quad \text{or} \quad \sigma_T^2/\sigma_R^2 \geq \delta_2 \tag{7.2.3}$$
$$\text{vs} \quad H_a: \delta_1 < \sigma_T^2/\sigma_R^2 < \delta_2.$$

where θ_1 and θ_2 are equivalent limits for the difference in variances and δ_1 and δ_2 are equivalent limits for the ratio of variances. The choices of θ_1, θ_2 and δ_1, δ_2 are rather subjective. Note that hypotheses (7.2.2) reduce to hypotheses (7.2.1) if $\theta_2 = -\theta_1 = \Delta\sigma_R^2$, which depend upon the unknown parameter σ_R^2.

7.3. POINT AND INTERVAL ESTIMATES

7.3.1 Point Estimates

The assumptions of Model (3.1.1) in Chapter 3 for a standard 2×2 crossover design, as stated in (3.1.4), are:

(i) $\{S_{ik}\}$ are i.i.d. normal with mean 0 and variance σ_S^2;
(ii) $\{e_{ijk}\}$ are i.i.d. normal with mean 0 and variance σ_e^2;
(iii) $\{S_{ik}\}$ and $\{e_{ijk}\}$ are mutually independent.

Under model (3.1.1), σ_S^2 and σ_e^2 are indicative of the inter-subject variability and the intra-subject variability, respectively. The information of σ_S^2 and σ_e^2 are useful in bioavailability studies. For example, prior information of σ_S^2 and σ_e^2 can be used for sample size determination in the planning stage of a bioavailability study. On the other hand, estimates of σ_S^2 and σ_e^2, which can be obtained from the data, are usually used to interpret the inter-subject and the intra-subject variabilities. Note that the sum of inter-subject and intra-subject variabilities (i.e., $\sigma_S^2 + \sigma_e^2$) is the variance of the marginal distribution of Y_{ijk}. As discussed in previous chapters, under a crossover design, the comparison of average bioavailabilities between formulations can be made using period differences based only upon the intra-subject variability. Therefore, the gain in precision for using the intra-subject variability alone can be expressed by

$$\rho_I = \frac{\sigma_s^2}{\sigma_s^2 + \sigma_e^2},$$

which is known as the intraclass correlation between responses of the two formulations. Although σ_s^2, σ_e^2, and ρ_I are indicative of the inter-subject variability, the intra-subject variability, and the intraclass correlation, respectively, in practice, they are unknown population parameters and must be estimated from the data. Therefore, in this section, we will investigate point and interval estimates for σ_s^2, σ_e^2 and ρ_I.

Under model (3.1.1), the structure of the crossover design with random components $\{S_{ik}\}$ and $\{e_{ijk}\}$ is, in fact, a balanced two-stage nested design. Therefore, standard estimation procedures for a balanced two-stage nested design can be used to obtain point and interval estimates for σ_s^2, σ_e^2, and ρ_I. Recall that from the analysis of variance table (Table 3.5.1), under the assumptions of (3.1.4), distributions of sum of squares of inter-subject residuals and intra-subject residuals are:

$$SS_{Inter} \sim (\sigma_e^2 + 2\sigma_s^2)\chi^2(n_1 + n_2 - 2), \quad \text{and}$$
$$SS_{Intra} \sim \sigma_e^2\chi^2(n_1 + n_2 - 2), \tag{7.3.1}$$

where SS_{Inter} and SS_{Intra} were defined in Section 3.5. The expected values of the mean squares for the inter-subject and intra-subject variabilities are then given by

$$E[MS_{Inter}] = E\left(\frac{SS_{Inter}}{n_1 + n_2 - 2}\right) = \sigma_e^2 + 2\sigma_s^2;$$

$$E[MS_{Intra}] = E\left(\frac{SS_{Intra}}{n_1 + n_2 - 2}\right) = \sigma_e^2 \tag{7.3.2}$$

Therefore, the analysis of variance estimates of σ_e^2 and σ_s^2 can be obtained by equating the observed mean squares and their expected values as follows:

$$\hat{\sigma}_e^2 = MS_{Intra}, \quad \text{and}$$

$$\hat{\sigma}_s^2 = \frac{MS_{Inter} - MS_{Intra}}{2}, \tag{7.3.3}$$

Note that, under model (3.1.1) with assumptions of (3.1.4), the estimators, $\hat{\sigma}_e^2$ and $\hat{\sigma}_s^2$, are minimum variance unbiased quadratic estimators for σ_e^2 and σ_s^2, respectively (Searle, 1971; Anderson, 1982). Furthermore, it can be verified that the distribution of $\hat{\sigma}_e^2$ is

$$\frac{\sigma_e^2}{n_1 + n_2 - 2}\chi^2(n_1 + n_2 - 2),$$

and $\hat{\sigma}_s^2$ is distributed as

$$\hat{\sigma}_s^2 \sim \frac{\sigma_e^2 + 2\sigma_s^2}{2(n_1 + n_2 - 2)} \chi^2(n_1 + n_2 - 2)$$

$$- \frac{\sigma_e^2}{2(n_1 + n_2 - 2)} \chi^2(n_1 + n_2 - 2). \quad (7.3.4)$$

From (7.3.4), it can be seen that there exists no closed-form for the distribution of $\hat{\sigma}_s^2$ because the coefficients of (7.3.4) involve unknown parameters σ_e^2 and σ_s^2 and the coefficient of the second term in (7.3.4) is negative (Searle, 1971). However, since SS_{Inter} and SS_{Intra} are independent, the variance of $\hat{\sigma}_s^2$ and $\hat{\sigma}_e^2$ can be easily obtained as follows:

$$Var(\hat{\sigma}_e^2) = \frac{2\sigma_e^4}{n_1 + n_2 - 2}, \quad \text{and}$$

$$Var(\hat{\sigma}_s^2) = \frac{1}{2(n_1 + n_2 - 2)} [(\sigma_e^2 + 2\sigma_s^2)^2 + \sigma_e^4]. \quad (7.3.5)$$

Since $\hat{\sigma}_s^2$ is based upon the difference in mean squares between inter-subject and intra-subject residuals, it is possible to obtain a negative estimate if $MS_{Inter} < MS_{Intra}$. Searle (1971) provided a formula for calculation of the probability for obtaining a negative estimate of σ_s^2, which is given by

$$P\{\hat{\sigma}_s^2 < 0\} = P\left\{F(n_1 + n_2 - 2, n_1 + n_2 - 2) < \frac{\sigma_e^2}{\sigma_e^2 + 2\sigma_s^2}\right\}, \quad (7.3.6)$$

where $F(n_1 + n_2 - 2, n_1 + n_2 - 2)$ is a central F distribution with degrees of freedom $n_1 + n_2 - 2$ and $n_1 + n_2 - 2$. Since σ_e^2 and σ_s^2 in (7.3.6) are unknown, an estimate for $\sigma_e^2/(\sigma_e^2 + 2\sigma_s^2)$ is necessary to evaluate the probability. $\sigma_e^2/(\sigma_e^2 + 2\sigma_s^2)$ can be estimated by $1/F_v$, where

$$F_v = \frac{MS_{Inter}}{MS_{Intra}}.$$

Note that probability (7.3.6) could be substantial when σ_s^2/σ_e^2 is small. A negative estimate may indicate that model (3.1.1) is incorrect or sample size is too small. More details regarding negative estimates in analysis of variance components can be found in Hocking (1985).

To avoid negative estimates, a typical approach is to consider the following estimator:

$$\tilde{\sigma}_s^2 = \max\{0, \hat{\sigma}_s^2\}$$

$$= \begin{cases} \hat{\sigma}_s^2 & \text{if } MS_{Inter} \geq MS_{Intra} \\ 0 & \text{if } MS_{Inter} < MS_{Intra} \end{cases} \quad (7.3.7)$$

and

$$\bar{\sigma}_e^2 = \begin{cases} \hat{\sigma}_e^2 \text{ if } MS_{Inter} \geq MS_{Intra} \\ \hat{\sigma}^2 \text{ if } MS_{Inter} < MS_{Intra}, \end{cases}$$

where

$$\hat{\sigma}^2 = \frac{SS_{Inter} + SS_{Intra}}{2(n_1 + n_2)}.$$

The above estimators are known as the restricted maximum likelihood estimators. It should be noted that maximum likelihood estimators for σ_e^2 and σ_s^2 can also be derived from the likelihood function. The MLE's are usually biased, though their mean squared errors are smaller than the variances of $\hat{\sigma}_e^2$ and $\hat{\sigma}_s^2$, respectively, (for details, see Anderson (1982) and Hocking (1985)).

For estimation of the intraclass correlation ρ_I, Snedecor and Cochran (1980) suggested the following estimator:

$$\hat{\rho}_I = \frac{MS_{Inter} - MS_{Intra}}{MS_{Inter} + MS_{Intra}}. \tag{7.3.8}$$

It can be seen that $\hat{\rho}_I$, again, could be negative because its numerator is the difference of mean squares between inter-subject and intra-subject variabilities. A negative estimate of ρ_I indicates that two responses on the same subjects are negatively correlated.

7.3.2. Confidence Intervals

To obtain interval estimates for σ_e^2, σ_s^2, and ρ_I, it is helpful to define the following quantiles for a chi-square and an F distribution. Let

$$\chi_L^2 = \chi^2(\alpha/2, n_1 + n_2 - 2),$$

$$\chi_U^2 = \chi^2(1 - \alpha/2, n_1 + n_2 - 2),$$

$$F_L = F(\alpha/2, n_1 + n_2 - 2, n_1 + n_2 - 2),$$

$$F_U = F(1 - \alpha/2, n_1 + n_2 - 2, n_1 + n_2 - 2).$$

$(1 - \alpha) \times 100\%$ confidence intervals for σ_e^2 and ρ_I are then given by (L_e, U_e) and (L_ρ, U_ρ), respectively, where

$$L_e = \frac{SS_{Intra}}{\chi_U^2} \quad \text{and} \quad U_e = \frac{SS_{Intra}}{\chi_L^2}, \tag{7.3.9}$$

and

$$L_\rho = \frac{F_v/F_U - 1}{F_v/F_U + 1} \quad \text{and} \quad U_\rho = \frac{F_v/F_L - 1}{F_v/F_L + 1}. \tag{7.3.10}$$

The $(1 - \alpha) \times 100\%$ confidence interval for ρ_I in (7.3.10) was derived by Graybill (1961). Confidence intervals (7.3.9) and (7.3.10), however, are not the shortest confidence intervals because we assign equal probability of $\alpha/2$ to the upper and lower tails of χ^2 or F distributions. These confidence intervals can provide at least $(1 - \alpha) \times 100\%$ coverage probability for the respective parameters.

For interval estimates of σ_s^2, there exists no exact $(1 - \alpha) \times 100\%$ confidence interval (Graybill, 1976). Tukey (1951) and Williams (1962), however, derived a confidence interval which has a confidence level between $(1 - 2\alpha) \times 100\%$ and $(1 - \alpha) \times 100\%$. We will refer to this confidence interval as William-Tukey's confidence interval. William-Tukey's confidence interval, denoted by (L_s, U_s), is given below:

$$L_s = \frac{(SS_{Inter})(1 - F_U/F_v)}{2\chi_U^2} \quad \text{and} \quad U_s = \frac{(SS_{Inter})(1 - F_L/F_v)}{2\chi_L^2}. \quad (7.3.11)$$

In a simulation study comparing (7.3.11) with eight other approximate confidence intervals, Boardman (1974) recommended that William-Tukey's confidence interval be used for obtaining a $(1 - \alpha) \times 100\%$ confidence interval for σ_s^2 because it yields approximately $(1 - \alpha) \times 100\%$ coverage probability. Wang (1990), further, showed that the lower bound of the confidence interval of the William-Tukey intervals is indeed $1 - \alpha$.

Example 7.3.1 To illustrate the use of these point and interval estimates for σ_e^2, σ_s^2, and ρ_I, we, once again, use AUC data given in Table 3.6.1. From the analysis of variance table (Table 3.6.3), we have

$$SS_{Inter} = 16211.49 \quad \text{and} \quad SS_{Intra} = 3679.43.$$

Since $n_1 + n_2 - 2 = 22$, $\hat{\sigma}_e^2$, $\hat{\sigma}_s^2$, and $\hat{\rho}_I$ are given by

$$\hat{\sigma}_e^2 = MS_{Intra} = \frac{SS_{Intra}}{n_1 + n_2 - 2} = \frac{3679.43}{22} = 167.25,$$

$$\hat{\sigma}_s^2 = \frac{MS_{Inter} - MS_{Intra}}{2} = \frac{736.886 - 167.25}{2} = 284.82,$$

$$\hat{\rho}_I = \frac{MS_{Inter} - MS_{Intra}}{MS_{Inter} + MS_{Intra}} = \frac{736.886 - 167.25}{736.886 + 167.25} = 0.63,$$

For interval estimation, since

$$\chi^2(0.025,22) = 10.982, \qquad \chi^2(0.975,22) = 36.781,$$
$$F(0.025,22,22) = 0.424, \qquad F(0.975,22,22) = 2.358, \quad \text{and}$$
$$F_v = MS_{Inter}/MS_{Intra} = 4.406,$$

The 95% confidence intervals for σ_e^2 and ρ_I are

$$L_e = \frac{3679.43}{36.781} = 100.037, \qquad U_e = \frac{3679.43}{10.982} = 335.032, \quad \text{and}$$

$$L_\rho = \frac{(4.406/2.358) - 1}{(4.406/2.358) + 1} = 0.303, \qquad U_\rho = \frac{(4.406/0.424) - 1}{(4.406/0.424) + 1} = 0.824,$$

respectively. In addition, the 95% confidence interval for σ_s^2 based on William-Tukey's confidence interval is given by

$$L_s = \frac{(16211.49)[1 - (2.358/4.406)]}{2(36.781)} = 102.443,$$

$$U_s = \frac{(16211.49)[1 - (0.424/4.406)]}{2(10.982)} = 667.027.$$

The probability of obtaining a negative estimate of σ_s^2 is also evaluating using (7.3.6). Since $F_v = 4.406$, which is an estimate of $\{\sigma_e^2/(\sigma_e^2 + 2\sigma_s)\}^{-1}$, the probability can be obtained as follows:

$$P\{\hat{\sigma}_s^2 < 0\} = P\{F(22,22) < 1/4.406\} = 0.00048.$$

Therefore, the chance of getting a negative estimate for σ_s^2 based upon the AUC data in Table 3.6.1 is negligible.

The above results are summarized in Table 7.3.1.

Example 7.3.2 For another example, point and interval estimates for σ_e^2, σ_s^2, and ρ_I for Clayton-Leslie's example given in Table 6.9.1 were also obtained based upon log-transformed data. The results are also summarized in Table 7.3.1.

In Clayton-Leslie's example, although we obtained positive estimates for σ_e^2, σ_s^2, and ρ_I, the probability for obtaining a negative estimate for σ_s^2 is about 15% which is non-negligible. This may be in part due to the two completely

Table 7.3.1 Summary of Point and Interval Estimation of Inter- and Intra-Subject Variability for Data Sets in Example 3.6.1 and 6.9.1[a]

Data set	Parameter	Point estimate	95% Confidence interval
Example 3.6.1	σ_e^2	167.25	(100.04, 335.03)
	σ_s^2	284.82	(102.44, 667.03)
	ρ_I	0.63	(0.30, 0.82)
	$P\{\hat{\sigma}_s^2 < 0\}$	0.00048	—
Example 6.9.1	σ_e^2	0.31	(0.17, 0.72)
	σ_s^2	0.11	(−0.09, 0.47)
	ρ_I	0.26	(−0.24, 0.65)
	$P\{\hat{\sigma}_s^2 < 0\}$	0.15	—

[a]Based upon log-transformed AUC data.

opposite responses observed on subject 7 for test and reference formulations. A high probability of obtaining a negative estimate may lead to a confidence interval with a negative lower confidence limit. It is clear that the negative portion of the interval is meaningless since σ_s^2 is always positive. Although one may ignore the negative portion and simply use the positive portion as the confidence interval for σ_s^2, whether the truncated confidence interval is still of the same confidence level is questionable. More discussion regarding this issue is given in Section 7.7.

7.4. TESTS FOR EQUALITY OF VARIABILITIES

In Chapter 4, we have introduced several methods for assessment of equivalence in average bioavailabilities between formulations. These methods were derived under model (4.1.1.) (or equivalently model (4.1.2)) for a standard 2×2 crossover design with assumptions described in (3.1.4). Under these assumptions, it is assumed that the intra-subject variabilities for the test and reference formulations are the same (i.e., $\sigma_T^2 = \sigma_R^2 = \sigma_e^2$). As indicated in previous sections, if the intra-subject variabilities between formulations are different, then equivalence in average bioavailabilities between formulations does not imply that the two formulations are therapeutically equivalent and are, thus, interchangeable. In this situation, the effectiveness and safety of the test formulation may be of great concern. Therefore, it is important to examine whether the intra-subject variability of the test formulation is the same as that of the reference formulation.

In the following section, we will introduce both parametric and nonparametric test procedures for testing equality of intra-subject variabilities between formulations under the assumption of no period effects. In the presence of period effects, these test procedures can still be applied with some modifications.

7.4.1. Pitman-Morgan's Adjusted F Test

In bioavailability studies, the most commonly used test procedure for equality of variabilities is probably the so-called Pitman-Morgan's adjusted F test, which will be derived below:

Under model (4.1.2), let $\mathbf{X}_{ik}$ be a bivariate random vector of the two responses observed on subject i of sequence k, i.e.,

$$\mathbf{X}_{ik} = \begin{bmatrix} Y_{iRk} \\ Y_{iTk} \end{bmatrix}, \qquad i = 1, 2, \ldots, n_k; \quad k = 1, 2. \tag{7.4.1}$$

If the intra-subject variabilities for the test and reference formulations are different, i.e., $\sigma_T^2 \neq \sigma_R^2$, then, according to (2.5.5), the covariance matrix of $\mathbf{X}_{ik}$ is

$$\Sigma = \begin{bmatrix} \sigma_R^2 + \sigma_s^2 & \sigma_s^2 \\ \sigma_s^2 & \sigma_T^2 + \sigma_s^2 \end{bmatrix}. \tag{7.4.2}$$

Note that

$$\begin{aligned} \text{Var}(Y_{iTk}) - \text{Var}(Y_{iRk}) &= (\sigma_T^2 + \sigma_s^2) - (\sigma_R^2 + \sigma_s^2) \\ &= \sigma_T^2 - \sigma_R^2. \end{aligned} \tag{7.4.3}$$

Therefore, under (7.4.2), the difference in variances between the marginal distributions of the two formulations is, in fact, the difference in intra-subject variabilities between the two formulations. Haynes (1981) first introduced Pitman-Morgan's test for equality of the variances of the marginal distributions of two correlated variables to bioavailability/bioequivalence studies. Under (7.4.2) and (7.4.3), the hypotheses for testing equality of variance of marginal distributions between the test and reference formulations becomes

$$H_0: \sigma_T^2 = \sigma_R^2 \quad \text{vs} \quad H_a: \sigma_T^2 \neq \sigma_R^2. \tag{7.4.4}$$

Applying the idea proposed by Pitman (1939) and Morgan (1939), a test statistic can be obtained based upon the correlation between the crossover differences defined in (3.4.1) and subject totals defined in (3.2.1).

Under the assumption of no carry-over effects, we first consider the situation where there are no period effects. It can be seen that X_{ik} are i.i.d. bivariate random vectors with mean

$$\mu = \begin{bmatrix} \mu + F_R \\ \mu + F_T \end{bmatrix},$$

and covariance matrix as defined in (7.4.2).

Let V_{ik} be twice of the crossover differences O_{ik} defined in (3.4.1), i.e.,

$$V_{ik} = 2(O_{ik}) = Y_{iTk} - Y_{iRk}, \quad i = 1,2, \ldots, n_k; \quad k = 1,2.$$

Also, let U_{ik} be subject totals as defined in (3.2.1), i.e.,

$$U_{ik} = Y_{iTk} + Y_{iRk}.$$

The bivariate random vectors $B_{ik} = (V_{ik}, U_{ik})$ are the i.i.d. with mean vector

$$\mu_B = \begin{bmatrix} F_T - F_R \\ 2\mu \end{bmatrix}, \tag{7.4.5}$$

and covariance matrix

$$\Sigma_B = \begin{bmatrix} \sigma_T^2 + \sigma_R^2 & \sigma_T^2 - \sigma_R^2 \\ \sigma_T^2 - \sigma_R^2 & 4\sigma_s^2 + \sigma_T^2 + \sigma_R^2 \end{bmatrix}. \tag{7.4.6}$$

From (7.4.6), it can be seen that the covariance between V_{ik} and U_{ik} is just the difference in intra-subject variabilities between the test and reference formulations. Therefore, the hypotheses (7.4.4) are equivalent to the hypotheses for testing the presence of correlation between the crossover differences V_{ik} and subject totals U_{ik}, i.e.,

$$H_0: \rho_{VU} = 0 \quad \text{vs} \quad H_a: \rho_{VU} \neq 0. \tag{7.4.7}$$

Hence, the following standard F test for correlation between two random variables can be applied to test hypotheses (7.4.4).

Under model (4.1.2) with assumptions of (3.1.4), the Pearson correlation coefficient between V_{ik} and U_{ik} is given by

$$r_{VU} = \frac{S_{VU}}{\sqrt{S_{VV}^2 S_{UU}^2}}, \tag{7.4.8}$$

where

$$S_{VV}^2 = \frac{1}{n_1 + n_2 - 1} \sum_{k=1}^{2} \sum_{i=1}^{n_k} (V_{ik} - \overline{V})^2,$$

$$S_{UU}^2 = \frac{1}{n_1 + n_2 - 1} \sum_{k=1}^{2} \sum_{i=1}^{n_k} (U_{ik} - \overline{U})^2,$$

$$S_{VU} = \frac{1}{n_1 + n_2 - 1} \sum_{k=1}^{2} \sum_{i=1}^{n_k} (V_{ik} - \overline{V})(U_{ik} - \overline{U}),$$

$$\overline{V} = \frac{1}{n_1 + n_2} \sum_{k=1}^{2} \sum_{i=1}^{n_k} V_{ik}, \quad \text{and}$$

$$\overline{U} = \frac{1}{n_1 + n_2} \sum_{k=1}^{2} \sum_{i=1}^{n_k} U_{ik}.$$

We then reject H_0 at the α level, if

$$F_{VU} = \frac{(n - 2) r_{VU}^2}{1 - r_{VU}^2} > F(\alpha, 1, n_1 + n_2 - 2), \tag{7.4.9}$$

where $F(\alpha, 1, n_1 + n_2 - 2)$ is the upper αth quantile $(0 < \alpha < 1)$ of an F distribution with degrees of freedom 1 and $n_1 + n_2 - 2$.

From (7.4.9), it can be seen that test statistic F_{VU} is obtained based on the transformed data V_{ik} and U_{ik}. To relate (7.4.9) to Y_{ijk}, let S_{tt}^2, S_{rr}^2, and S_{tr} be the sample variances and covariances of Y_{iTk} and Y_{iRk}, i.e.,

$$S_{tt}^2 = \frac{1}{n_1 + n_2 - 1} \sum_{k=1}^{2} \sum_{i=1}^{n_k} (Y_{iTk} - \overline{Y}_t)^2,$$

$$S_{rr}^2 = \frac{1}{n_1 + n_2 - 1} \sum_{k=1}^{2} \sum_{i=1}^{n_k} (Y_{iRk} - \overline{Y}_r)^2,$$

$$S_{tr} = \frac{1}{n_1 + n_2 - 1} \sum_{k=1}^{2} \sum_{i=1}^{n_k} (Y_{iTk} - \overline{Y}_t)(Y_{iRk} - \overline{Y}_r),$$

$$\overline{Y}_t = \frac{1}{n_1 + n_2} \sum_{k=1}^{2} \sum_{i=1}^{n_k} Y_{iTk}, \quad \text{and}$$

$$\overline{Y}_r = \frac{1}{n_1 + n_2} \sum_{k=1}^{2} \sum_{i=1}^{n_k} Y_{iRk}.$$

It can then be easily verified that

$$S_{VV}^2 = S_{tt}^2 + S_{rr}^2 - 2S_{tr},$$
$$S_{UU}^2 = S_{tt}^2 + S_{rr}^2 + 2S_{tr},$$
$$S_{VU} = S_{tt}^2 - S_{rr}^2. \tag{7.4.10}$$

Therefore,

$$r_{VU}^2 = \frac{(S_{tt} - S_{rr})^2}{(S_{tt}^2 + S_{rr}^2)^2 - 4S_{tr}^2},$$

and

$$1 - r_{VU}^2 = \frac{4(S_{tt}^2 S_{rr}^2 - S_{tr}^2)}{(S_{tt}^2 + S_{rr}^2)^2 - 4S_{tr}^2}.$$

It follows that

$$\begin{aligned} F_{VU} &= \frac{(n_1 + n_2 - 2)r_{VU}^2}{1 - r_{VU}^2} \\ &= \frac{(n_1 + n_2 - 2)(S_{tt}^2 - S_{rr}^2)^2}{4(S_{tt}^2 S_{rr}^2 - S_{tr}^2)} \\ &= \frac{(n_1 + n_2 - 2)\left[\dfrac{S_{tt}^2}{S_{rr}^2} - 1\right]^2}{4\left[\dfrac{S_{tt}^2}{S_{rr}^2}\right]\left[1 - \dfrac{S_{tr}^2}{S_{tt}^2 S_{rr}^2}\right]}. \end{aligned}$$

If we let $F_{tr} = S_{tt}^2/S_{rr}^2$ and $r_{tr} = S_{tr}/(S_{tt}S_{rr})$, then F_{VU} becomes the familiar Pitman-Morgan's adjusted F test, as introduced by Haynes (1981), i.e.,

$$F_{PM} = \frac{(n_1 + n_2 - 2)[F_{tr} - 1]^2}{4F_{tr}(1 - r_{tr}^2)}. \tag{7.4.11}$$

Note that Pitman-Morgan's adjusted F test is the uniformly most powerful test for hypotheses (7.4.4). We then reject H_0 in (7.4.4) if

$$F_{PM} > F(\alpha, 1, n_1 + n_2 - 2).$$

Example 7.4.1 For AUC data in Table 3.6.1, assuming that there are no period effects, we have

$$S_{VV}^2 = 323.08, \qquad S_{UU}^2 = 1433.69, \qquad S_{VU} = 13.32,$$
$$S_{tt}^2 = 445.85, \qquad S_{rr}^2 = 432.53, \qquad S_{tr} = 277.65.$$

The Pearson correlation coefficient between crossover differences and subject totals is then given by

$$r_{VU} = \frac{13.32}{\sqrt{(323.03)(1433.69)}} = 0.0196.$$

This gives

$$F_{VU} = \frac{(22)(0.0196)^2}{1 - (0.0196)^2} = 0.00843,$$

with a p-value of 0.93. Therefore, we fail to reject H_0: $\sigma_T^2 = \sigma_R^2$ at the 5% level of significance. On the other hand, since

$$F_{tr} = \frac{445.85}{432.53} = 1.0308, \quad \text{and}$$

$$r_{tr} = \frac{277.65}{\sqrt{(445.85)(432.53)}} = 0.6323,$$

Pitman-Morgan's test statistic (7.4.11) is then given by

$$F_{PM} = \frac{(22)(1.0308 - 1)^2}{(4)(1.0308)[1 - (0.6323)^2]} = 0.00843,$$

which agrees with the value obtained based upon F_{VU}.

Similarly, the above tests can also be applied to Clayton and Leslie's example (Table 6.9.1). It can be verified that, based upon the log-transformed data, $r_{VU} = 0.3035$ and $F_{PM} = 1.6231$ which yields a p-value of 0.22. Therefore, we fail to reject the null hypothesis of equality in intra-subject variabilities between the two erythromycin formulations.

7.4.2. A Distribution-Free Test Based on Spearman's Rank Correlation Coefficient

In the previous section, it should be noted that if normality assumptions on $\{S_{ik}\}$ and $\{e_{ijk}\}$ are seriously violated, neither test statistic (7.4.9) nor (7.4.11) follows an F distribution. In other words, Pitman-Morgan's adjusted F test for the hypothesis of equality in intra-subject variabilities between formulations is not valid. In this situation, alternatively, the following distribution-free test based upon Spearman's rank correlation coefficient is useful.

Let $R(V_{ik})$ and $R(U_{ik})$ be the rank of V_{ik} and U_{ik} in the combined sequence of $\{V_{ik}\}$ and $\{U_{ik}\}$, $k = 1,2; i = 1,2, \ldots ,n_k$, respectively. The Spearman's rank correlation coefficient r_S can be obtained by replacing V_{ik} and U_{ik} in (7.4.8) with their corresponding ranks $R(V_{ik})$ and $R(U_{ik})$, $k = 1,2; i = 1,2, \ldots ,n_k$. If there are no ties, then r_S is given by

$$r_S = \frac{12 \sum_{k=1}^{2} \sum_{i=1}^{n_k} \left[R(V_{ik}) - \frac{n_1 + n_2 + 1}{2} \right]\left[R(U_{ik}) - \frac{n_1 + n_2 + 1}{2} \right]}{(n_1 + n_2)\,[(n_1 + n_2)^2 - 1]}. \tag{7.4.12}$$

If there are ties, then r_S should be calculated as follows:

$$r_S = \frac{\sum_{k=1}^{2} \sum_{i=1}^{n_k} R(V_{ik})R(U_{ik}) - K_S}{\left\{ \left[\sum_{k=1}^{2} \sum_{i=1}^{n_k} R^2(V_{ik}) - K_S \right]\left[\sum_{k=1}^{2} \sum_{i=1}^{n_k} R^2(U_{ik}) - K_S \right] \right\}^{1/2}}, \tag{7.4.13}$$

where $K_S = (n_1 + n_2)[(n_1 + n_2)/2]^2$.

We then reject H_0 in (7.4.4) at the α level if

$$|r_S| > r_S(\alpha/2, n_1 + n_2),$$

where $r_S(\alpha/2, n_1 + n_2)$ is the αth quantile of the distribution of Spearman's rank correlation coefficient based upon the $n_1 + n_2$ observations. When $n_1 + n_2 > 30$, the αth upper quantile of r_S can be approximated by

$$r_S(\alpha, n_1 + n_2) \approx \frac{Z_\alpha}{\sqrt{n_1 + n_2 - 1}}, \tag{7.4.14}$$

where Z_α is the αth upper quantile of a standard normal variable.

As it can be seen, Pitman-Morgan's adjusted F test is derived based upon Pearson's correlation coefficient, r_{VU}, under normality assumptions. In most bioavailability/bioequivalence studies, however, the normality assumption is not easy to verify. Therefore, the above distribution-free test based upon Spearman's rank correlation coefficient is usually considered as an alternative to Pitman-Morgan's adjusted F test for testing the equality of intra-subject variabilities if the normality assumption is in doubt. When the normality assumption is violated, McCulloch (1987) showed that the asymptotic variance of Pearson's correlation coefficient r_{VU} depends upon the common kurtosis between the crossover differences and subject totals. The actual size of the test (7.4.8) is larger (smaller) than the nominal level if the common kurtosis is larger (smaller) than 0. Based on a simulation study, McCulloch (1987) pointed out that Spearman's test is the only test among five tests under study including test (7.4.8) that controls the actual size at the nominal level with a very competitive power for a variety of different distributions. Hence, Spearman's test ((7.4.12) or (7.4.13)) is recommended for testing the equality between intra-subject variabilities.

Example 7.4.2 To illustrate the use of Spearman's test for equality of variabilities, similarly, we consider the two examples given in Section 3.6 and Section 6.9.

For AUC data in Table 3.6.1, it can be verified that Spearman's rank correlation coefficient between crossover differences and subject totals based upon the raw data is 0.1165 with a p-value of 0.59. Like the Pearson correlation coefficient and Pitman-Morgan's test, we fail to reject H_0 at the 5% level of significance. However, it should be noted that the p-value of r_S is much smaller than that of the Pearson correlation coefficient.

For Clayton and Leslie's example, based upon log-transformed AUC, we have $r_S = 0.3684$ which yields a p-value of 0.13. Thus, we fail to reject H_0 at the 5% level of significance. This result agrees with that from the Pitman-Morgan test.

7.4.3. Tests in the Presence of Period Effects

In previous sections, in addition to the assumption of no carry-over effects, we assume that there are no period effects to derive Pitman-Morgan's adjusted F test. This result does not generally hold because Pitman-Morgan's adjusted F test depends upon S_{tt}^2, S_{rr}^2, and S_{tr} whose expected values include the period effects (Ho and Patel, 1988). In this situation, Pitman-Morgan's adjusted F test can be modified as follows.

Assuming that there are no unequal carry-over effects, under model (4.1.2) with normality assumptions, we have

$$X_{ik} \sim N(\mu_k, \Sigma), \qquad i = 1, 2, \ldots, n_k; \quad k = 1, 2,$$

where Σ is the covariance matrix as defined in (7.4.2) and

$$\mu_1 = \begin{bmatrix} \mu + F_R + P_1 \\ \mu + F_T + P_2 \end{bmatrix} \quad \text{and} \quad \mu_2 = \begin{bmatrix} \mu + F_R + P_2 \\ \mu + F_T + P_1 \end{bmatrix}.$$

The pooled sample covariance matrix $\hat{\Sigma}$ is an unbiased estimator of Σ which is given below

$$
\begin{aligned}
\hat{\Sigma} &= \frac{1}{n_1 + n_2 - 2} (S_1 + S_2) \\
&= \begin{bmatrix} S_{RR}^2 & S_{TR} \\ S_{TR} & S_{TT}^2 \end{bmatrix},
\end{aligned}
$$

where S_1 and S_2 are defined in Section 4.2.4 and S_{RR}^2, S_{TT}^2, and S_{TR} are defined in Section 4.2.3, which were used to construct a $(1 - 2\alpha) \times 100\%$ confidence interval for μ_T/μ_R based on Fieller's theorem. Since there are $n_1 + n_2 - 2$ degrees of freedom for $\hat{\Sigma}$, the adjusted F statistic for the Pitman-Morgan's test becomes

$$
\tilde{F}_{PM} = \frac{(n_1 + n_2 - 3)(F_{TR} - 1)^2}{4F_{TR}(1 - r_{TR}^2)}, \tag{7.4.15}
$$

where $F_{TR} = S_{TT}^2/S_{RR}^2$ and $r_{TR}^2 = S_{TR}^2/(S_{TT}^2 S_{RR}^2)$.

We then reject H_0 at the α level if

$$
\tilde{F}_{PM} > F(\alpha, 1, n_1 + n_2 - 3). \tag{7.4.16}
$$

Note that the above test is equivalent to Pearson's correlation coefficient between the crossover differences and subject totals based upon the residuals from the sequence-by-period means, i.e.,

$$
R_{ik} = X_{ik} - \overline{X}_k = \begin{bmatrix} X_{iRk} - \overline{X}_{\cdot Rk} \\ X_{iTk} - \overline{X}_{\cdot Tk} \end{bmatrix}, \qquad i = 1, 2, \dots, n_k; \quad k = 1, 2,
$$

where

$$
\overline{X}_k = \frac{1}{n_k} X_{ik} = \begin{bmatrix} \dfrac{1}{n_k} \displaystyle\sum_{i=1}^{n_k} X_{iRk} \\ \dfrac{1}{n_k} \displaystyle\sum_{i=1}^{n_k} X_{iTk} \end{bmatrix}, \qquad k = 1, 2.
$$

Similarly, Spearman's test can also be computed based upon the ranks of the residuals from sequence-by-period means using either (7.4.12) or (7.4.13). We will denote this test statistic by $\tilde{r}_s$ and reject H_0 at the α level if

$$
|\tilde{r}_s| > r_S(\alpha/2, n_1 + n_2 - 2). \tag{7.4.17}
$$

Example 7.4.3 To illustrate use of Pitman-Morgan's test and Spearman's test in the presence of period effects, again, we consider AUC data in Table 3.6.1 and log-transformed AUC data given in Table 6.9.1.

When there are period effects, the Pitman-Morgan test should be applied to the residuals from the sequence-by-period means. From Table 3.6.1, it can be found that

$$S_{TT}^2 = 463.557, \qquad S_{RR}^2 = 440.576, \quad \text{and} \quad S_{TR} = 284.82.$$

This leads to

$$F_{TR} = \frac{463.557}{440.576} = 1.0522,$$

and

$$r_{TR} = \frac{284.82}{\sqrt{(463.557)(440.576)}} = 0.6302.$$

The adjusted Pitman-Morgan test statistic is then given by

$$\tilde{F}_{PM} = \frac{(21)(1.0522 - 1)^2}{(4)(1.0522)[1 - (0.6302)^2]} = 0.0225,$$

which gives a p-value of 0.8821.

Hence, we also fail to reject the hypothesis of no difference in intra-subject variabilities between formulations.

One can also easily verify that the Spearman's test based upon the residuals from the sequence-by-period is given by 0.1017 with a p-value of 0.65.

For log-transformed AUC in Clayton and Leslie's example, it can be verified that $\tilde{F}_{PM} = 1.0345$ (p-value $= 0.33$) and $\tilde{r}_S = 0.3560$ (p-value $= 0.15$). Both tests fail to reject the null hypothesis of no difference in intra-subject variabilities between the two erythromycin formulations.

Test results for the two examples (one from Section 3.6 and the other one from Section 6.9) with and without period effects are summarized in Table 7.4.1.

Table 7.4.1 Test Results for Equality of Variabilities

Example	Method	Test[a] (P-Value)	Test[b] (P-Value)
3.6.1	Pearson	0.0196 (0.93)	0.0327 (0.88)
	Pitman-Morgan	0.0084 (0.93)	0.0225 (0.88)
	Spearman	0.1165 (0.59)	0.1017 (0.65)
6.9.1[c]	Pearson	0.3035 (0.22)	0.2540 (0.33)
	Pitman-Morgan	1.6231 (0.22)	1.0345 (0.33)
	Spearman	0.3684 (0.13)	0.3560 (0.15)

[a]Based upon raw data.
[b]Based upon residuals from sequence-by-period means.
[c]Based upon log-transformed AUC data.

7.5. EQUIVALENCE IN VARIABILITY OF BIOAVAILABILITY

Under normality assumptions, Pitman-Morgan's adjusted F test is the uniformly most powerful test for the hypothesis of equality of variances between the marginal distributions of the test and reference formulations. However, it is not an appropriate test for equivalence in variability between formulations. Similar to the interval hypotheses (4.3.1) (for the difference in means) and (4.3.2) (for the ratio of means) for assessment of average bioavailability, equivalence in variability of bioavailability can be assessed by testing interval hypotheses (7.2.2) or (7.2.3). Hypotheses (7.2.3) for the ratio of variances can be decomposed into two one-sided hypotheses as follows:

$$H_{01}: \sigma_T^2/\sigma_R^2 \leq \delta_1 \quad vs \quad H_{a1}: \sigma_T^2/\sigma_R^2 > \delta_1,$$

and

$$H_{02}: \sigma_T^2/\sigma_R^2 \geq \delta_2 \quad vs \quad H_{a2}: \sigma_T^2/\sigma_R^2 < \delta_2. \tag{7.5.1}$$

On the other hand, hypotheses (7.2.2) for the difference in variances can be decomposed into the following two one-sided hypotheses:

$$H_{01}: \sigma_T^2 - \sigma_R^2 \leq \theta_1 \quad vs \quad H_{a1}: \sigma_T^2 - \sigma_R^2 > \theta_1,$$

and

$$H_{02}: \sigma_T^2 - \sigma_R^2 \geq \theta_2 \quad vs \quad H_{a2}: \sigma_T^2 - \sigma_R^2 < \theta_2. \tag{7.5.2}$$

For testing hypotheses (7.5.1), the idea in Section 7.4.1 can be extended to derive a test (Liu and Chow, 1992c). Define

$$V_{ik} = Y_{iTk} - Y_{iRk}, \quad and$$

$$U_{ik} = Y_{iTk} + \delta Y_{iRk},$$

where $\delta > 0$. It can be easily verified that, in the absence of period effects,

$$E(V_{ik}) = F_T - F_R, \quad E(U_{ik}) = (1 + \delta)\mu + (\delta - 1)F_T,$$

and

$$Var(V_{ik}) = \sigma_T^2 + \sigma_R^2, \quad Var(U_{ik}) = \sigma_T^2 + \delta^2\sigma_R^2 + (1 + \delta)^2\sigma_s^2.$$

The covariance between V_{ik} and U_{ik} is then given by

$$\begin{aligned} Cov(V_{ik}, U_{ik}) &= Cov(Y_{iTk} - Y_{iRk}, Y_{iTk} + \delta Y_{iRk}) \\ &= Var(Y_{iTk}) + (\delta - 1)Cov(Y_{iTk}, Y_{iRk}) - \delta Var(Y_{iRk}) \\ &= \sigma_T^2 - \delta\sigma_R^2. \end{aligned}$$

Therefore, $Cov(V_{ik}, U_{ik})$ is the difference between σ_T^2 and $\delta\sigma_R^2$. Hypotheses (7.5.1) are equivalent to the following two one-sided hypotheses:

$$H_{01}: \sigma_T^2 - \delta_1\sigma_R^2 \leq 0 \quad \text{vs} \quad H_{a1}: \sigma_T^2 - \delta_1\sigma_R^2 > 0, \tag{7.5.3}$$

and

$$H_{02}: \sigma_T^2 - \delta_2\sigma_R^2 \geq 0 \quad \text{vs} \quad H_{a2}: \sigma_T^2 - \delta_2\sigma_R^2 < 0,$$

which are equivalent to the hypotheses for testing the presence of a positive correlation between V_{ik} and U_{Lik} and the presence of a negative correlation between V_{ik} and U_{Uik}, respectively, where

$$U_{Lik} = Y_{iTk} + \delta_1 Y_{iRk}, \quad \text{and}$$

$$U_{Uik} = Y_{iTk} + \delta_2 Y_{iRk}.$$

In other words, hypotheses (7.5.3) are equivalent to the following two one-sided hypotheses:

$$H_{01}: \rho_L \leq 0 \quad \text{vs} \quad H_{a1}: \rho_L > 0, \tag{7.5.4}$$

and

$$H_{02}: \rho_U \geq 0 \quad \text{vs} \quad H_{a2}: \rho_U < 0,$$

where ρ_L and ρ_U are correlation coefficients between V_{ik}, U_{Lik} and V_{ik}, U_{Uik}, respectively.

Let r_L and r_U be the sample Pearson correlation coefficients between V_{ik} and U_{Lik}, and between V_{ik} and U_{Uik}, respectively. H_{01} is then rejected at the α level of significance if

$$t_L = \frac{r_L}{\left[\dfrac{1 - r_L^2}{n_1 + n_2 - 2}\right]^{1/2}} > t(\alpha, n_1 + n_2 - 2), \tag{7.5.5}$$

and H_{02} is rejected at the α level of significance if

$$t_U = \frac{r_U}{\left[\dfrac{1 - r_U^2}{n_1 + n_2 - 2}\right]^{1/2}} < -t(\alpha, n_1 + n_2 - 2). \tag{7.5.6}$$

We conclude that σ_T^2 and σ_R^2 are equivalent if both H_{01} and H_{02} are rejected.

Note that the above tests (7.5.5) and (7.5.6) can be directly applied to obtain a distribution-free test by simply replacing r_L and r_U with their corresponding Spearman's rank correlation coefficients when normality assumptions are in doubt.

In the presence of period effects, the above procedure can be carried out using residuals from sequence-by-period means as raw data with degrees of freedom $n_1 + n_2 - 3$. In other words, test statistics t_L and t_U can be expressed in terms of S_{TT}^2, S_{RR}^2 and S_{TR} as follows:

$$t_L = (F_L)^{1/2} \quad \text{and} \quad t_U = (F_U)^{1/2},$$

where

$$F_L = \frac{(n_1 + n_2 - 3)[S_{TT}^2 - \delta_1 S_{RR}^2 + (\delta_1 - 1)S_{TR}]^2}{(1 + \delta_1)^2(S_{TT}^2 S_{RR}^2 - S_{TR}^2)}$$

and

$$F_U = \frac{(n_1 + n_2 - 3)[S_{TT}^2 - \delta_2 S_{RR}^2 + (\delta_2 - 1)S_{TR}]^2}{(1 + \delta_2)^2(S_{TT}^2 S_{RR}^2 - S_{TR}^2)}$$

We then reject H_{01} at the α level if $S_{TT}^2 - \delta_1 S_{RR}^2 + (\delta_1 - 1)S_{TR} > 0$, and

$$|t_L| > t(\alpha, n_1 + n_2 - 3), \tag{7.5.7}$$

and reject H_{02} at the α level if $S_{TT}^2 - \delta_2 S_{RR}^2 + (\delta_2 - 1)S_{TR} < 0$, and

$$|t_U| > t(\alpha, n_1 + n_2 - 3). \tag{7.5.8}$$

Appendix B.4 provides SAS programs for hypotheses (7.4.4) and (7.5.1) based upon the residuals from sequence-by-period means.

For testing hypotheses (7.5.2), Liu (1991) proposed a two one-sided tests procedure based upon the idea of orthogonal transformations introduced by Cornell (1980). An orthogonal transformation is used to transform the $n_1 + n_2$ independent bivariate random vectors Y_{ik} into $n_1 + n_2 - 2$ independent bivariate random vectors Z_{gk} with mean 0 and covariance matrix Σ_k, where i = 1,2, . . . , n_k; g = 1,2, . . . , $n_k - 1$; k = 1,2. The differences of the squares of the two components of each Z_{gk} are then unbiased estimates of the difference of intra-subject variabilities between the two formulations. Consequently, the nonparametric two one-sided tests procedure for average bioavailability described in Chapter 4 can be directly applied to hypotheses (7.5.2) based upon the differences of the squares of two components of Z_{gk}.

Let $Y_{jk} = (Y_{1jk}, . . . ,Y_{n_k jk})'$, be the vector of n_k responses observed on subjects during period j in sequence k. Also let c_{gk} be a $n_k \times 1$ vector of coefficients of normalized linear orthogonal contrasts of degree n_k such that $1'c_{gk} = 0$, $c_{gk}'c_{gk} = 1$, and $c_{gk}'c_{g'k} = 0$, for $g \neq g'$, where g = 1,2, . . . , $n_k - 1$; j = 1,2; k = 1,2. If we define

$$Z_{gjk} = c_{gk}'Y_{jk}, \quad \text{and}$$
$$\mathbf{Z}_{gk} = (Z_{g1k}, Z_{g2k})', \quad g = 1,2, . . . ,n_k - 1; \quad j = 1,2; \quad k = 1,2, \tag{7.5.9}$$

then, $\mathbf{Z}_{gk}$, g = 1,2, . . . ,n_k; k = 1,2, are independent bivariate normal vectors with mean vector $\mathbf{O}$ and covariance matrix Σ_k, where Σ_k is defined in (2.5.5). Since orthogonal transformations are linear, the units of the original data are maintained in the transformed data. In addition, orthogonal transformation preserves the covariance structure of the original data. Define the period difference of the squares of Z_{gjk} as follows.

$$Q_{gk} = \tfrac{1}{2} (Z^2_{g2k} - Z^2_{g1k}), \qquad g = 1,2, \ldots ,n_k - 1, \quad k = 1,2. \qquad (7.5.10)$$

Q_{gk} then follow a distribution as a half of the difference of two correlated χ^2 random variables (each with one degree of freedom). Under the normality assumptions of $\{S_{ik}\}$ and $\{e_{ijk}\}$ as described in (2.5.5), Q_{gk} are independently distributed with mean

$$E(Q_{gk}) = \begin{cases} \tfrac{1}{2} (\sigma^2_T - \sigma^2_R) & \text{for } k = 1 \\ \tfrac{1}{2} (\sigma^2_R - \sigma^2_T) & \text{for } k = 2, \end{cases}$$

and common variance

$$Var(Q_{gk}) = \tfrac{1}{2} [(\sigma^4_T + \sigma^4_R) + 2\sigma^2_s(\sigma^2_T + \sigma^2_R)].$$

Since $\{Q_{g1}, g = 1,2. \ldots ,n_1 - 1\}$ and $\{Q_{g2}, g = 1,2, \ldots ,n_2 - 1\}$ are two independent samples from the distributions with a common variance and a location difference of $\sigma^2_T - \sigma^2_R$, the Wilcoxon-Mann-Whitney's two one-sided procedure and the distribution-free confidence interval based upon Hodges-Lehmann's estimator discussed in Chapter 4 can be directly applied to $\{Q_{gk}, g = 1,2, \ldots ,n_k - 1; k = 1,2\}$ to test interval hypotheses of equivalence in intra-subject variabilities. Similar to the nonparametric procedures for average bioavailability as discussed in Sections 4.5.1 and 4.5.2, the resultant nonparametric two one-sided tests procedure for hypotheses (7.5.2) is also operationally equivalent to the corresponding distribution-free confidence interval.

Note that, in computation of W_L and W_U, the $n_1 + n_2 - 2$ Q_{gk} values should be treated as raw data and quantiles $w(\alpha)$ and $w(1 - \alpha)$ are based upon sample sizes $n_1 - 1$ and $n_2 - 1$. The orthogonal transformations based upon (7.5.9) can be easily obtained using statistical software such as the ORPOL function of SAS IML.

Compared to tests for hypotheses (7.5.1), one disadvantage of Liu's test procedure is that the equivalent limits θ_1 and θ_2 depend upon the estimate of σ^2_R which can certainly introduce bias to the procedure.

Sometimes, as pointed out by Anderson and Hauck (1990), in order to maintain a comparable safety profile, it is desirable to test whether variability of the test formulation does not exceed that of the reference formulation by at least some specific amount. In this situation, one only needs to test the second set of hypotheses in (7.5.1) or (7.5.2), i.e., H_{02} vs H_{a2} which has the same form as the hypotheses for evaluation of equivalence of efficacy in clinical trials considered by others (e.g., Blackwelder, 1982). Note that the hypotheses H_{02} vs H_{a2} can be tested using test statistics, either t_U for (7.5.1) or W_U for (7.5.2).

For testing hypotheses (7.5.2), Esinhart and Chinchilli (1990) also proposed several test procedures based upon Wald's test, Rao's score test, and the likelihood ratio test. Esinhart and Chinchilli (1991) also proposed the use of generalized estimating equations (GEEs) (Liang and Zeger, 1986) for analyses of average and variability of bioavailability. However, these test procedures are based upon asymptotic results. The GEEs procedure was originally developed for analyses of epidemiology studies which involve a large number of patients (sometimes in the thousands). In practice, since the number of subjects involved in a bioavailability/bioequivalence study is usually small (e.g., 12–24), the application of these procedures needs further investigation. Although the asymptotic GEEs procedure only requires that distributions be expressed in an exponential form with the first two moments, unlike the nonparametric two one-sided procedure for average bioavailability, it is not general enough to include some distributions such as Cauchy which do not have any moments greater than or equal to 1. In addition, the asymptotic $(1 - 2\alpha) \times 100\%$ confidence intervals, based upon these tests, may not provide a true $(1 - 2\alpha) \times 100\%$ coverage probability in small samples. Furthermore, it is doubtful that these confidence intervals are operationally equivalent to their corresponding test procedures.

Example 7.5.1   We first use AUC data from Example 3.6.1 to illustrate the two one-sided tests procedure (7.5.5) and (7.5.6) for hypotheses (7.5.1) of equivalence in variability of bioavailability. For the purpose of illustration, we choose $\delta_1 = 0.8$ and $\delta_2 = 1.2$. The Pearson's correlation coefficients between V_{ik} and U_{Lik} and between V_{ik} and U_{Uik}, based upon the residuals from the sequence-by-period means, are given by 0.0854 and -0.0106, respectively. This gives

$$t_L = \frac{0.0854}{\left[\dfrac{1 - (0.0854)^2}{21} \right]^{1/2}} = 0.3928, \quad \text{and}$$

$$t_U = \frac{-0.0106}{\left[\dfrac{1 - (-0.0106)^2}{21} \right]^{1/2}} = -0.0485.$$

Therefore, we fail to reject H_{01} (p-value = 0.3492) and H_{02} (p-value = 0.4809) at the 5% level of significance. We then conclude that the two formulations are not bioequivalent in variability.

Similarly, it can be verified that the Spearman's rank correlation coefficients between V_{ik} and U_{Lik} and between V_{ik} and U_{Uik} are 0.1635 and 0.0217. This leads to the same conclusion that the two formulations are not equivalent in variability (p-values > 0.1). Note that t_L and t_U can also be obtained from the squared root of F_L and F_U with values of S_{TT}^2, S_{RR}^2, and S_{TR} given in Example 7.4.3.

The above results are summarized in Table 7.5.1. Table 7.5.1 also gives test results of Clayton and Leslie's example which were obtained based upon log-

Table 7.5.1 Summary of Test Results for Equivalence in Variability Based on the Ratio of Variances

Example	Method	Test[a] (P-Value)	
3.6.1	Pearson	r_L =	0.0854 (0.3492)
		r_U =	-0.0106 (0.4809)
	Spearman	r_L =	0.1635 (>0.1)
		r_U =	0.0217 (>0.5)
6.9.1[b]	Pearson	r_L =	0.3312 (0.0970)
		r_U =	0.1869 (0.7637)
	Spearman	r_L =	0.3829 ($0.05 < P < 0.1$)
		r_U =	0.2632 (>0.5)

[a]Based upon the residuals from the sequence-by-period means.
[b]Based upon the log-transformed AUC data.

transformed AUC. The results indicate that the two erythromycin formulations are not equivalent in variability.

For the nonparametric two one-sided tests procedure for hypotheses (7.5.2), for the purpose of illustration, we choose $\theta_2 = -\theta_1 = 168$, which is an estimate of the intra-subject variability obtained from the analysis of variance table (Table 3.6.3). Therefore, we will conclude equivalence in intra-subject variabilities if the 90% confidence interval for $\sigma_T^2 - \sigma_R^2$ is within ± 168.

Table 7.5.2 gives Z_{gjk} and Q_{gk} which are obtained after orthogonal transformations. Based upon these values, the Hodges-Lehmann's estimator yields an estimate of 9.66 and a distribution-free 90% confidence interval of (-216.76, 263.28). Since the 90% confidence interval is not within (-168, 168), we conclude that there is not enough evidence to support the equivalence in intra-subject variabilities between the two formulations. It can be verified that this result agrees with that obtained using nonparametric two one-sided tests procedure based upon Q_{gk}.

Test results for Clayton and Leslie's example based upon log-transformed AUC data are also obtained and summarized in Table 7.5.3. However, it should be noted that, in Clayton and Leslie's example, $\theta_2 = -\theta_1$ was chosen to be 0.30, which is the mean squared error obtained from the analysis of variance table given in Table 6.9.2. From Table 7.5.3, it can be seen that the two formulations from Example 3.6.1 are equivalent in average bioavailability, but not in variability of bioavailability. On the other hand, the two erythromycin formulations are not equivalent in both average and variability of bioavailability.

It should be noted that in most bioavailability studies, sample sizes are chosen so that there is sufficient power for detection of significant difference in average

Table 7.5.2 Results of Orthogonal Transformation for Data in Table 3.6.1

Sequence	Z_{g1k}	Z_{g2k}	Q_{gk}
1	43.4344	−11.0164	−1765.19
	−35.7584	−13.2207	−1103.88
	23.4740	2.7537	−543.45
	35.2714	26.8370	−523.85
	31.3209	54.0996	1945.77
	4.9200	8.4439	47.09
	−7.0244	−25.3875	595.18
	15.1894	0.3349	−230.61
	2.1556	23.7542	559.61
	−22.0535	−23.3809	60.31
	−15.7381	−12.1105	−101.03
2	3.6063	−5.0457	12.45
	4.3842	−13.8096	171.48
	−19.3074	−20.0137	27.77
	−40.5814	−39.4527	−90.33
	22.4712	6.1321	−467.35
	26.5541	3.3575	−693.85
	22.6315	5.7194	−479.43
	16.6862	1.0559	−277.32
	−14.8249	−20.4558	198.66
	1.2681	−2.8886	6.94
	0.2745	5.0672	25.60

Table 7.5.3 Summary of Test Results for Equivalence in Variability Based on the Difference in Variances

Example	Point estimate	Distribution-free 90% confidence intervals
3.6.1	9.660	(−216.76, 263.28)
6.9.1[a]	0.035	(−0.26, 0.32)

[a]Based upon log-transformed AUC data.

bioavailabilities between formulations, which may not be large enough for assessment of equivalence in variability of bioavailability. Therefore, a nonsignificant test result for interval hypotheses of variability may imply that the sample size is too small to declare equivalence in variability if the two formulations are indeed equivalent.

7.6. CV ASSESSMENT

As indicated in previous sections, it is important to compare the inter-subject and the intra-subject variabilities between formulations in addition to average bioavailability. Therefore, it is helpful to estimate the inter-subject and the intra-subject variabilities for each formulation. In practice, if the drug has a short half-life, it is possible that each subject has replicates for each formulation, i.e., each subject receives several administrations of the sequence of formulations. Based upon these replicates, the inter-subject and the intra-subject variabilities can then be estimated using the methods introduced in Section 7.3. However, in many cases, each subject may have a different intra-subject mean and intra-subject variability for the parameter under study. In this case, assessment of the inter-subject variability and the intra-subject variability is not possible because standard statistical methods assume that intra-subject variability is the same from subject to subject. However, if we assume that the coefficient of variation (CV) for each subject is the same from subject to subject, then intra-subject variability can be assessed in terms of the common CV. Assuming that there are no period effects, Chow and Tse (1988, 1990b) provided several estimators for the common CV under a conditional one-way random effects model. The conditional one-way random effects will be described in the next section.

7.6.1. Conditional One-Way Random Effects Model

Let Y_{ij} be the response of a bioavailability parameter, say AUC, of the jth replicate of the ith subject for a given formulation of a drug. To estimate the inter-subject and the intra-subject variabilities, the following one-way random effects model is usually considered:

$$Y_{ij} = \mu + S_i + e_{ij}, \qquad i = 1,2, \ldots ,k; \quad j = 1,2, \ldots ,n, \qquad (7.6.1)$$

where μ is the overall mean, $S_i's$ and $e_{ij}'s$ are independent $N(0,\sigma_s^2)$ and $N(0,\sigma_e^2)$, respectively. Under model (7.6.1) the inter-subject variability and the intra-subject variability of the AUC are described by σ_s^2 and σ_e^2, respectively. However, this model is based upon the assumption that the intra-subject variability is the same from subject to subject. In order to account for the different intra-subject means and variances, Chow and Tse (1988, 1990b) considered the following conditional one-way random effects model:

$$Y_{ij} = \mu + S_i + e_{ij}, \qquad i = 1,2, \ldots ,k; \quad j = 1,2, \ldots ,n, \qquad (7.6.2)$$

where μ is the overall mean, S_i's are independent $N(0,\sigma_s^2)$ and given $S_i = a_i$, the e_{ij}'s are independent $N(0,\sigma_i^2)$, i.e.,

$$e_{ij} \mid S_i = a_i \sim N(0,\sigma_i^2).$$

The difference between model (7.6.1) and (7.6.2) is that the variance of the e_{ij}'s in the latter model differs from subject to subject.

Suppose μ_i and σ_i^2 are the mean and variance for the ith subject. The CV of the ith subject is then given by

$$\lambda_i = \sigma_i/\mu_i.$$

Under the assumption that the CV is the same from subject to subject, we have

$$\lambda = \lambda_i, \qquad i = 1, \ldots ,k.$$

Therefore, the expected value and variance of Y_{ij} given $S_i = a_i$ are given below:

$$E(Y_{ij}|S_i = a_i) = \mu_i = \mu + a_i, \quad \text{and}$$

$$Var(Y_{ij}|S_i = a_i) = \sigma_i^2 = \lambda^2\mu_i^2 = \lambda^2(\mu + a_i)^2.$$

It appears that model (7.6.2) is useful because it can account for the different intra-subject means and variances. Note that, instead, one could suggest using a log-transformed model to stabilize the intra-subject variabilities. However, this method may not be effective in a typical bioavailability/bioequivalence study, which often involves the comparison of bioavailability among several formulations. In this case, the attempt to stabilize the variabilities fails if the CV's are different from formulation to formulation. However, model (7.6.2) can be extended to the case where there are several formulations.

7.6.2. Estimators for the Common CV

Let $\overline{Y}_i$ and $\hat{\sigma}_i$ be the sample mean and sample standard deviation of the ith subject. Under model (7.6.2), the following estimators of the common CV are considered by Chow and Tse (1990b):

(i) $\hat{\lambda}_1 = \dfrac{1}{k} \sum_{i=1}^{k} (\hat{\sigma}_i/\overline{Y}_i);$

(ii) $\hat{\lambda}_2 = \left[\dfrac{1}{k} \sum_{i=1}^{k} (\hat{\sigma}_i/\overline{Y}_i)^2 \right]^{1/2};$

(iii) $\quad \hat{\lambda}_3 = \left[\sum_{i=1}^{k} \hat{\sigma}_i \overline{Y}_i\right] \bigg/ \left[\sum_{i=1}^{k} \overline{Y}_i^2\right];$

(iv) $\quad \hat{\lambda}_4 = \left\{\left[\sum_{i=1}^{k} \hat{\sigma}_i^2 \overline{Y}_i^2\right] \bigg/ \left[\sum_{i=1}^{k} \overline{Y}_i^4\right]\right\}^{1/2};$

(v) $\quad \hat{\lambda}_5 = \left[\dfrac{M_2}{\overline{Y}^2 + \hat{\sigma}_s^2}\right]^{1/2}, \qquad$ where $\hat{\sigma}_s^2 = (M_1 - M_2)/n$ and

$$M_1 = \frac{n}{k-1} \sum_{i=1}^{k} (\overline{Y}_i - \overline{Y})^2,$$

$$M_2 = \frac{1}{k(n-1)} \sum_{i=1}^{k} \sum_{j=1}^{n} (Y_{ij} - \overline{Y}_i)^2.$$

The first two estimators are derived by taking the average of the estimates of λ_i and λ_i^2 (i.e., $\hat{\sigma}_i/\overline{Y}_i$ and $\hat{\sigma}_i^2/\overline{Y}_i^2$), respectively. The third and fourth estimators are obtained by fitting a least regression function (through the origin) of $\hat{\sigma}_i$ and $\hat{\sigma}_i^2$ on $\overline{Y}$ and $\overline{Y}_i^2$, respectively, based upon the fact that $\sigma_i = \lambda\mu_i$ and $\sigma_i^2 = \lambda^2\mu_i^2$. The fifth estimator is the analysis of variance estimator which is obtained by equating M_1 and M_2 to their expectations and solving for λ, i.e.,

$$M_1 = E(M_1) = n\sigma_s^2 + \mu^2\lambda^2 + \sigma_s^2\lambda^2, \quad \text{and}$$

$$M_2 = E(M_2) = \mu^2\lambda^2 + \sigma_s^2\lambda^2.$$

Chow and Tse (1990b) compared the above estimators in terms of the asymptotic order terms for their biases and mean squared errors. The results indicate that, for large n and k, the biases can be approximated by

$\text{Bias}(\hat{\lambda}_1) \approx -5\lambda c^4/2;$

$\text{Bias}(\hat{\lambda}_2) \approx 13\lambda c^4/8;$

$\text{Bias}(\hat{\lambda}_3) \approx -5\lambda c^4/8;$

$\text{Bias}(\hat{\lambda}_4) \approx -19\lambda c^4/8;$

$\text{Bias}(\hat{\lambda}_5) \approx -[\lambda c^2(1 + c^2)^{-1}]/2k,$

where $c = \sigma_s/\mu$.

The results show that these estimators, except for $\hat{\lambda}_2$, underestimate λ. In addition, $\hat{\lambda}_5$ is asymptotically unbiased. All the other four estimators are with biases of order $0(1)$. Based on a simulation study, Chow and Tse (1990b) indicated that $\hat{\lambda}_5$ gives the best performance in most cases. When λ is small, $\hat{\lambda}_2$, $\hat{\lambda}_3$, and $\hat{\lambda}_4$ may be possible contenders. However, they perform miserably for large λ.

Table 7.6.1 AUCs for an Intravenous Formulation

Subject	AUC (mμ × min/ml)			Mean	S.D.	C.V.
1	7462.7	6212.8	5782.5	6486.0	872.78	0.1346
2	7533.9	5337.7	7371.4	6747.7	1223.76	0.1814
3	6509.3	10084.6	6665.4	7753.1	2020.67	0.2606
4	6184.8	6042.9	8026.9	6751.5	1106.76	0.1639
5	5693.1	7139.3	8597.7	7263.4	1635.81	0.2252
6	9551.9	6550.8	8987.1	8363.3	1594.82	0.1907
7	8643.5	6153.3	8800.0	7865.6	1484.93	0.1888
8	9167.8	6756.3	6192.9	7372.3	1580.22	0.2143
9	7900.5	6767.8	9123.3	7930.5	1178.01	0.1485
10	6522.4	8040.5	6236.0	6932.9	969.77	0.1399
11	8178.9	6329.5	9211.0	7906.5	1459.93	0.1847
12	6835.5	7002.1	8906.8	7581.4	1150.78	0.1518
13	6533.7	6948.9	9669.3	7717.3	1703.26	0.2207
14	5876.9	8066.4	6511.7	6818.3	1126.46	0.1652

Source: Chow and Tse (1990b).

Example 7.6.2 Chow and Tse (1990b) considered an example concerning the comparison of bioavailability between two intranasal formulations and one intravenous formulation of a drug product. To assess the intra-subject variability in terms of the common CV, each subject was administered the same formulation three times. For the purpose of illustration, let's consider the intravenous formulation. Table 7.6.1 gives AUC's of 14 subjects for the intravenous formulation. The results indicate that the subjects have different intra-subject means and intra-subject variances. Therefore, the usual one-way random effects model is not an appropriate model because the heterogeneity of intra-subject variability among subjects. Since the CV's are similar from subject to subject, we then consider model (7.6.2) to assess the intra-subject variability in terms of the common CV. Estimates of the common CV for the intravenous formulation using $\hat{\lambda}_1$ through $\hat{\lambda}_5$ are

$$\hat{\lambda}_1 = 18.36\%, \quad \hat{\lambda}_2 = 18.69\%, \quad \hat{\lambda}_3 = 18.56\%,$$
$$\hat{\lambda}_4 = 19.05\%, \quad \text{and} \quad \hat{\lambda}_5 = 18.94\%.$$

These results indicate that intra-subject variability for the AUC of the formulation is less than 20%.

7.7. DISCUSSION

In Section 7.2, we briefly discussed the decision rule for assessing equivalence in variability of bioavailability. In the current FDA guidelines, however, there is no discussion of how much difference in variability would be of clinical significance. The question How much difference in variability of bioavailability is of clinical significance? generally depends upon the characteristic and the therapeutic range of the study drug and/or routes of administration. The decision rule for equivalence in variability as suggested in Section 7.2 may depend upon variance of the estimated variability of the reference formulation. To obtain an estimate of σ_R^2 and its estimated variance, each subject will have to receive several administrations of the reference formulation at different periods. The variance of $\hat{\sigma}_R^2$, which usually depends upon sample size, could be substantial. Therefore, it may require more subjects to establish equivalence in variability of bioavailability with a desired power than equivalence in average bioavailability for detection of clinical significance difference.

In Section 7.3, several estimators for σ_s^2 are considered to assess the inter-subject variability. The restricted maximum likelihood estimator and maximum likelihood estimator are commonly used to avoid negative estimates by truncating the estimates at 0. However, these estimators are likely to yield a zero estimate when σ_s^2 is small. This is because there is a high probability of obtaining a negative estimate for small σ_s^2. To avoid negative and zero estimates in variance components models, Chow and Shao (1988) proposed an estimation procedure for variance components and their ratios which is shown to have lower mean squared error than the customary estimators over a large range of parameters space. In the interest of obtaining a positive estimate for σ_s^2, this estimation procedure can be applied with some modification under model (3.1.1). For interval estimate of σ_s^2, it should be noted that the lower confidence limit for σ_s^2 could be less than 0, especially for small σ_s^2. The portion $(L_s, 0)$ would not be meaningful since σ_s^2 is always positive. In this case, a modification by truncating at 0 is usually considered, i.e., L_s is set to be $\max(0, L_s)$. However, with this modification, the confidence level may no longer be $1 - \alpha$. In addition, the expected width of (L_s, U_s) may be too wide to be useful. As an alternative approach, Chow (1985) considered a nonparametric procedure using the jackknife technique in conjunction with the ideas of grouping, truncation, and shifting to construct an approximate $(1 - \alpha) \times 100\%$ confidence interval for σ_s^2. Based upon a simulation study, the resultant confidence interval is shown to have approximately $(1 - \alpha) \times 100\%$ coverage probability with a much narrower confidence interval. This procedure is useful yet complicated and hence is not included.

The two one-sided tests procedure for average bioavailability such as Schuirmann's procedure together with the two one-sided tests procedure for variability of bioavailability can be used to assess population bioequivalence between test and reference formulations if the responses (or logarithmic transformation) are approximately normally distributed. However, when data are not normal, we can only establish equivalence in average and variability, not population equivalence, using nonparametric methods described in Chapter 4 and this chapter. For more discussion of population bioequivalence using nonparametric procedures, see Cornell (1990), and Liu and Chow (1992c).

From Chapter 4, it can be seen that both the parametric and nonparametric two one-sided tests procedure for assessing equivalence in average bioavailability are based upon the mean period differences which are linear combinations of the sequence-by-period means. For hypotheses (7.5.3), it is suggested that the test statistics be calculated based upon the residuals from the sequence-by-period means because the resultant statistics do not involve fixed period effects and they are independent of the sequence-by-period means. Furthermore, the test statistics, proposed by Liu (1991), for hypotheses (7.5.2) are based upon the orthogonal transformations which are also independent of the sequence-by-period means. Therefore, the two one-sided tests procedure for variability of bioavailability are independent of the two one-sided tests procedure for average bioavailability. For assessment of bioequivalence between formulations, some researchers (e.g., Metzler and Huang, 1983) suggested that the equivalence in variability of bioavailability be tested before the assessment of equivalence in average bioavailability is performed. It should, however, be noted that testing for equivalence between variances generally requires much larger sample sizes than does testing for difference in averages. In addition, either testing for equivalence in variability of bioavailability and testing for equivalence in average bioavailability are done separately or simultaneously, an appropriate justification on the α level should be considered in order to have a pre-determined overall type I error rate.

Instead of testing for equivalence in both average bioavailability and variability of bioavailability, testing for equivalence in CV may be useful if the CV is the same from subject to subject. Let μ_T and μ_R be the means for test and reference formulation, respectively. Also, let λ_T and λ_R be the corresponding CV, i.e., $\lambda_T = \sigma_T/\mu_T$ and $\lambda_R = \sigma_R/\mu_R$. Bioequivalence between formulations may then be determined by testing the following hypotheses:

$$H_0: \lambda_T - \lambda_R \leq \lambda_L \quad \text{or} \quad \lambda_T - \lambda_R \geq \lambda_U \qquad (7.7.1)$$

$$\text{vs} \quad H_a: \lambda_L < \lambda_T - \lambda_R < \lambda_U,$$

where λ_L and λ_U are equivalence limits. It can be seen that if $\sigma_T^2 = \sigma_R^2$, then hypotheses (7.7.1) reduces to hypotheses (4.3.1) for average bioavailability. On the other hand, if $\mu_T = \mu_R$, then testing for hypotheses (7.7.1) is equivalent to testing for hypotheses (7.5.2).

8

Assumptions and Outliers Detection

8.1. INTRODUCTION

For a standard 2×2 crossover design, as indicated in Chapter 2, the following model is usually considered for assessing bioequivalence between formulations:

$$Y_{ijk} = \mu + S_{ik} + F_{(j,k)} + C_{(j-1,k)} + P_j + e_{ijk}, \qquad (8.1.1)$$

where $i = 1,2, \ldots , n_k; j = 1,2; k = 1,2$.

Several statistical methods for assessment of bioequivalence in average bioavailability between formulations were discussed under model (8.1.1) with the following normality assumptions:

(i) $\{S_{ik}\}$ are i.i.d. normal with mean 0 and variance σ_s^2;
(ii) $\{e_{ijk}\}$ are i.i.d. normal with mean 0 and variance σ_e^2;
(iii) $\{S_{ik}\}$ and $\{e_{ijk}\}$ are mutually independent.   (8.1.2)

In most bioavailability studies, the above assumptions may not hold. The validation of assumptions (8.1.2) has an impact on the assessment of bioequivalence in average bioavailability in the following ways:

8.1.1. Model Selection

In a bioavailability study, since the methods introduced in Chapter 4 can be applied to either raw data or log-transformed data, it is important to check the

217

assumptions before an appropriate statistical model is chosen. A different model may result in a different conclusion regarding bioequivalence (Chow, 1990).

8.1.2. Drug Safety and Exchangeability

As indicated in the previous chapter, a significant difference in intra-subject variability between formulations may raise a great concern in drug safety and exchangeability of the test formulations (Liu, 1991; Liu and Chow, 1991). Assumptions (8.1.2) basically require that the intra-subject variabilities are the same from subject to subject and from formulation to formulation. In practice, this is often not true since individual subjects may differ widely in their responses to a drug (Wagner, 1971; Chow and Tse 1988, 1990b). Therefore, assumptions (8.1.2) should be examined prior to assessment of bioequivalence.

8.1.3. Outlying Data

One of the problems commonly encountered in bioavailability studies is that the data set sometimes contains either some extremely large and/or small observations. We will refer to these extreme values as outlying data. These outlying data may have dramatic effects on the bioequivalence test. Results with and without the outlying observations could be totally opposite in a marginal case (Chow and Tse, 1990a). It is therefore important to examine the outlying data under model (8.1.1) carefully.

The objectives of this chapter are then to (i) provide some tests for assumptions which can be used for model selection between the raw data model and the log-transformation model, and (ii) introduce several statistical test procedures for detection of outlying data under mode (8.1.1).

In the next section, several tests for assumptions using the inter-subject and intra-subject residuals are discussed. We will then describe three different types of outlying observations that often occur in bioavailability studies in Section 8.3. In Sections 8.4 and 8.5, statistical tests for detection of outlying subjects and outlying observations for an individual subject will be presented. A brief discussion is given in Section 8.6.

8.2. TESTS FOR ASSUMPTIONS

For a standard 2×2 crossover design, under model (8.1.1) with assumptions (8.1.2), as indicated in Chapter 3, the total sum of squares can be partitioned into the between-subject sum of squares ($SS_{Between}$) and the within-subject sum of squares (SS_{Within}). $SS_{Between}$ can be further partitioned into the sum of squares of carry-over effects and the sum of squares of inter-subject error (SS_{Inter}). The within-subject sum of squares can be further decomposed into sum of squares of formulation effects (SS_{Drug}), sum of squares of period effects (SS_{period}), and sum of squares of intra-subject residuals (SS_{Intra}). For testing assumptions (8.1.2), Jones and Kenward (1989) suggested the use of the inter-subject and intra-subject residuals. In the following, tests for assumptions using the intra-subject and inter-subject residuals will be outlined. We will first consider the use of intra-subject residuals.

8.2.1. Intra-Subject Residuals

The intra-subject residual for subject i within sequence k during period j, denoted by $\hat{e}_{ijk}$, is defined as the difference between the observed response Y_{ijk} and its predicted value $\hat{Y}_{ijk}$, which can be obtained based upon model (8.1.1), i.e.,

$$\hat{e}_{ijk} = Y_{ijk} - \hat{Y}_{ijk}$$
$$= Y_{ijk} - \overline{Y}_{i \cdot k} - \overline{Y}_{\cdot jk} + \overline{Y}_{\cdot \cdot k}, \tag{8.2.1}$$

where $\hat{Y}_{ijk} = \overline{Y}_{i \cdot k} + \overline{Y}_{\cdot jk} - \overline{Y}_{\cdot \cdot k}$, and

$$\overline{Y}_{i \cdot k} = \frac{1}{2} \sum_{j=1}^{2} Y_{ijk} = \frac{Y_{i \cdot k}}{2}, \tag{8.2.2}$$

$$\overline{Y}_{\cdot jk} = \frac{1}{n_k} \sum_{i=1}^{n_k} Y_{ijk} = \frac{Y_{\cdot jk}}{n_k},$$

$$\overline{Y}_{\cdot \cdot k} = \frac{1}{2n_k} \sum_{j=1}^{2} \sum_{i=1}^{n_k} Y_{ijk} = \frac{Y_{\cdot \cdot k}}{2n_k} = \frac{1}{2} \sum_{j=1}^{2} \overline{Y}_{\cdot jk},$$

where $i = 1,2, \ldots, n_k$; $j, k = 1,2$.

From (8.2.2), it can be verified that

$$Y_{i \cdot k} = 2\overline{Y}_{i \cdot k} \quad \text{and} \quad 2\overline{Y}_{\cdot \cdot k} = \overline{Y}_{\cdot 1k} + \overline{Y}_{\cdot 2k}.$$

It follows that

$$\hat{e}_{i1k} + \hat{e}_{i2k} = 0, \tag{8.2.3}$$

or equivalently,

$$\hat{e}_{i1k} = -\hat{e}_{i2k}, \quad \text{for } i = 1,2, \ldots, n_k \quad \text{and} \quad k = 1,2.$$

Also, for $k = 1,2$,

$$\sum_{i=1}^{n_k} \hat{e}_{ijk} = \sum_{i=1}^{n_k} (Y_{ijk} - \overline{Y}_{i \cdot k} - \overline{Y}_{\cdot jk} + \overline{Y}_{\cdot \cdot k})$$
$$= n_k (\overline{Y}_{\cdot jk} - \overline{Y}_{\cdot \cdot k} - \overline{Y}_{\cdot jk} + \overline{Y}_{\cdot \cdot k}) = 0 \tag{8.2.4}$$

Since there are $n_1 + n_2$ constraints in (8.2.3) and two constraints in (8.2.4), the degrees of freedom for the sum of squares of the intra-subject residuals (SS_{Intra}) is $2(n_1 + n_2) - (n_1 + n_2) - 2 = n_1 + n_2 - 2$.

The variances and covariances of $\hat{e}_{ijk}$ under assumption (ii) of (8.1.2) are given by

$$\text{Var}(\hat{e}_{ijk}) = \frac{(n_k - 1)}{2n_k} \sigma_e^2, \tag{8.2.5}$$

and

$$\text{Cov}(\hat{e}_{ijk}, \hat{e}_{i'j'k'}) = \begin{cases} -\dfrac{n_k - 1}{2n_k} \sigma_e^2 & \text{if } i = i', j \neq j', k = k' \\ -\sigma_e^2/2n_k & \text{if } i \neq i', j = j', k = k' \\ \sigma_e^2/2n_k & \text{if } i \neq i', j \neq j', k = k' \\ 0 & \text{if } k \neq k' \end{cases} \tag{8.2.6}$$

From (8.2.3) and (8.2.6), it can be seen that there is a perfect negative correlation (i.e., $\rho = -1$) between the two intra-subject residuals within the same subject. Therefore, it is sufficient to examine intra-subject residuals at the first period, i.e.,

$$\hat{e}_{i1k}, i = 1, 2, \ldots, n_k, \qquad k = 1, 2.$$

If σ_e^2 is known, we may consider the following standardized intra-subject residuals

$$e_{i1k}^* = \frac{\hat{e}_{i1k}}{\left[\dfrac{n_k - 1}{2n_k} \sigma_e^2 \right]^{1/2}}, \qquad i = 1, 2, \ldots, n_k; \quad k = 1, 2. \tag{8.2.7}$$

The marginal distribution of e_{i1k}^* is the standard normal distribution with mean 0 and variance 1. Due to the constraints of (8.2.4), the joint distribution of $\{e_{i1k}^*\}$ is an $n_1 + n_2$ dimensional singular multivariate distribution with mean vector **0** and covariance matrix

$$\text{Cov}(e_{i1k}^*, e_{i1k'}^*) = \begin{cases} 1 & \text{if } i = i', k = k' \\ -\dfrac{1}{n_k} & \text{if } i \neq i', k = k' \\ 0 & \text{if } k \neq k' \end{cases} \tag{8.2.8}$$

In practice, however, σ_e^2 is usually unknown but can be estimated unbiasedly by MS_{Intra}. Substituting σ_e^2 in (8.2.7) with MS_{Intra} yields the following studentized intra-subject residuals

$$\bar{e}_{i1k} = \frac{\hat{e}_{i1k}}{\left[\dfrac{n_k - 1}{2n_k} MS_{\text{Intra}} \right]^{1/2}}, \qquad i = 1, 2, \ldots, n_k; \quad k = 1, 2. \tag{8.2.9}$$

Although $\bar{e}_{i1k}$ also has mean 0 and variance 1, the joint distribution of the studentized intra-subject residuals is quite complicated. In practice, as indicated by Jones and Kenward (1989), we can treat $\bar{e}_{i1k}$ as standard normal random variables to evaluate assumptions (8.1.2).

Based upon $\{\bar{e}_{i1k}\}$ and $\{\hat{Y}_{i1k}\}$ where $i = 1,2, \ldots ,n_k$ and $k = 1,2$, assumptions (8.1.2) and the adequacy of model (8.1.1) can be examined in terms of the normal probability plot of $\{\bar{e}_{i1k}\}$ or the residual plot between $\{\bar{e}_{i1k}\}$ and $\{\hat{Y}_{i1k}\}$. The normal probability plot of $\{\bar{e}_{i1k}\}$ is used to examine the normality assumption on the intra-subject variability of e_{ijk}, while the residual plot between $\{\bar{e}_{i1k}\}$ and $\{\hat{Y}_{i1k}\}$ is used to examine whether model (8.1.1) is adequate. Note that the residual plot and normal probability plot can also provide preliminary information of potential outlying data.

The normal probability plot is obtained by plotting $\bar{e}_{i1k}$ against the expected normal order statistics (or normal scores), which can be easily computed using some statistical software programs such as PROC RANK in SAS. If e_{ijk} is, indeed, normally distributed, then the normal probability plot should be a straight line. Therefore, any marked deviation from a straight line may indicate that the normality assumption of e_{ijk} does not hold for the data set under study. Alternatively, the normality of e_{ijk} can also be examined using the test of Shapiro and Wilk (1965) on $\bar{e}_{i1k}$. This can be accomplished using PROC UNIVARIATE in SAS. It, however, should be noted that Shapiro and Wilk's test requires the observations be independent. Therefore, the p-value for the normality test of Shapiro and Wilk based upon the studentized intra-subject residuals should be used with caution, especially with a small sample size, because they are not independent.

For the residual plot, if model (8.1.1) is an appropriate model, then $\bar{e}_{i1k}$ should be randomly scattered between -2 and 2 on the plot. Any distinct pattern observed in the plot may indicate that the model is not adequate to describe the data. In this case, additional terms or covariates may be added to the model. In addition, if there are some extremely large or small unusual values (i.e., outside interval $(-2,2)$) observed on the plot, the corresponding subjects may possibly be outlying subjects. These subjects should be carefully examined in terms of their demographic information, plasma concentration, and other important factors.

8.2.2. Inter-Subject Residuals

Similarly, the inter-subject residuals can be used to evaluate the normality assumption imposed on the inter-subject variability of S_{ik}. The inter-subject residuals, denoted by $\hat{S}_{ik}$, are given below:

$$\hat{S}_{ik} = Y_{i \cdot k} - \overline{Y}_{\cdot \cdot k}, \qquad i = 1,2, \ldots ,n_k, \quad k = 1,2. \tag{8.2.10}$$

Under assumption (i) of (8.1.2), the variances and covariances of $\hat{S}_{ik}$ are given by

$$\text{Var}(\hat{S}_{ik}) = \frac{2(n_k - 1)}{n_k}(2\sigma_s^2 + \sigma_e^2), \qquad (8.2.11)$$

and

$$\text{Cov}(\hat{S}_{ik}, \hat{S}_{i'k'}) = \begin{cases} -\dfrac{2(2\sigma_s^2 + \sigma_e^2)}{n_k} & \text{if } i \neq i', k = k' \\ 0 & \text{if } k \neq k' \end{cases} \qquad (8.2.12)$$

The studentized inter-subject residuals are then given by

$$\tilde{S}_{ik} = \frac{\hat{S}_{ik}}{\left[\dfrac{2(n_k - 1)}{n_k} MS_{\text{Inter}}\right]^{1/2}}, \quad i = 1,2, \ldots, n_k, \quad k = 1,2. \quad (8.2.13)$$

Therefore, the normality assumption of S_{ik} can be checked by either examining the normal probability plot of $\{\tilde{S}_{ik}\}$ or performing the Shapiro-Wilk's test on $\{\tilde{S}_{ik}\}$.

Note that the assumption (iii) of independence between $\{S_{ik}\}$ and $\{e_{ijk}\}$ in (8.1.2) can also be checked by examining either Pearson's correlation coefficient or Spearman's rank correlation coefficient between $\{\tilde{e}_{i1k}\}$ and $\{\tilde{S}_{ik}\}$.

Example 8.2.1   We will use AUC data (both raw data and log-transformed data) of the two erythromycin formulations in Clayton and Leslie's study to illustrate procedures for examination of assumptions (8.1.2) through the inter-subject and intra-subject residuals.

Raw Data

Table 8.2.1 provides studentized intra-subject and inter-subject residuals of the raw AUC data, which were obtained by plugging the values of $MS_{\text{Intra}} (= 5.992)$ and $MS_{\text{Inter}} (= 8.911)$ into (8.2.9) and (8.2.13), respectively. Note that only studentized intra-subject residuals at the first period are given in Table 8.2.1. The residual plot of predicted values vs studentized intra-subject residuals is given in Figure 8.2.1. Figure 8.2.1 exhibits a slightly downward pattern of distribution of the studentized intra-subject residuals. It can be seen that studentized intra-subject residuals corresponding to approximately the same value of $\hat{Y}_{i1k}$ for three subjects are quite different. These three subjects are subject 2 (6.137 vs 1.675), subject 7 (5.972 vs −3.053), and subject 14 (6.046 vs −1.101). As a matter of fact, these three subjects yield the largest three studentized intra-

Table 8.2.1 Intra-Subject and Inter-Subject Residuals of Raw AUC for Clayton and Leslie's Study at the First Period

Subj.	Y_{i1k}	$\hat{Y}_{i1k}$	$\hat{e}_{i1k}$	$\tilde{e}_{i1k}$	$Y_{i\cdot k}$	$\overline{Y}_{\cdot\cdot k}$	$\hat{S}_{ik}$	$\tilde{S}_{ik}$
1	2.52	3.277	−0.757	−0.464	7.99	8.461	−0.471	−0.118
2	8.87	6.137	2.733	1.675	13.71	8.461	5.249	1.319
3	0.79	0.802	−0.012	−0.007	3.04	8.461	−5.421	−1.362
4	1.68	1.032	0.648	0.397	3.50	8.461	−4.961	−1.247
5	6.95	6.692	0.258	0.158	14.82	8.461	6.359	1.598
6	1.05	1.432	−0.382	−0.234	4.30	8.461	−4.161	−1.046
7	0.99	5.972	−4.982	−3.053	13.38	8.461	4.919	1.236
8	5.60	4.467	1.133	0.695	10.37	8.461	1.909	0.480
9	3.16	1.802	1.358	0.832	5.04	8.461	−3.421	−0.860
10	4.98	4.481	0.499	0.306	8.17	10.234	−2.064	−0.519
11	7.14	8.881	−1.741	−1.069	16.97	10.234	6.736	1.692
12	1.81	2.756	−0.946	−0.580	4.72	10.234	−5.514	−1.386
13	7.34	6.356	0.984	0.603	11.92	10.234	1.686	0.424
14	4.25	6.046	−1.796	−1.101	11.30	10.234	1.066	0.268
15	6.66	5.431	1.229	0.753	10.07	10.234	−0.164	−0.041
16	4.76	4.021	0.739	0.453	7.25	10.234	−2.984	−0.750
17	7.16	7.066	0.094	0.058	13.34	10.234	3.106	0.780
18	5.52	4.581	0.939	0.575	8.37	10.234	−1.864	−0.469

subject residuals in absolute magnitude, though they are in the opposite direction. In addition, subject 7 of sequence 1 has a very large studentized intra-subject residual (-3.053). Therefore, the first normality assumption of (8.1.2) regarding the intra-subject variability is questionable. From Table 6.9.1, which gives the raw AUC data, it can be seen that the AUC of erythromycin stearate for subject 7 is 0.99, while the AUC of erythromycin base is 12.39. Therefore, subject 7 has an extremely low relative bioavailability which indicates that subject 7 is a potential outlying subject. For a further investigation, Table 8.2.2 presents hourly plasma erythromycin concentrations of the two formulations for subject 7. The extremely large AUC of the base formulation may be caused by extremely high plasma concentrations occurring at 4, 6, and 8 hours after administration of the drug. It is, however, not discussed in Clayton and Leslie (1981) regarding the possible causes of this extreme difference.

For evaluation of assumption (ii) of (8.1.2), the normal probability plot of studentized intra-subject residuals is used, which is given in Figure 8.2.2. The plot shows a quadratic trend which certainly argues against the normality assumption. Departure from linearity may be explained by the unusual studentized

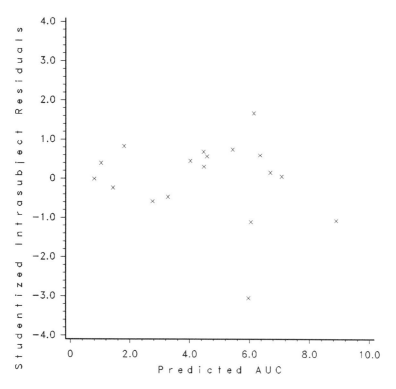

Figure 8.2.1   Predicted AUC versus studentized intra-subject residuals.

Table 8.2.2   Hourly Plasma Erythromycin
Concentrations (μg/ml) for Subject 7 in Clayton
and Leslie's Study

	Erythromycin	
Hours	Stearate	Base
0.0	0.00	0.00
0.5	0.09	0.07
1.0	0.10	0.07
1.5	0.13	0.07
2.0	0.18	0.07
3.0	0.28	0.34
4.0	0.11	3.00
6.0	0.09	2.96
8.0	0.07	1.47
AUC (μg $\cdot$ hr/ml)	0.99	12.39

Source: Clayton and Leslie (1981).

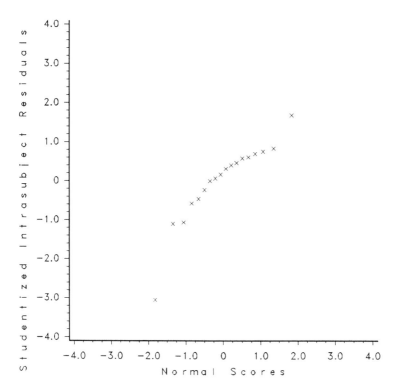

Figure 8.2.2 Normal probability plot of studentized intra-subject residuals. (From Liu and Weng (1992).)

intra-subject residuals occurring in subjects 2, 7, and 14. The Shapiro-Wilk's test gives a p-value of 0.028 which also leads to rejection of the normality assumption hypothesis of e_{i1k}.

Studentized inter-subject residuals can be used to examine the first normality assumption given in (8.1.2). From Table 8.2.1, no substantial studentized inter-subject residuals were observed. The normal probability plot (Figure 8.2.3) exhibits a linear relationship between residuals and their corresponding normal scores, which is in favor of the normality assumption of S_{ik}. The Shapiro-Wilk's test yields a p-value of 0.267. Therefore, we fail to reject the normality assumption hypothesis of S_{ik}. Since the normality assumption for e_{ijk} may not hold, Spearman's rank correlation coefficient between $\tilde{e}_{i1k}$ and $\tilde{S}_{ik}$ is calculated to evaluate the independence between S_{ik} and e_{ijk}. Spearman's rank correlation coefficient gives a value of -0.013 with a p-value of 0.958. Thus, there is no evidence to suspect that assumption (iii) of (8.1.2) is not true. Therefore, based upon raw AUC data, we are comfortable with all the assumptions in (8.1.2) except for the normality assumption of e_{ijk}.

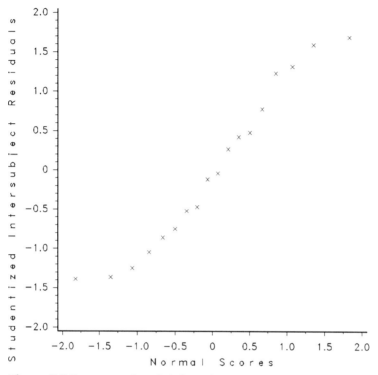

Figure 8.2.3 Normal probability plot of studentized inter-subject residuals. (From Liu and Weng (1992).)

Log-Transformed Data

Similarly, we can examine the assumptions of (8.1.2) using intra-subject and inter-subject residuals based upon log-transformed AUCs. Studentized intra-subject and inter-subject residuals are given in Table 8.2.3. The scatter plot of $\hat{Y}_{ilk}$ versus $\tilde{e}_{ilk}$ for log-transformed data is presented in Figure 8.2.4. From Figure 8.2.4, it can be seen that the pattern observed in the raw data disappears after log-transformation. This may indicate that the log-transformed model is more appropriate than the raw data model. However, it should be noted that subject 7 still has an unusually large studentized intra-subject residual (-2.750) which is outside the range of ($-2,2$).

For evaluation of the first assumption of (8.1.2), both the normal probability plot for inter-subject residuals given in Figure 8.2.5 and the Shapiro-Wilk's test (p-value = 0.903) suggest that S_{ik} follows a normal distribution. For assumption (ii), the normal probability plot (Figure 8.2.6) for intra-subject studentized residuals shows a slight departure from a straight line, which is probably due to

Table 8.2.3 Intra-Subject and Inter-Subject Residuals of Log (AUC) for Clayton and Leslie's Study at the First Period

Subj.	Y_{i1k}	$\hat{Y}_{i1k}$	$\hat{e}_{i1k}$	$\tilde{e}_{i1k}$	$Y_{i\cdot k}$	$\overline{Y}_{\cdot\cdot k}$	$\hat{S}_{ik}$	$\tilde{S}_{ik}$
1	0.92	1.067	− 0.143	− 0.386	2.62	2.320	− 0.304	− 0.315
2	2.18	1.636	0.547	1.477	3.76	2.320	1.440	1.495
3	− 0.24	0.043	− 0.279	− 0.753	0.58	2.320	− 1.745	− 1.812
4	0.52	0.315	0.204	0.551	1.12	2.320	− 1.202	− 1.249
5	1.94	1.757	0.182	0.491	4.00	2.320	1.682	1.747
6	0.05	0.369	− 0.321	− 0.865	1.23	2.320	− 1.092	− 1.135
7	− 0.01	1.009	− 1.019	− 2.750	2.51	2.320	0.187	0.194
8	1.72	1.398	0.325	0.876	3.29	2.320	0.965	1.002
9	1.15	0.647	0.504	1.360	1.78	2.320	− 0.538	− 0.559
10	1.61	1.480	0.126	0.340	2.77	3.082	− 0.316	− 0.328
11	1.97	2.222	− 0.257	− 0.693	4.25	3.082	1.170	1.215
12	0.59	0.928	− 0.334	− 0.902	1.66	3.082	− 1.420	− 1.475
13	1.99	1.854	0.139	0.375	3.52	3.082	0.433	0.450
14	1.45	1.797	− 0.350	− 0.944	3.40	3.082	0.318	0.331
15	1.90	1.658	0.238	0.642	3.12	3.082	0.041	0.043
16	1.56	1.333	0.227	0.613	2.47	3.082	− 0.609	− 0.633
17	1.97	1.992	− 0.023	− 0.063	3.79	3.082	0.708	0.736
18	1.71	1.475	0.234	0.631	2.76	3.082	− 0.326	− 0.338

the unusual residual of subject 7. The p-value from the Shapiro-Wilk's test is given by 0.105, which also indicates a slight deviation from the normality assumption of e_{ijk}. Independence between S_{ik} and e_{ijk} can be examined using either Pearson's correlation coefficient or Spearman's rank correlation coefficient. Both Pearson's correlation coefficient (0.211) and Spearman's rank correlation coefficient (0.214) failed to reject the hypothesis of independence (p-values = 0.40 and 0.39, respectively).

Note that (i) the signs of both correlation coefficients for the log-transformed data are positive, while correlation coefficients between $\tilde{S}_{ik}$ and $\tilde{e}_{i1k}$ are negative for the raw data; (ii) the magnitudes of the two correlation coefficients for the log-transformed data are very close, while the magnitude of Spearman's rank correlation coefficient is much smaller than that of Pearson's correlation coefficient for raw data ($= -0.204$). This is because e_{ijk} of log-transformed AUCs may be normally distributed while e_{ijk} of the raw AUCs are not.

It is possible that violation of the normality assumption of intra-subject variability is caused by the unusual AUCs for both stearate and base formulations observed on subject 7 in sequence 1. Therefore, we should repeat the same analyses with subject 7 deleted from the data set. The results are also summarized

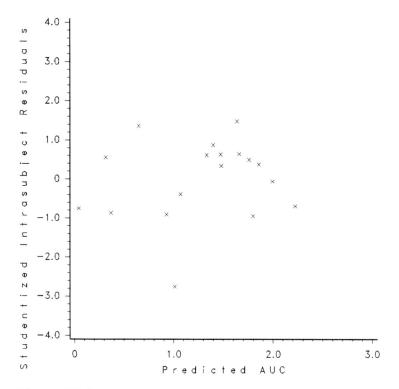

Figure 8.2.4 Predicted AUC vs studentized intra-subject residuals (log-transformed data).

in Table 8.2.4 along with those obtained based upon the complete data set. Since results of the inter-subject residuals remain almost unchanged, only the normal probability plots of studentized intra-subject residuals are presented which are given in Figure 8.2.7 (for raw data) and Figure 8.2.8 (for log-transformed data). Figure 8.2.7 indicates little evidence of a normality assumption violation of e_{ijk} if subject 7 is excluded from the analysis. Shapiro-Wilk's test for normality also confirms this with a p-value of 0.657. For the log-transformed AUC, with and without subject 7, the normal probability plots are similar. The Shapiro-Wilk's test also gives similar p-values, 0.117 (without subject 7) and 0.105 (with subject 7).

For raw data, with subject 7 deleted, both Pearson's and Spearman's correlation coefficients between studentized intra-subject and inter-subject residuals give p-values greater than 0.70. Although the independence assumption between e_{ijk} and S_{ik} seems to hold for log-transformed data without subject 7, the cor-

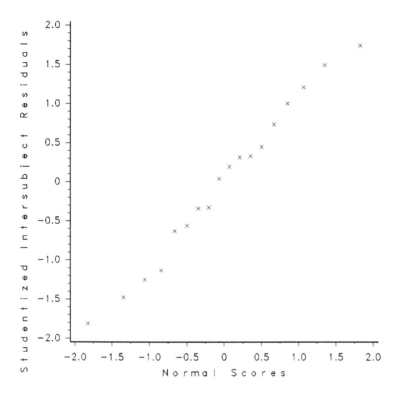

Figure 8.2.5 Normal probability plot of studentized inter-subject residuals (log-transformed data). (From Liu and Weng (1992).)

relation coefficient magnitudes are much larger than those obtained based upon the raw data. As a result, if subject 7 is excluded, assumptions (8.1.2) hold for the raw data. In this case, the raw data model is an appropriate model for assessment of bioequivalence.

Table 8.2.5 provides estimates for inter-subject and intra-subject variabilities with and without subject 7. For the raw data, the estimate of intra-subject variability with subject 7 deleted reduces more than 55% from the estimate with subject 7 included, while the estimate of inter-subject variability without subject 7 increases more than 107% from the estimate including subject 7. The null hypothesis $H_0: \sigma_s^2 = 0$ is rejected at the 5% level if subject 7 is excluded from the analysis. Similar results are obtained for the log-transformed AUC. Consequently, inclusion and exclusion of the possible outlying subject has a tremendous impact on assessment of bioequivalence. Therefore, for the remainder

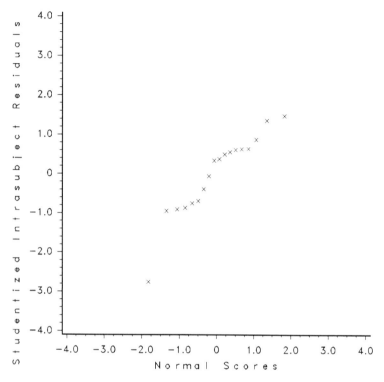

Figure 8.2.6   Normal probability plot of studentized intra-subject residuals (log-transformed data). (From Liu and Weng (1992).)

Table 8.2.4   Summary of Test Results for Normality Assumptions

| | Shapiro-Wilk[a] | | Pearson[b] | | Spearman[c] | |
Data set	$\bar{e}_{i1k}$	$\tilde{S}_{i1k}$	Test	P	Test	P
Raw AUC	0.028	0.267	−0.204	0.417	−0.013	0.958
Log (AUC)	0.105	0.903	0.211	0.400	0.214	0.395
Raw AUC[d]	0.657	0.276	0.054	0.836	0.098	0.708
Log (AUC)[d]	0.117	0.876	0.339	0.183	0.275	0.286

[a]P-value of Shapiro-Wilk's test for normality.

[b]Pearson correlation coefficient.

[c]Spearman's rank correlation coefficient.

[d]The results are obtained with deletion of subject 7.

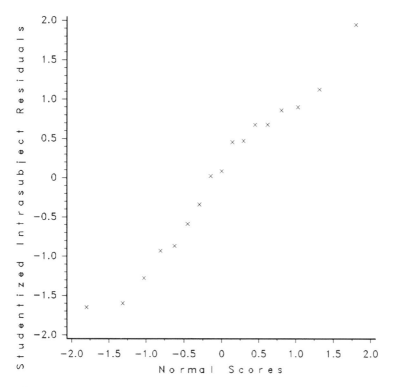

Figure 8.2.7   Normal probability plot of studentized intra-subject residuals without subject 7.

of this chapter efforts will be directed toward detection of potential outlying data.

8.3.   THE DEFINITION OF OUTLYING OBSERVATIONS

As indicated earlier, one of the problems commonly encountered in bioavailability studies is that the data set may contain some extremely large or small (i.e., outlying) observations. These observations may have an impact on the conclusion of bioequivalence. Basically, there are three different kinds of outliers, which are described below:

(1)   Unexpected observations in the blood or plasma concentration-time curve;
(2)   Extremely large or small observations within a given formulation;
(3)   Unusual subjects who exhibit extremely high or low bioavailability with respect to the reference formulation.

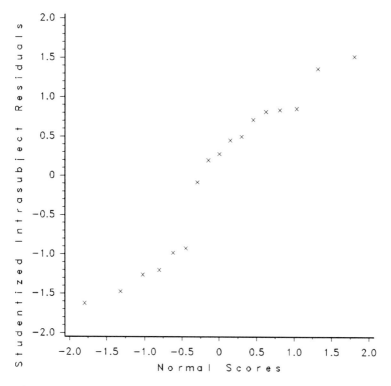

Figure 8.2.8   Normal probability plot of studentized intra-subject residuals (log-transformed data) without subject 7.

Table 8.2.5   Summary of Inter-Subject and Intra-Subject Variability of AUC Data from Clayton and Leslie's Study

Data set	$\hat{\sigma}_e^2$	$\hat{\sigma}_s^2$	P-Value
Raw AUC	5.992	1.459	0.218
Log (AUC)	0.309	0.106	0.153
Raw AUC[a]	2.669	2.964	0.015
Log (AUC)[a]	0.174	0.191	0.016

[a]The results are obtained with deletion of subject 7.

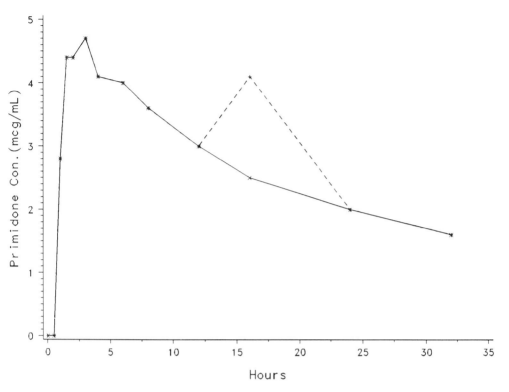

Figure 8.3.1   Primidone concentration-time curve.   (X: raw data, +: un-expected primidone concentration of 16 hours.)

For the first kind of outlier, Rodda (1986) indicated that unexpected obser-vations in the plasma concentration-time curve usually have little effect on cal-culation of AUC and consequently have little effect on the comparison of bio-availability. As an example, let's consider the hourly plasma samples in Table 1.3.1. Suppose we observe an extremely high concentration at hour 16 (say 4.1). The concentration-time curve is plotted in Figure 8.3.1 which indicates that the concentration at hour 16 is a potential outlying data point. The AUC is recal-culated in Table 8.3.1, which is given by 95.55. It can be seen that there is about 11.2% increase in AUC. However, contribution of this outlying concen-tration to the overall average is only 0.5% of 85.95 (original AUC) (= 11.2/24), if there are 24 subjects in the study.

Table 8.3.1 Calculation of AUC Using the Trapezoidal Rule[a]

Blood sample i	t_i	C_i	$(C_i + C_{i-1})/2$	$t_i + t_{i-1}$	$\dfrac{(C_i + C_{i-1})}{(t_i + t_{i-1})/2}$
1	0.0	0.0	—	—	—
2	0.5	0.0	0.00	0.5	0.00
3	1.0	2.8	1.40	0.5	0.70
4	1.5	4.4	3.60	0.5	1.80
5	2.0	4.4	4.40	0.5	2.20
6	3.0	4.7	4.55	1.0	4.55
7	4.0	4.1	4.40	1.0	4.40
8	6.0	4.0	4.05	2.0	8.10
9	8.0	3.6	3.80	2.0	7.60
10	12.0	3.0	3.30	4.0	13.20
11	16.0	4.1	3.55	4.0	14.20
12	24.0	2.0	3.05	8.0	24.40
13	32.0	1.6	1.80	8.0	14.40

[a]AUC $(0 - 32) = 95.55$.

To illustrate outlying observations of a given formulation, let us again consider Clayton and Leslie's study. Figure 8.3.2 plots AUC vs subject for stearate and base formulations. It can be seen that the AUC of subject 7 has a potential outlying observation within the base formulation. This outlying observation within the base formulation certainly has an impact on the assessment of bioequivalence in average bioavailability because the average bioavailability is very sensitive to extreme values.

For the third kind of outlier, as indicated by Chow and Tse (1990a), an outlying subject may negate the conclusion of a bioequivalence study. For a given set of data, a potential outlying subject can be examined by plotting the relative AUCs. For Clayton and Leslie's study, Figure 8.3.3 plots AUCs of the stearate formulation vs the base formulation. An outlying subject can be detected by a relatively large deviation from the straight line $y = x$.

It can be seen that Figures 8.3.2 and 8.3.3 are useful for a preliminary evaluation of potential outlying data. In the following, statistical tests for detection of outlying subjects and observations within a given formulation will be outlined.

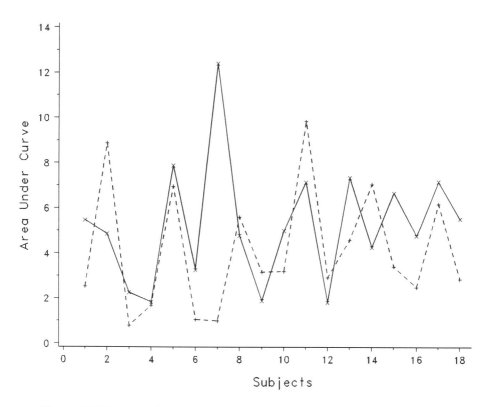

Figure 8.3.2 Distribution of AUC of stearate and base formulations. (+: stearate formulation, X: base formulation.)

8.4. DETECTION OF OUTLYING SUBJECTS

8.4.1. Likelihood Distance and Estimates Distance

For a $k \times k$ crossover design comparing f formulations of a drug product, model (8.1.1) can be rewritten as follows:

$$Y_{ijl} = \mu + S_i + F_j + P_l + e_{ijl}, \qquad j,l = 1, \ldots ,f; \quad i = 1, \ldots ,N, \tag{8.4.1}$$

where Y_{ijl} is the response variable on the ith subject in the lth period under the jth formulation; μ is the overall mean; F_j is the fixed effect of the jth formulation with $\Sigma_j F_j = 0$; P_l is the fixed effect of the lth period. With $\Sigma_l P_l = 0$; S_i is the

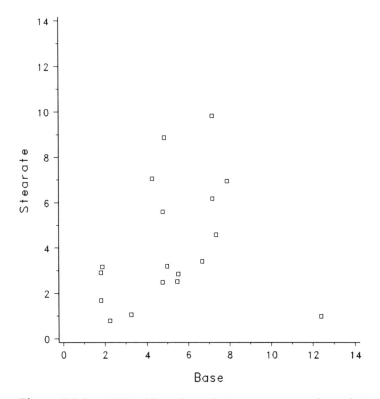

Figure 8.3.3   AUC of base formulations vs stearate formulations.

random effect of the ith subject, e_{ijl} is the error term. In the above model, it is assumed that $\{S_i\}$ and $\{e_{ijl}\}$ are independent and normally distributed with means 0 and variances σ_s^2 and σ_e^2, respectively.

For detection of an outlying subject, Chow and Tse (1990a) proposed two test procedures, namely, the likelihood distance (LD) and the estimates distance (ED) under the assumption that there are no period effects and formulation effects in (8.4.1). Under the assumption that there are no period and formulation effects, model (8.4.1) reduces to

$$Y_{ij} = \mu + S_i + e_{ij}, \qquad j = 1, \ldots, f; \quad i = 1, \ldots, N. \qquad (8.4.2)$$

In this case, the parameters of interest are μ, σ_s^2, and σ_e^2. Let $\theta = (\theta_1, \theta_2, \theta_3)'$ where $\theta_1 = \mu$, $\theta_2 = \sigma_e^2$, and $\theta_3 = \sigma_e^2 + f\sigma_s^2$. The log-likelihood function is then given by

$$L(\boldsymbol{\theta}) = \frac{-Nf}{2} \log 2\pi - \frac{N}{2} \log(\theta_2 \theta_3^{f-1}) - \frac{1}{2\theta_3} \sum_{i=1}^{N} \sum_{j=1}^{f} (Y_{ij} - \theta_1)^2$$

$$- \frac{f}{2} \left(\frac{1}{\theta_2} - \frac{1}{\theta_3} \right) \sum_{i=1}^{N} (\overline{Y}_i - \theta_1)^2. \qquad (8.4.3)$$

The maximum likelihood estimator (MLE) $\hat{\boldsymbol{\theta}}$ of $\boldsymbol{\theta}$ are given by

$$\hat{\theta}_1 = \overline{Y} = \frac{1}{Nf} \sum_{i=1}^{N} \sum_{j=1}^{f} Y_{ij}$$

$$\hat{\theta}_2 = m_1$$

$$\hat{\theta}_3 = (N - 1)m_2/N,$$

where $m_1 = \dfrac{1}{N(f-1)} \sum\limits_{i=1}^{N} \sum\limits_{j=1}^{f} (Y_{ij} - \overline{Y}_i)^2$ and $m_2 = \dfrac{f}{N-1} \sum\limits_{i=1}^{N} (\overline{Y}_i - \overline{Y})^2$.

It should be noted that although $\theta_3 \geq \theta_2$, it is possible that $\hat{\theta}_3 < \hat{\theta}_2$ (i.e., $(N - 1)m_2 < Nm_1$). In this case, the maximum likelihood estimators of θ_2 and θ_3 are modified as follows:

$$\hat{\theta}_2 = \hat{\theta}_3 = \frac{1}{Nf} \sum_{i=1}^{N} \sum_{j=1}^{f} (Y_{ij} - \overline{Y})^2,$$

which are obtained by maximizing $L(\boldsymbol{\theta})$ under the condition that $\theta_2 = \theta_3$. The LD test procedure is given by

$$LD_i(\hat{\boldsymbol{\theta}}) = 2 [L(\hat{\boldsymbol{\theta}}) - L(\hat{\boldsymbol{\theta}}_{(i)})], \qquad (8.4.4)$$

where $\hat{\boldsymbol{\theta}}_{(i)}$ denotes the MLE of $\boldsymbol{\theta}$ with deletion of the ith subject. As N tends to infinity, it can be verified that $LD_i(\hat{\boldsymbol{\theta}})$ is asymptotically distributed as a chi-square variable with three degrees of freedom. Thus we consider the ith subject as an outlying subject if

$$LD_i(\hat{\boldsymbol{\theta}}) > \chi_3^2(\alpha),$$

where $\sigma_3^2(\alpha)$ is the αth upper percentile point of χ_3^2.

A similar idea can also be applied to the estimates distance. We thus have

$$ED_i(\hat{\boldsymbol{\theta}}) = f^2 (\hat{\boldsymbol{\theta}}_{(i)} - \hat{\boldsymbol{\theta}}) \hat{\Sigma}^{-1} (\hat{\boldsymbol{\theta}}_{(i)} - \hat{\boldsymbol{\theta}}), \qquad (8.4.5)$$

where $\hat{\Sigma}$ is the estimate of

$$\Sigma = \begin{bmatrix} \dfrac{\theta_3}{N} & 0 & 0 \\ 0 & \dfrac{2\theta_2^2}{N-1} & 0 \\ 0 & 0 & 2\theta_3^2 \end{bmatrix},$$

which can be obtained by substituting $\boldsymbol{\theta}$ with $\hat{\boldsymbol{\theta}}$.

Chow and Tse (1990a) showed that $ED_i(\hat{\boldsymbol{\theta}})$ is distributed as a chi-square variable with three degrees of freedom as N tends to infinity. Therefore, we consider the ith subject as an outlying subject if

$$ED_i(\hat{\boldsymbol{\theta}}) > \chi_3^2(\alpha).$$

Example 8.4.1. To illustrate the use of LD and ED test procedures, let's consider AUC data from Clayton and Leslie's study. Table 8.4.1 summarizes the MLEs of θ_1, θ_2, and θ_3 with each subject deleted and the corresponding LD and ED test results for both raw data and log-transformed data.

For raw data, it is obvious that estimates with subject 7 deleted differ substantially from the MLEs. This indicates that subject 7 is very influential in the estimation process. Furthermore, $LD_7(\hat{\theta}) = 9.203$ and $ED_7(\hat{\theta}) = 55.273$, and $ED_{11}(\hat{\theta}) = 8.091$ which are greater than $\chi_3^2(0.05) = 7.815$. Both LD and ED methods indicate that subject 7 is an outlier under model (8.4.1) at the 5% level of significance. However, ED method also indicates that subject 11 is an outlier based upon the raw data.

For log-transformed data, the LD test procedure did not detect any outlying subject. The ED test procedure, however, indicates that subjects 3 and 7 are potential outlying subjects ($ED_3(\hat{\theta}) = 11.640$ and $ED_7(\hat{\theta}) = 39.072$, respectively).

A SAS program for LD and ED procedures can be found in Lin, Chow and Tse (1991).

8.4.2. Hotelling T²

Liu and Weng (1991) proposed a procedure based upon the order statistics of the two-sample Hotelling T^2 to identify possible outlying subjects. A sequential step-down closed testing procedure (Hochberg and Tamhane, 1987) is also introduced below to detect multiple outlying subjects.

Under the assumption that there are no period effects and with further relaxation of the compound symmetry assumption for covariance structure of f responses observed on subject i, model (8.4.1) becomes

$$Y_{ij} = \alpha_j + \varepsilon_{ij}, \qquad i = 1,2, \ldots ,N; \quad j = 1,2, \ldots ,f. \tag{8.4.6}$$

Let $\mathbf{Y}_i = (Y_{i1}, \ldots ,Y_{if})'$ be $f \times 1$ vector of the responses to f formulations observed on subject i. Thus, $\mathbf{Y}_i$ are f-dimensional multivariate normal (MVN) random vectors with mean vector $\boldsymbol{\alpha}$ and covariance matrix Λ, where

Table 8.4.1 Summary of Estimation Results (Raw Data)

Subject[a]	$\hat{\theta}_{1(i)}$	$\hat{\theta}_{2(i)}$	$\hat{\theta}_{3(i)}$	$LD_i(\hat{\theta})$	P	$ED_i(\hat{\theta})$	P
1	4.714	6.096	8.745	0.032	0.999	0.603	0.896
2	4.546	5.875	8.210	0.078	0.994	1.378	0.711
3	4.859	6.290	7.564	0.267	0.966	4.380	0.223
4	4.846	6.352	7.738	0.214	0.975	3.642	0.303
5	4.513	6.327	7.870	0.171	0.982	2.964	0.397
6	4.822	6.210	8.009	0.122	0.989	2.135	0.545
7	4.555	2.530	8.296	9.203	0.027	55.273	0.000[b]
8	4.644	6.332	8.770	0.055	0.997	1.057	0.787
9	4.801	6.304	8.225	0.093	0.993	1.688	0.640
10	4.709	6.258	8.759	0.045	0.998	0.861	0.835
11	4.450	6.140	6.993	0.549	0.908	8.091	0.044
12	4.810	6.317	8.136	0.110	0.991	1.971	0.578
13	4.598	6.128	8.597	0.038	0.998	0.709	0.871
14	4.616	6.122	8.684	0.034	0.998	0.646	0.886
15	4.653	6.042	8.786	0.029	0.999	0.567	0.904
16	4.736	6.201	8.666	0.041	0.998	0.770	0.857
17	4.556	6.324	8.306	0.084	0.994	1.549	0.671
18	4.703	6.143	8.773	0.034	0.998	0.651	0.885

(Log-Transformed Data)

Subject	$\hat{\theta}_{1(i)}$	$\hat{\theta}_{2(i)}$	$\hat{\theta}_{3(i)}$	$LD_i(\hat{\theta})$	P	$ED_i(\hat{\theta})$	P
1	1.353	0.346	0.567	0.029	0.999	0.569	0.903
2	1.319	0.353	0.533	0.072	0.995	1.301	0.729
3	1.413	0.332	0.427	0.845	0.839	11.640	0.009
4	1.397	0.364	0.490	0.264	0.967	4.391	0.222
5	1.312	0.363	0.515	0.145	0.986	2.559	0.465
6	1.394	0.326	0.500	0.204	0.977	3.415	0.332
7	1.356	0.176	0.566	5.121	0.163	39.072	0.000[b]
8	1.333	0.363	0.557	0.059	0.996	1.123	0.772
9	1.377	0.356	0.541	0.060	0.996	1.105	0.776
10	1.348	0.358	0.567	0.044	0.998	0.844	0.839
11	1.305	0.361	0.493	0.239	0.971	3.978	0.264
12	1.381	0.357	0.534	0.076	0.995	1.385	0.709
13	1.326	0.357	0.547	0.055	0.997	1.017	0.797
14	1.330	0.356	0.552	0.047	0.997	0.882	0.830
15	1.338	0.351	0.562	0.033	0.998	0.635	0.888
16	1.357	0.352	0.566	0.034	0.998	0.644	0.886
17	1.318	0.363	0.531	0.098	0.992	1.783	0.619
18	1.349	0.351	0.567	0.033	0.998	0.635	0.888

[a]Deleted subject.
[b]P-value < 0.001.

239

$\boldsymbol{\alpha} = (\alpha_1, \ldots, \alpha_f)'$, and

$$\Lambda = \text{Cov}(Y_{ij}, Y_{i'j'}) = \begin{cases} \sigma_j^2, & \text{if } i = i' \text{ and } j = j' \\ \sigma_{jj'}, & \text{if } i = i' \text{ and } j \neq j' \\ 0, & \text{otherwise.} \end{cases}$$

The hypotheses for outlying subjects due to a location shift can be formulated as follows:

$$H_0: \mathbf{Y}_i \sim \text{MVN}(\boldsymbol{\alpha}, \Lambda), \qquad \text{for all } i = 1, 2, \ldots, N$$

$$\text{vs} \quad H_a: \mathbf{Y}_i \sim \text{MVN}(\boldsymbol{\alpha} + \boldsymbol{\Delta}_i, \Lambda), \qquad \text{for at least one } i. \tag{8.4.7}$$

The above hypotheses can be further decomposed into the following N subhypotheses:

$$H_{0i}: \mathbf{Y}_i \sim \text{MVN}(\boldsymbol{\alpha}, \Lambda) \quad \text{vs} \quad H_{ai}: \mathbf{Y}_i \sim \text{MVN}(\boldsymbol{\alpha} + \boldsymbol{\Delta}_i, \Lambda), \tag{8.4.8}$$

where $i = 1, 2, \ldots, N$.

Note that $H_0 = \cap_i H_{0i}$ and $H_a = \cup_i H_{ai}$. Therefore, the sequential step-down closed testing procedure introduced by Hochberg and Tamhane (1987) can be applied to subhypotheses (8.4.8) for detection of possible multiple outlying subjects. Hypotheses (8.4.8) can be tested using the two-sample Hotelling's T^2 statistics by comparing a sample that consists only of the ith subject with the sample of other $N - 1$ subjects (Hawkins, 1980).

Let $\overline{\mathbf{Y}}$ and $\mathbf{A}$ be the sample mean and matrix of the sums of squares and cross products computed from $\mathbf{Y}_1, \ldots, \mathbf{Y}_N$, respectively. Also, let $\overline{\mathbf{Y}}_{(-i)}$ and $\mathbf{A}_{(-i)}$ be the sample mean and matrix of the sum of squares and cross products computed from $N - 1$ subjects with deletion of the ith subject. The two-sample Hotelling T^2 statistic for the ith subject is then given by

$$T_i^2 = \frac{(N - 1)(N - 2)}{N} (\mathbf{Y}_i - \overline{\mathbf{Y}}_{(-i)})' \mathbf{A}_{(-i)}^{-1} (\mathbf{Y}_i - \overline{\mathbf{Y}}_{(-i)}), \tag{8.4.9}$$

where $i = 1, 2, \ldots, N$.

Computation of T_i^2 can be simplified as

$$T_i^2 = \frac{(N - 2)D_i^2}{\left[\dfrac{N - 1}{N} - D_i^2\right]}, \tag{8.4.10}$$

where

$$D_i^2 = (\mathbf{Y}_i - \overline{\mathbf{Y}})' A^{-1} (\mathbf{Y}_i - \overline{\mathbf{Y}}), \qquad i = 1, 2, \ldots, N.$$

Note that Hotelling T^2 is invariant under any full-rank linear transformation. Consequently, the joint distribution of $\{T_i^2, i = 1, 2, \ldots, N\}$ is independent of the unknown parameters α and Λ. The testing procedure proposed by Liu and Weng (1991) is outlined below.

Let $T_{(1)}^2, \ldots, T_{(N)}^2$ be the order statistics of $T_1^2, \ldots, T_N^2$, and $H_{0(i)}$ be the corresponding subhypotheses based upon $T_{(i)}^2$. Also, let $(W_1^2, \ldots, W_N^2)$ be a vector of N Hotelling T^2 statistics computed from a sample of size N from an f-dimensional multivariate normal with mean $\mathbf{0}$ and covariance matrix I_f. We start with the order subhypothesis $H_{0(N)}$. Hypothesis $H_{0(i)}$ is rejected if

$$P\left\{ \max_{1 \le j \le i} W_j^2 > T_{(i)}^2 \right\} < \alpha, \tag{8.4.11}$$

provided that $H_{0(N)}, H_{0(N-1)}, \ldots, H_{0(i+1)}$ are rejected at the α level of significance, where $0 < \alpha < 1$.

Note that the joint distribution of order statistics of $\{T_i^2, i = 1, 2, \ldots, N\}$ is rather complicated. However, the sampling distribution of $T_{(i)}^2$ under $H_{0(i)}$ can be empirically evaluated by the Monte Carlo simulation using standard multivariate normal vectors because it is independent of α and Λ. Tables 8.4.2 and 8.4.3 give the 5% and 1% upper quantiles of the distribution of ordered T^2 statistics for $f = 2, 3$ and $N = 10(1)20$ and $20(5)50$, which were obtained based upon 3000 simulation samples.

The procedure for detection of a potential outlying subject proposed by Chow and Tse (1990a) is based upon the asymptotic distributions of LD and ED. The disadvantage of Chow and Tse's procedure is that the sample size for a bioavailability study is sometimes too small to apply asymptotic distributions of LD and ED. In addition, evaluation of LD or ED involves simultaneous estimation of formulation average, inter-subject and intra-subject variabilities. As an alternative, Liu and Weng (1991) suggested the use of the parametric bootstrap technique (Efron, 1982) for empirical evaluation of the sampling distribution of the order statistics of LD_i because the joint distribution of LD_i is not independent of the unknown parameters θ_1, θ_2, and θ_3. Chow and Tse's method can be further extended to detect possible multiple outlying subjects which is stated below.

At each iteration, a sample of size N for f-dimensional multivariate normal random vectors with mean $\hat{\theta}_1 \mathbf{1}_f$ and covariance matrix $\hat{\theta}_2 I_f + (\hat{\theta}_3 - \hat{\theta}_2) f^{-1} J_f$ is generated, where $\mathbf{1}_f$ is a $f \times 1$ vector of 1s, I_f is a $f \times f$ identity matrix, and J_f is a $f \times f$ matrix of 1s. We can compute LD_i (or ED_i) according to (8.4.4) (or (8.4.5)). Sampling distribution of the order statistics of LD_i (or ED_i) can then be obtained by iterating the process a large number of times (e.g., B times)

Table 8.4.2 $\alpha\%$ Upper Quantiles of the Ordered T^2 Statistics for $f = 2$ and $\alpha = 0.05$

Sample size	$T^2_{(N)}$	$T^2_{(N-1)}$	$T^2_{(N-2)}$	$T^2_{(N-3)}$
10	22.67	9.01	5.25	3.72
11	21.80	8.97	5.46	3.88
12	20.16	9.03	5.55	4.05
13	18.83	9.12	5.70	4.32
14	18.67	9.03	6.03	4.46
15	18.55	9.30	6.09	4.63
16	17.92	9.11	6.22	4.73
17	17.15	9.34	6.34	4.90
18	17.32	9.25	6.43	5.04
19	16.88	9.33	6.48	5.18
20	16.47	9.49	6.71	5.20
25	16.21	9.61	6.96	5.70
30	15.79	9.83	7.39	6.09
35	16.33	10.05	7.77	6.43
40	16.29	10.27	8.00	6.74
45	16.12	10.37	8.29	6.95
50	15.86	10.57	8.40	7.17

for $f = 2$ and $\alpha = 0.01$

Sample size	$T^2_{(N)}$	$T^2_{(N-1)}$	$T^2_{(N-2)}$	$T^2_{(N-3)}$
10	40.26	12.04	6.44	4.35
11	36.38	12.23	6.64	4.59
12	32.37	11.67	6.79	4.69
13	29.65	11.98	7.11	4.97
14	28.68	11.37	7.16	5.20
15	28.32	11.63	7.17	5.25
16	25.96	11.49	7.30	5.50
17	26.07	12.47	7.53	5.57
18	24.99	11.79	7.58	5.82
19	22.73	12.01	7.74	5.83
20	25.14	11.59	7.87	5.98
25	22.51	11.98	8.03	6.49
30	21.96	11.89	8.69	6.82
35	21.90	12.17	8.97	7.14
40	20.84	12.21	9.20	7.65
45	20.73	12.67	9.31	7.70
50	19.67	12.17	9.61	7.90

Table 8.4.3 $\alpha\%$ Upper Quantiles of the Ordered T^2 Statistics
for f = 3 and α = 0.05

Sample size	$T^2_{(N)}$	$T^2_{(N-1)}$	$T^2_{(N-2)}$	$T^2_{(N-3)}$
10	39.91	15.64	9.73	6.61
11	34.28	15.42	9.63	6.84
12	30.63	14.78	9.51	6.92
13	28.41	14.47	9.76	7.19
14	27.22	13.99	9.56	7.20
15	26.09	13.92	9.70	7.37
16	25.19	13.38	9.65	7.51
17	24.62	13.55	9.52	7.59
18	24.50	13.37	9.82	7.65
19	23.02	13.46	9.93	7.79
20	22.73	13.48	9.86	7.59
25	21.32	13.17	10.18	8.27
30	20.49	13.11	10.34	8.64
35	20.57	13.47	10.57	9.03
40	19.86	13.55	10.71	9.24
45	19.59	13.33	10.96	9.41
50	19.42	13.72	11.07	9.77

for f = 3 and α = 0.01

Sample size	$T^2_{(N)}$	$T^2_{(N-1)}$	$T^2_{(N-2)}$	$T^2_{(N-3)}$
10	63.13	22.04	12.06	7.67
11	52.28	21.40	12.16	8.20
12	50.43	19.38	11.65	7.97
13	40.41	18.90	11.65	8.30
14	40.90	18.04	11.55	8.29
15	38.14	18.21	11.46	8.46
16	36.59	17.17	11.58	8.50
17	37.93	16.87	11.36	8.65
18	34.60	16.41	11.52	8.81
19	31.54	17.01	11.76	8.76
20	30.51	16.98	11.96	8.86
25	28.02	16.22	11.72	9.37
30	26.18	15.31	11.96	9.79
35	26.61	15.90	11.88	10.17
40	25.36	15.93	12.33	10.24
45	24.73	15.98	12.34	10.51
50	23.90	15.68	12.66	10.66

Source: Liu and Weng (1991).

and the same procedure in (8.4.11) can be applied by replacing T^2 with LD_i (or ED_i).

Example 8.4.2   We will again use AUC data from Clayton-Leslie's study to illustrate the above procedure for detection of possible outlying subjects. Table 8.4.4 gives likelihood distances, which were calculated based upon model (8.4.2), and Hotelling T^2 statistics, which were obtained under model (8.4.6), for both raw and log-transformed AUC data.

For raw AUC data, as expected, subject 7 gives the maximum T^2 value of 28.738 with a p-value of 0.004. Therefore, we reject the order subhypothesis $H_{0(18)}$ at the 5% level of significance. We then proceed to test the subsequent order subhypothesis $H_{0(17)}$. Subject 11 yields the second largest T^2 ($T^2_{(17)}$ = 6.237) with a corresponding p-value of 0.38. Thus, we fail to reject $H_{0(17)}$ at the 5% level and the sequential testing procedure stops. Based upon the empirical

Table 8.4.4   Hotelling T^2 and Likelihood Distances for AUC Data from Clayton and Leslie's Study

Subject	Raw data		Log-transformed data	
	T^2	LD	T^2	LD
1	0.403	0.032	0.327	0.029
2	4.222	0.078	2.163	0.072
3	2.629	0.267	5.452	0.845
4	2.329	0.214	3.511	0.264
5	1.898	0.171	1.650	0.145
6	1.695	0.122	2.678	0.204
7	28.738	9.203	17.545	5.121
8	0.395	0.055	0.571	0.059
9	1.749	0.093	3.388	0.060
10	0.118	0.045	0.029	0.044
11	6.237	0.549	2.677	0.239
12	1.860	0.110	3.593	0.076
13	0.647	0.038	0.795	0.055
14	1.703	0.034	1.414	0.047
15	0.444	0.029	0.494	0.033
16	0.369	0.041	0.167	0.034
17	0.957	0.084	1.119	0.098
18	0.267	0.034	0.206	0.033

sampling distribution of the order statistics of Hotelling T^2, we conclude that subject 7 is an outlying subject. For use of likelihood distance, subject 7 also gives the maximum LD with a value of 9.203. Since $9.203 > \chi_3^2(0.05) = 7.815$, according to Chow and Tse (1990a), subject 7 is also identified as an outlying subject. However, the p-value based on the empirical sampling distribution of $LD_{(18)}$, which was obtained from 3000 bootstrap samples, is 0.002. Therefore, $H_{0(18)}$ is rejected at the 5% level. Similarly, $H_{0(17)}$ is not rejected because the p-value corresponding to the $LD_{(17)} = 0.549$ of subject 11 is 0.42. Hence, the sequential testing procedure stops and we conclude that subject 7 is an outlying subject.

For log-transformed AUC data, again subject 7 yields the maximum T^2 with a value of 17.545 which is greater than 17.32, the 5% upper quantile of $T_{(18)}^2$ (see Table 8.4.2). Therefore, $H_{0(18)}$ is rejected at the 5% level. Since $T_{(17)}^2 = 5.452$ of subject 3 is less than 9.25, the 5% quantile of $T_{(17)}^2$, we fail to reject $H_{0(17)}$ at the 5% level. Consequently, the sequential testing procedure stops and we conclude that subject 7 is an outlying subject. For likelihood distance, subject 7 also gives the largest LD of 5.121. Since 5.121 is less than $\chi_3^2(0.05) = 7.815$, we cannot declare that subject 7 is an outlying subject according to the procedure proposed by Chow and Tse (1990a). However, the p-value evaluated from the empirical sampling distribution of $LD_{(18)}$ is 0.023, which suggests a rejection of $H_{0(18)}$ at the 5% level. The p-value corresponding to $LD_{(17)} = 0.845$ of subject 3 is 0.179. Therefore, the sequential testing procedure stops and we conclude that subject 7 is an outlying subject.

Inconsistency between the empirical sampling distribution of the order statistics and the asymptotic distribution of LD for log-transformed AUC data may be due to the fact that the number of subjects in this study is too small. A comparison between the 5% upper quantile of a χ^2 with 3 degrees of freedom and that of the empirical sampling distribution of LD strongly suggests that the empirical sampling distribution of LD is stochastically smaller than that of χ_3^2. This suggests that the empirical sampling distribution of LD should be used if the LD test procedure is considered for detection of a possible outlying subject.

8.5. DETECTION OF OUTLYING OBSERVATIONS

Liu and Weng (1991) proposed a method for detection of potential outlying observations. Their procedure is based upon residuals from the formulation means which can be estimated under the following model:

$$Y_{ij} = \mu + F_j + S_i + e_{ij} \tag{8.5.1}$$
$$= \alpha_j + S_i + e_{ij},$$

where $\alpha_j = \mu + F_j$, $i = 1,2, \ldots , N$ (number of subjects) and $j = 1,2, \ldots ,$ f (number of formulations). Model (8.5.1) is a reduced model of (8.4.1) under the assumption of no period effects.

Let $\mathbf{Y}_i = (Y_{i1}, \ldots ,Y_{if})'$, $\qquad i = 1, \ldots ,N$; and $\mathbf{Y} = (\mathbf{Y}_1', \ldots ,\mathbf{Y}_N')'$.

Under the assumptions of model (8.4.1), $\mathbf{Y}$ is then an M dimensional multivariate normal vector with mean vector $\mathbf{1}_N \otimes \boldsymbol{\alpha}$ and covariance matrix $I_N \otimes \Sigma$, where $M = Nf$, $\otimes$ is the kronecker product between two matrices $\boldsymbol{\alpha} = (\alpha_1, \ldots ,\alpha_f)'$ and

$$\Sigma = \sigma_e^2 I_f + \sigma_s^2 J_f.$$

Define

$$MSF = \frac{N}{f - 1} \sum_{j=1}^{f} (\overline{Y}_{\cdot j} - \overline{Y}_{\cdot \cdot})^2,$$

$$MSB = \frac{f}{N - 1} \sum_{i=1}^{N} (\overline{Y}_{i \cdot} - \overline{Y}_{\cdot \cdot})^2,$$

and

$$MSE = \frac{1}{(N - 1)(f - 1)} [SST - (f - 1)MSF - (N - 1)MSB],$$

$$(8.5.2)$$

where

$$SST = \sum_{i=1}^{N} \sum_{j=1}^{f} (Y_{ij} - \overline{Y}_{\cdot \cdot})^2,$$

and

$$\overline{Y}_{\cdot j} = \frac{1}{N} \sum_{i=1}^{N} Y_{ij}, \qquad \overline{Y}_{i \cdot} = \frac{1}{f} \sum_{j=1}^{f} Y_{ij}, \quad \text{and} \quad \overline{Y}_{\cdot \cdot} = \frac{1}{Nf} \sum_{i=1}^{N} \sum_{j=1}^{f} Y_{ij}.$$

The restricted maximum likelihood estimators (Hocking, 1985) for α_j, σ_s^2, and σ_e^2 under model (8.5.1) are then given by

$$\hat{\alpha}_j = \overline{Y}_{\cdot j},$$
$$\hat{\sigma}_e^2 = MSE,$$
$$\hat{\sigma}_s^2 = \frac{(MSB - MSE)}{f}.$$

$$(8.5.3)$$

When MSB < MSE, $\hat{\sigma}_s^2 = 0$, and

$$\hat{\sigma}_e^2 = \frac{1}{f(N-1)} \sum_{i=1}^{N} \sum_{j=1}^{f} (Y_{ij} - \overline{Y}_{\cdot j})^2.$$

Therefore, consistent estimators for α and Σ can be obtained as follows:

$$\hat{\alpha} = (\overline{Y}_{\cdot 1}, \ldots, \overline{Y}_{\cdot f})', \quad \text{and} \tag{8.5.4}$$

$$\hat{\Sigma} = \hat{\sigma}_e^2 I_f + \hat{\sigma}_s^2 J_f.$$

The vector of residuals from the formulation means is then given by

$$\mathbf{R} = \mathbf{Y} - \mathbf{1}_N \otimes \hat{\alpha}$$

$$= \left[\left(I_N - \frac{1}{N} J_N \right) \otimes I_f \right] \mathbf{Y}. \tag{8.5.5}$$

$\mathbf{R}$ is distributed as an M-dimensional multivariate normal (singular) with mean $\mathbf{0}$ and covariance matrix

$$\text{Cov}(\mathbf{R}) = \left[\left(I_N - \frac{1}{N} J_N \right) \otimes \Sigma \right]. \tag{8.5.6}$$

The studentized residuals from the formulation means are then given by

$$r_{ij} = \frac{Y_{ij} - \overline{Y}_{\cdot j}}{\left[\dfrac{N-1}{N} (\hat{\sigma}_e^2 + \hat{\sigma}_s^2) \right]^{1/2}}, \quad i = 1, 2, \ldots, N; \quad j = 1, 2, \ldots, f. \tag{8.5.7}$$

Define $z_m = r_{ij}$, where $m = f(i-1) + j$, where $i = 1, \ldots, N; j = 1, \ldots, f$, and $m = 1, 2, \ldots, M$. $\mathbf{Z} = (z_1, \ldots, z_M)'$ is then approximately distributed as a multivariate normal vector (singular, $M - f$) with mean vector $\mathbf{0}$ and covariance V, where V is the M by M correlation matrix of $\mathbf{R}$ obtained from the estimated covariance matrix $\hat{\Sigma}$. If the possible outlying observations are due to a location shift and not due to dispersion, we can then express the hypotheses of outlying observations in terms of the means, Δ_m, of the distribution of z_m, i.e.,

$$H_0: \Delta_m = 0 \quad \text{for all } m = 1, 2, \ldots, M \tag{8.5.8}$$
$$\text{vs} \quad H_a: \Delta_m \neq 0 \quad \text{for at least one } m.$$

Under model (8.5.1) and H_0 in (8.5.8), Z_m are identically distributed as a standard normal variable. Hypotheses (8.5.8) can be further decomposed into M sub-hypotheses given below:

$$H_{0m}: \Delta_m \neq 0 \quad \text{vs} \quad H_{am}: \Delta_m \neq 0, \quad m = 1, \ldots, M. \tag{8.5.9}$$

We can apply the sequential step-down closed testing procedure to these M subhypotheses for detection of possible outlying observations as presented below.

Let $|z|_{(1)}, \ldots, |z|_{(M)}$ be the order statistics of $|z_1|, \ldots, |z_M|$ and $H_{0(m)}$ be the corresponding ordered null subhypotheses, where $|z_m|$ is the absolute value of z_m. Also, let $(T_1, \ldots, T_M)'$ be an M-dimensional vector of the studentized residuals generated from model (8.5.1). We then start with the ordered null subhypothesis $H_{0(M)}$. We reject $H_{0(m)}$, $m = M, M - 1, \ldots, 1$, if

$$P\left\{ \max_{1 \leq j \leq m} |T_j| > |z|_{(m)} \right\} < \alpha, \tag{8.5.10}$$

provided that $H_{0(M)}, \ldots, H_{0(m+1)}$ are all rejected at the α level of significance.

Since the exact joint distribution of the order statistics for $|z_1|, \ldots, |z_M|$ is somewhat complicated to obtain, Liu and Weng (1991) suggested the following parametric bootstrap procedure to evaluate the empirical sampling distribution of $|z|_{(m)}$ under $H_{0(m)}$, $m = M, M - 1, \ldots, 1$:

At each iteration, a sample of size N for f-dimensional normal random vectors with mean $\hat{\alpha}$ and covariance matrix $\hat{\Sigma}$ are generated and studentized residuals $T_1, \ldots, T_M$ are calculated according to (8.5.7). The empirical sampling distribution of $|z|_{(m)}$ can then be obtained by repeating the above process a large number (e.g., B) of times. The probability (8.5.10) can then be evaluated by computing the proportion of the number of events that satisfy

$$\left\{ \max_{1 \leq j \leq m} |T_j| > z_{(m)} \right\}, \qquad m = M, M - 1, \ldots, 1.$$

Example 8.5.1 Under model (8.5.1), the restricted MLE of σ_e^2 and σ_s^2 based upon the AUC data (or Log(AUC)) from Clayton and Leslie's study are given by 5.695 and 1.554 (0.302 and 0.133), respectively. Studentized residuals from the formulation means are given in Table 8.5.1 (for raw AUC) and Table 8.5.2 (for log(AUC)).

From Table 8.5.1, it can be seen that the maximum absolute studentized residual $|z|_{(36)}$ occurred at the AUC value of the erythromycin base formulation in subject 7 within sequence 1 with a value of 2.736. The p-value based upon the empirical sampling distribution of $|z|_{(36)}$ from 3000 bootstrap samples is 0.15. Therefore, we fail to reject $H_{0(36)}$ at the 5% level of significance. Thus, the sequential testing procedure stops and we conclude that there are no outlying observations for the raw AUC data. Similarly, from Table 8.5.2, the empirical sampling distribution of $|z|_{(36)}$ for log-transformed AUC gives a p-value of 0.681 which leads to the same conclusion of no outlying observations.

Note that we are unable to identify any outlying observations according to the sequential testing procedure discussed in this section, though subject 7 in sequence 1 was identified as an outlying subject by either the two-sample

Table 8.5.1 Residuals from the Formulation Means: Raw Data

Sequence	Subject	Formulation	AUC	Formulation mean	Residual	Absolute studentized residual
1	1	Base	5.47	5.2311	0.2389	0.0913
1	1	Stearate	2.52	4.1167	−1.5967	0.6102
1	2	Base	4.84	5.2311	−0.3911	0.1495
1	2	Stearate	8.87	4.1167	4.7533	1.8167
1	3	Base	2.25	5.2311	−2.9811	1.1394
1	3	Stearate	0.79	4.1167	−3.3267	1.2714
1	4	Base	1.82	5.2311	−3.4111	1.3037
1	4	Stearate	1.68	4.1167	−2.4367	0.9313
1	5	Base	7.87	5.2311	2.6389	1.0086
1	5	Stearate	6.95	4.1167	2.8333	1.0829
1	6	Base	3.25	5.2311	−1.9811	0.7572
1	6	Stearate	1.05	4.1167	−3.0667	1.1721
1	7	Base	12.39	5.2311	7.1589	2.7361
1	7	Stearate	0.99	4.1167	−3.1267	1.1950
1	8	Base	4.77	5.2311	−0.4611	0.1762
1	8	Stearate	5.60	4.1167	1.4833	0.5669
1	9	Base	1.88	5.2311	−3.3511	1.2808
1	9	Stearate	3.16	4.1167	−0.9567	0.3656
2	10	Base	4.98	5.2311	−0.2511	0.0960
2	10	Stearate	3.19	4.1167	−0.9267	0.3542
2	11	Base	7.14	5.2311	1.9089	0.7296
2	11	Stearate	9.83	4.1167	5.7133	2.1836
2	12	Base	1.81	5.2311	−3.4211	1.3075
2	12	Stearate	2.91	4.1167	−1.2067	0.4612
2	13	Base	7.34	5.2311	2.1089	0.8060
2	13	Stearate	4.58	4.1167	0.4633	0.1771
2	14	Base	4.25	5.2311	−0.9811	0.3750
2	14	Stearate	7.05	4.1167	2.9333	1.1211
2	15	Base	6.66	5.2311	1.4289	0.5461
2	15	Stearate	3.41	4.1167	−0.7067	0.2701
2	16	Base	4.76	5.2311	−0.4711	0.1801
2	16	Stearate	2.49	4.1167	−1.6267	0.6217
2	17	Base	7.16	5.2311	1.9289	0.7372
2	17	Stearate	6.18	4.1167	2.0633	0.7886
2	18	Base	5.52	5.2311	0.2889	0.1104
2	18	Stearate	2.85	4.1167	−1.2667	0.4841

Source: Clayton and Leslie (1981).

Table 8.5.2 Residuals from the Formulation Means: Log-Transformed Data

Sequence	Subject	Formulation	AUC	Formulation mean	Residual	Absolute studentized residual
1	1	Base	1.6993	1.5209	0.1784	0.2783
1	1	Stearate	0.9243	1.1798	−0.2556	0.3987
1	2	Base	1.5769	1.5209	0.0560	0.0874
1	2	Stearate	2.1827	1.1798	1.0029	1.5646
1	3	Base	0.8109	1.5209	−0.7100	1.1077
1	3	Stearate	−0.2357	1.1798	−1.4155	2.2085
1	4	Base	0.5988	1.5209	−0.9221	1.4386
1	4	Stearate	0.5188	1.1798	−0.6610	1.0313
1	5	Base	2.0631	1.5209	0.5422	0.8459
1	5	Stearate	1.9387	1.1798	0.7589	1.1840
1	6	Base	1.1787	1.5209	−0.3422	0.5340
1	6	Stearate	0.0488	1.1798	−1.1310	1.7646
1	7	Base	2.5169	1.5209	0.9960	1.5539
1	7	Stearate	−0.0101	1.1798	−1.1899	1.8564
1	8	Base	1.5623	1.5209	0.0414	0.0647
1	8	Stearate	1.7228	1.1798	0.5429	0.8471
1	9	Base	0.6313	1.5209	−0.8896	1.3880
1	9	Stearate	1.1506	1.1798	−0.0293	0.0456
2	10	Base	1.6054	1.5209	0.0845	0.1319
2	10	Stearate	1.1600	1.1798	−0.0198	0.0309
2	11	Base	1.9657	1.5209	0.4448	0.6940
2	11	Stearate	2.2854	1.1798	1.1056	1.7249
2	12	Base	0.5933	1.5209	−0.9276	1.4472
2	12	Stearate	1.0682	1.1798	−0.1117	0.1742
2	13	Base	1.9933	1.5209	0.4724	0.7371
2	13	Stearate	1.5217	1.1798	0.3419	0.5334
2	14	Base	1.4469	1.5209	−0.0740	0.1154
2	14	Stearate	1.9530	1.1798	0.7732	1.2063
2	15	Base	1.8961	1.5209	0.3752	0.5854
2	15	Stearate	1.2267	1.1798	0.0469	0.0732
2	16	Base	1.5602	1.5209	0.0393	0.0614
2	16	Stearate	0.9123	1.1798	−0.2675	0.4174
2	17	Base	1.9685	1.5209	0.4476	0.6983
2	17	Stearate	1.8213	1.1798	0.6415	1.0008
2	18	Base	1.7084	1.5209	0.1875	0.2925
2	18	Stearate	1.0473	1.1798	−0.1325	0.2067

Source: Clayton and Leslie (1981).

Hotelling T^2 or likelihood distance. This inconsistency may be explained by the following reasons. First of all, model (8.5.1) proposed by Liu and Weng (1991) may not be adequate for Clayton and Leslie's data set. Secondly, the procedure proposed by Liu and Weng (1991) may not be able to detect outlying observations due to differences in variabilities. Recall in Example 8.2.1 that intra-subject and inter-subject variabilities change more than 50% with deletion of subject 7. This indicates that subject 7 may be drawn from a population with different intra-subject and/or inter-subject variabilities than the rest of subjects in the study. Finally, the AUC of the two erythromycin formulations for subject 7 may differ, individually, from their respective formulation means. However, they are not extreme enough to be declared as outlying observations when considered alone. On the other hand, the two-sample Hotelling T^2 and likelihood distance depend upon the joint distribution of AUCs for both formulations. In other words, they take into account the magnitude of relative change for each subject. Therefore, subject 7 is identified as an outlying subject by both Hotelling T^2 and likelihood distance because the relative bioavailability or erythromycin stearate to erythromycin base for subject 7 is only 0.08.

8.6. DISCUSSION

In Section 8.2, we introduced the use of intra-subject and inter-subject residuals to verify the normality assumptions for model (8.1.1). Although the residual plot and the normal probability plot based upon studentized residuals can provide useful information about the normality assumption, they are not rigorous statistical tests for normality. Further research on this topic is worthwhile.

For the model section between the raw data model and the log-transformation model, one may compare residual plots for the raw AUC and the log-transformed AUC. When the two residual plots exhibit a similar pattern, it is not easy to determine which model is more appropriate. In this case, as an alternative, we may consider the following method. Suppose there are f formulations. We may apply Shapiro-Wilk's test to test normality for each formulation and the difference between any two formulations. In other words, a total of $f + f(f - 1)/2$ tests are to be performed for each raw AUC and log(AUC). Based upon the test results (i.e., p-values), we then select the model with the higher percentage of failure in rejection of the normality assumption hypothesis (p-values > 0.05). Note that the above alternative provides a quick review for model selection between the raw data model and the log-transformed model which, however, does not directly address assumptions of S_{ik} and e_{ijk}.

In Sections 8.4 and 8.5, statistical tests for detection of possible outlying subjects and outlying observations for an individual subject were discussed. However, it should be noted that outliers are determined under an assumed statistical model. An outlier may indicate that the model is not correct. In other words, an extreme value may be identified as an outlier under model A but may not be concluded as an outlier under model B. However, model B may produce different outliers. Therefore, the treatment of a possible outlier in statistical analysis is critical. In bioavailability/bioequivalence studies, statistical analyses with and without the possible outlier may lead to a totally opposite conclusion for bioequivalence.

A report by a bioequivalence task force, which resulted from bioequivalence hearings conducted by the FDA in 1986, indicated that removal of certain subjects in a study because their data do not conform with the remaining data may affect validation of the study. One cannot determine whether the apparently non-conforming data result from laboratory error, data transcription, or other causes unrelated to bioequivalence. The above concerns certainly have built a case against data removal.

9

Optimal Crossover Designs for Two Formulations

9.1. INTRODUCTION

In previous chapters, most of our efforts were directed at the assessment of bioequivalence in average bioavailability for a standard 2×2 crossover design for comparing two formulations of a drug product. The standard two-sequence, two-period crossover, however, is not useful in the presence of carry-over effects. In addition, it does not provide independent estimates of intra-subject variabilities. To account for these disadvantages, in practice, it is of interest to consider a higher-order crossover design. A higher-order crossover design is defined as a crossover design in which either the number of periods or the number of sequences is greater than the number of formulations to be compared. The most commonly used higher-order designs for comparing two formulations are a four-sequence, two-period design (or Balaam's design), a two-sequence, three-period design, and a four-period design with two or four sequences. Some of these designs were briefly described as designs A, B, and C in Section 2.5. In this chapter, statistical methods for assessing bioequivalence of average bioavailability from these experimental designs will be discussed.

Consider the following general model for a higher-order crossover design:

$$Y_{ijk} = \mu + G_k + S_{ik} + P_j + F_{(j,k)} + C_{(j-1,k)} + e_{ijk}, \qquad (9.1.1)$$

where $i = 1, 2, \ldots, n_k$, $j = 1, \ldots, J$, $k = 1, \ldots, K$, and Y_{ijk}, μ, P_j, $F_{(j,k)}$, $C_{(j-1,k)}$, S_{ik} and e_{ijk} are defined as those in (2.5.1), and G_k is the fixed effect of sequence k. For Balaam's design, $K = 4$ and $J = 2$. For a two-sequence, three-period (or four-period) design, $K = 2$ and $J = 3$ (or $J = 4$). For a four-sequence, four-period design, $K = J = 4$. Note that, unlike the model (2.5.1) for a standard two-sequence, two-period crossover design, G_k was included in the above model. This is because a higher-order crossover design provides a statistical test for the sequence effect in the presence of the period effect, the direct formulation effect, and the carry-over effect. The test for the sequence effect can be used to examine the success or failure of the randomization.

For assessment of equivalence in average bioavailability, as indicated in Chapter 4, Schuirmann's two one-sided tests procedure, the classical confidence interval approach, and Bayesian methods proposed by Rodda and Davis and by Grieve will essentially reach the same conclusion regarding bioequivalence. Therefore, in this chapter, we will focus on Schuirmann's two one-sided tests procedure and the classical confidence interval approach for assessing bioequivalence under a higher-order crossover design. Since both approaches depend upon the direct formulation effect, F, our primary emphasis will be placed on estimation of F and its variance.

In Sections 9.2 through 9.4, statistical analyses for assessing bioequivalence under Balaam's design, a two-sequence, three-period design and a four-period design with two or four sequences are given. The use of log-transformation (or individual subject ratios) is outlined in Section 9.5. A brief discussion is given in the last section.

9.2. BALAAM'S DESIGN

In this section, we first consider the following four-sequence and two-period design:

Balaam's Design		
	Period	
Sequence	I	II
1	T	T
2	R	R
3	R	T
4	T	R

This design is usually referred to as Balaam's design (Balaam, 1968). It can be seen that if sequences 1 and 2 are omitted, then Balaam's design reduces to the

Table 9.2.1 Expected Values of the Sequence-by-Period Means for Balaam's Design

Sequence	Period	
	I	II
1	$\mu + G_1 + P_1 + F_T$	$\mu + G_1 + P_2 + F_T + C_T$
2	$\mu + G_2 + P_1 + F_R$	$\mu + G_2 + P_2 + F_R + C_R$
3	$\mu + G_3 + P_1 + F_R$	$\mu + G_3 + P_2 + F_T + C_R$
4	$\mu + G_4 + P_1 + F_T$	$\mu + G_4 + P_2 + F_R + C_T$

standard two-sequence, two-period crossover design. Patients in sequences 1 and 2 receive the same formulation twice, either the test formulation or the reference formulation, which allows us to estimate intra-subject variabilities.

Table 9.2.1 gives the expected values of the sequence-by-period means, i.e., $\overline{Y}_{\cdot jk}$, $j = 1,2,3,4$ and $k = 1,2$, where

$$\overline{Y}_{\cdot jk} = \frac{1}{n_k} \sum_{i=1}^{n_k} Y_{ijk}. \qquad (9.2.1)$$

There are 7 degrees of freedoms associated with the eight sequence-by-period means, which can be decomposed as follows:

Source	d.f.
Sequence effect	3
Formulation effect	1
Period effect	1
Carry-over effect	1
Formulation-by-carry-over interaction	1
Total	7

In model (9.1.1), however, we assume there is no formulation by carry-over interaction. Therefore, the one degree of freedom is then combined with that of intra-subject residuals. The analysis of variance table for Balaam's design in terms of degrees of freedom is given in Table 9.2.2. From Table 9.2.2, it can see that Balaam's design allows us to estimate the direct formulation effect in the presence of the carry-over effects. It also provides independent estimates of intra-subject variabilities, σ_T^2 and σ_R^2. Note that, unlike the standard two-

Table 9.2.2 Analysis of Variance Table for
Balaam's Design

Source of variation	Degrees of freedom
Inter-subject	$n_1 + n_2 + n_3 + n_4 - 1$
Sequence	3
Residual	$n_1 + n_2 + n_3 + n_4 - 4$
Intra-subject	$n_1 + n_2 + n_3 + n_4$
Period	1
Formulation	1
Carry-over	1
Residual	$n_1 + n_2 + n_3 + n_4 - 3$
Total	$2(n_1 + n_2 + n_3 + n_4) - 1$

sequence, two-period crossover design, the carry-over effect under Balaam's design is not confounded with the sequence effect.

9.2.1. Analysis of Average Bioavailability

Let d_{ik} be defined as (3.3.1), i.e.,

$$d_{ik} = \tfrac{1}{2}(Y_{i2k} - Y_{i1k}), \qquad i = 1, 2, \ldots, n_k;$$
$$k = 1, 2, 3, 4. \tag{9.2.2}$$

Then, under normality assumptions and $\sigma_T^2 = \sigma_R^2 = \sigma_e^2$, the expected values and variances of the sample means of d_{ik} are given by

$$E(\bar{d}_{\cdot k}) = \begin{cases} \tfrac{1}{2}[(P_2 - P_1) + C_T], & k = 1 \\ \tfrac{1}{2}[(P_2 - P_1) + C_R], & k = 2 \\ \tfrac{1}{2}[(P_2 - P_1) + (F_T - F_R) + C_R], & k = 3 \\ \tfrac{1}{2}[(P_2 - P_1) - (F_T - F_R) + C_T], & k = 4, \end{cases} \tag{9.2.3}$$

and

$$\mathrm{Var}(\bar{d}_{\cdot k}) = \frac{1}{2n_k}\sigma_e^2. \tag{9.2.4}$$

It follows that

$$E(\bar{d}_{\cdot 3} - \bar{d}_{\cdot 4}) = F_T - F_R + \tfrac{1}{2}(C_R - C_T)$$

and

$$E(\bar{d}_{\cdot 2} - \bar{d}_{\cdot 1}) = \tfrac{1}{2}(C_R - C_T).$$

Consequently,

$$E[(\overline{d}._3 - \overline{d}._4) - (\overline{d}._2 - \overline{d}._1)] = F_T - F_R = F.$$

Therefore, the best linear unbiased estimator for the direct formulation effect F after adjustment of the carry-over effect is given by

$$\hat{F}|C = (\overline{d}._3 - \overline{d}._4) - (\overline{d}._2 - \overline{d}._1)$$
$$= \tfrac{1}{2} [(\overline{Y}._{23} - \overline{Y}._{13}) - (\overline{Y}._{24} - \overline{Y}._{14}) - (\overline{Y}._{22} - \overline{Y}._{12}) \qquad (9.2.5)$$
$$+ (\overline{Y}._{21} - \overline{Y}._{11})].$$

Since $\overline{d}._k$, $k = 1,2,3,4$, are independent of each other, the variance of $\hat{F}|C$ can be obtained as follows

$$\text{Var}(\hat{F}|C) = \tfrac{1}{2} \sigma_e^2 \left(\frac{1}{n_1} + \frac{1}{n_2} + \frac{1}{n_3} + \frac{1}{n_4} \right). \qquad (9.2.6)$$

Furthermore, since $F = F_T - F_R$ and $F_T + F_R = 0$, we have

$$F_T = \tfrac{1}{2} F \quad \text{and} \quad F_R = -\tfrac{1}{2} F.$$

Therefore, unbiased estimates for $\mu_T = \mu + F_T$ and $\mu_R = \mu + F_R$ are given by

$$\hat{\mu}_T = \overline{Y}... + \tfrac{1}{2} (\hat{F}|C)$$
$$\hat{\mu}_R = \overline{Y}... - \tfrac{1}{2} (\hat{F}|C), \qquad (9.2.7)$$

where

$$\overline{Y}... = \frac{1}{8} \sum_{k=1}^{4} \sum_{j=1}^{2} \overline{Y}._{jk},$$

which is an unbiased estimator for the overall mean μ. Therefore, the classical $(1 - 2\alpha) \times 100\%$ confidence interval for F or equivalently $\mu_T - \mu_R$, denoted by (L,U), is given by

$$(L,U) = \hat{F}|C \pm t[\alpha, N - 3)] S \sqrt{\frac{1}{2} \left(\frac{1}{n_1} + \frac{1}{n_2} + \frac{1}{n_3} + \frac{1}{n_4} \right)}, \qquad (9.2.8)$$

where $N = n_1 + n_2 + n_3 + n_4$ and S^2 is the intra-subject mean squared error from the analysis of variance table which can be obtained by fitting model (9.1.1) using PROC GLM of SAS.

Thus, we claim the two formulations are bioequivalent if (L,U) is within the bioequivalent limits (θ_L, θ_U).

Similarly, Schuirmann's two one-sided tests procedure can be obtained as follows. We reject the interval hypotheses (4.3.1) and conclude bioequivalence at the α level of significance if

$$T_L = \frac{\hat{F}|C - \theta_L}{S\sqrt{\dfrac{1}{2}\left(\dfrac{1}{n_1} + \dfrac{1}{n_2} + \dfrac{1}{n_3} + \dfrac{1}{n_4}\right)}} > t[\alpha, N - 3],$$

and

$$T_U = \frac{\hat{F}|C - \theta_U}{S\sqrt{\dfrac{1}{2}\left(\dfrac{1}{n_1} + \dfrac{1}{n_2} + \dfrac{1}{n_3} + \dfrac{1}{n_4}\right)}} < - t[\alpha, N - 3]. \tag{9.29}$$

Note that the p-value of Anderson and Hauck's procedure for interval hypotheses (4.3.1) can also be obtained by simply plugging T_L and T_U into either (4.3.12) or (4.3.13).

9.2.2. Inference for the Carry-Over Effect

Since $E[2(\bar{d}._1 - \bar{d}._2)] = C_T - C_R = C$, the following estimator is the best linear unbiased estimator for the carry-over effect C after adjustment of formulation effect:

$$\hat{C}|F = 2(\bar{d}._1 - \bar{d}._2)$$
$$= [(\bar{Y}._{21} - \bar{Y}._{11}) - (\bar{Y}._{22} - \bar{Y}._{12})]. \tag{9.2.10}$$

The variance of $\hat{C}|F$ is given by

$$\text{Var}(\hat{C}|F) = 2 \sigma_e^2 \left(\frac{1}{n_1} + \frac{1}{n_2}\right). \tag{9.2.11}$$

Hence, unlike the standard two-sequence, two-period crossover design, the variance of the estimate of the carry-over effect is a function which only involves the intra-subject variability. Therefore, we reject the null hypothesis (3.2.4) of equal carry-over effects at the α level of significance if

$$|T_C| > t(\alpha/2, n_1 + n_2 + n_3 + n_4 - 3), \tag{9.2.12}$$

where

$$T_C = \frac{\hat{C}|F}{S\sqrt{2\left(\dfrac{1}{n_1} + \dfrac{1}{n_2}\right)}}. \tag{9.2.13}$$

If we fail to reject the null hypothesis H_0: $C_T = C_R$, we may drop the carry-over effect from the model. In this case, the best linear unbiased estimator of F can be obtained based only upon sequence-by-period means from the last two sequences, i.e.,

$$\hat{F} = \bar{d}._3 - \bar{d}._4$$

$$= \tfrac{1}{2} [(\bar{Y}._{23} - \bar{Y}._{13}) - (\bar{Y}._{24} - \bar{Y}._{14})].$$

The variance of $\hat{F}$ under the assumption of equal carry-over effects is given by

$$Var(\hat{F}) = \tfrac{1}{2} \sigma_c^2 \left(\frac{1}{n_3} + \frac{1}{n_4} \right).$$

Therefore, bioequivalence in average bioavailability can be assessed by using either $(1 - 2\alpha) \times 100\%$ confidence interval or Schuirmann's two one-sided tests procedure, which are given below.

$$(L,U) = \hat{F}|C \pm t[\alpha, N - 2)] S \sqrt{\frac{1}{2} \left(\frac{1}{n_3} + \frac{1}{n_4} \right)},$$

$$T_L = \frac{\hat{F} - \theta_L}{S \sqrt{\frac{1}{2} \left(\frac{1}{n_3} + \frac{1}{n_4} \right)}} > t[\alpha, N - 2], \quad \text{and}$$

$$T_U = \frac{\hat{F} - \theta_U}{S \sqrt{\frac{1}{2} \left(\frac{1}{n_1} + \frac{1}{n_2} \right)}} < -t[\alpha, N - 2],$$

where S^2 is the intra-subject mean squared error obtained from the model without the carry-over effect.

9.2.3. The Assessment of Intra-Subject Variabilities

In Section 9.2.1, for assessment of bioequivalence of average bioavailability, we assume the intra-subject variability is the same from formulation to formulation, i.e., $\sigma_T^2 = \sigma_R^2 = \sigma_e^2$. Under the assumption of equal carry-over effects, statistical methods for assessing bioequivalence were derived based upon data from sequences 3 and 4. In practice, the intra-subject variabilities may differ from formulation to formulation, i.e.,

$$Var(e_{ijk}) = \begin{cases} \sigma_T^2 & \text{for the test formulation} \\ \sigma_R^2 & \text{for the reference formulation.} \end{cases} \qquad (9.2.14)$$

In this case, the confidence interval and the two one-sided tests procedure are still valid. This can be seen by the fact that the estimate of the formulation effect F depends only upon the period differences from sequences 3 and 4, which has a variance of $\frac{1}{4}(\sigma_T^2 + \sigma_R^2)$ for both sequences. One advantage of Balaam's design is that, under assumption of equal carry-over effects, data collected from sequences 1 and 2 can be used to assess equivalence of intra-subject variabilities by testing hypotheses (7.2.3) for equivalence in variances.

Let SSD_1 and SSD_2 be sums of squares for period differences of sequences 1 and 2, respectively, i.e.,

$$SSD_k = \sum_{i=1}^{n_k} (d_{ik} - \bar{d}_{\cdot k})^2, \qquad k = 1,2, \tag{9.2.15}$$

where

$$\bar{d}_{\cdot k} = \frac{1}{n_k} \sum_{i=1}^{n_k} d_{ik}, \qquad k = 1,2.$$

Note that SSD_1 and SSD_2 are independent, and

$$2SSD_1 \sim \sigma_T^2 \chi^2(n_1 - 1),$$
$$2SSD_2 \sim \sigma_R^2 \chi^2(n_2 - 1).$$

Therefore,

$$\hat{\sigma}_T^2 = \frac{2SSD_1}{n_1 - 1} \quad \text{and} \quad \hat{\sigma}_R^2 = \frac{2SSD_2}{n_2 - 1}$$

are independent and unbiased estimators of σ_T^2 and σ_R^2, respectively. We reject the interval hypotheses (7.2.3) and conclude equivalence of intra-subject variabilities at the α level of significance if

$$F = \hat{\sigma}_T^2/\hat{\sigma}_R^2 > \delta_1 F(\alpha, n_1 - 1, n_2 - 1),$$

and

$$F = \hat{\sigma}_T^2/\hat{\sigma}_R^2 < \delta_2 F(1 - \alpha, n_1 - 1, n_2 - 1), \tag{9.2.16}$$

where $0 < \delta_1 < 1 < \delta_2$ are lower and upper equivalence limits and $F(a, \nu_1, \nu_2)$ is the upper ath quantile of an F distribution with ν_1 and ν_2 degrees of freedom.

The lower and upper limits of the classical $(1 - 2\alpha) \times 100\%$ confidence interval for σ_T^2/σ_R^2 are given by

$$L = \frac{\hat{\sigma}_T^2/\hat{\sigma}_R^2}{F(\alpha, n_1 - 1, n_2 - 1)}, \quad \text{and}$$
$$U = (\hat{\sigma}_T^2/\hat{\sigma}_R^2)F(\alpha, n_2 - 1, n_1 - 1). \tag{9.2.17}$$

Since $F(1 - \alpha, n_1 - 1, n_2 - 1) = [F(\alpha, n_2 - 1, n_1 - 1)]^{-1}$, (L, U) is within (δ_1, δ_2) if and only if (9.2.16) holds. Thus, the confidence interval approach is operationally equivalent to the two one-sided tests procedure for determination of equivalence of intra-subject variabilities.

Example 9.2.1 To illustrate Balaam's design, we consider AUC data given in Table 9.2.3. This study was designed to compare a test formulation with a reference formulation using Balaam's design with 24 normal subjects, who were assigned to each of the four sequences (TT, RR, RT, and TR) at random. The sequence-by-period means are given in Table 9.2.4. The intra-subject mean squared error S^2 from the analysis of variance table, which is obtained by fitting model (9.1.1) using PROC GLM of SAS, is 3827.79. Table 9.2.5 summarizes the results of Schuirmann's two one-sided tests procedure and the 90% confidence

Table 9.2.3 AUC Data for Balaam's Design

Sequence	Subject	Period I	Period II
1: TT	1	280	482
	2	219	161
	3	230	99
	4	229	260
	5	494	274
	6	112	171
2: RR	7	205	221
	8	349	420
	9	285	288
	10	266	247
	11	161	175
	12	240	248
3: RT	13	325	225
	14	374	439
	15	416	372
	16	243	119
	17	248	269
	18	345	334
4: TR	19	177	290
	20	174	224
	21	235	271
	22	380	340
	23	308	270
	24	269	249

Table 9.2.4 Sequence-by-
Period Means for AUC Data in
Table 9.2.3

Sequence	Period	
	I	II
1	260.67	241.17
2	251.00	266.50
3	325.17	293.00
4	257.17	274.00

$S^2 = 16.67$.

interval based upon the ± 20 rule with $\hat{\mu}_R$ assumed to be the true μ_R. In the presence of the carry-over effect, from (9.2.5), we have

$$\hat{\mu} = \overline{Y}... = 271.08 \quad \text{and} \quad \hat{F}|C = -42.0.$$

Thus, estimates for the test and reference formulations means are given by

$$\hat{\mu}_T = 250.08 \quad \text{and} \quad \hat{\mu}_R = 292.08.$$

The estimate of the carry-over effect, $\hat{C}|F$ is 35 with a standard error of 50.52 which leads to a p-value of 0.496. Hence, we conclude there is no carry-over effect. Furthermore, it can be verified that estimates of F, μ_T, and μ_R in the absence of unequal carry-over effects are -24.5, 258.83, and 283.33, respectively. From Table 9.2.5, it can be seen that the interval approach and the two one-sided tests procedure lead to the same conclusion of bioinequivalence regardless of the presence of the carry-over effect.

Table 9.2.5 Summary of Results for AUC Data in Table 9.2.3

Carry-over	$\hat{F}$	Test statistic[a]		P-Value	90% C.I.[b]
Yes	-42.0	$T_L =$	0.46	0.325	$(-103.5, 19.5)$
		$T_U =$	-2.81	0.005	$(64.6\%, 106.7\%)$
No	-24.5	$T_L =$	1.29	0.105	$(-67.4, 18.4)$
		$T_U =$	-3.25	0.002	$(76.2\%, 106.5\%)$

[a]The ± 20 rule was applied by assuming $\mu_R = \hat{\mu}_R = 292.08$.
[b]C.I. = confidence interval which was expressed in terms of $\hat{\mu}_R = \hat{\mu} - \frac{1}{2}\hat{F} = 292.08$ in presence of carry-over effect, and of $\hat{\mu}_R = 283.3$ in absence of carry-over effect.

For testing the equivalence of intra-subject variabilities, estimates of σ_T^2 and σ_R^2 are given by 11188.94 and 448.55. The 90% confidence interval for σ_T^2/σ_R^2 based upon (9.2.17) is (494%, 12597%) which indicates that the two formulations are not equivalent in intra-subject variabilities.

9.3. THE TWO-SEQUENCE DUAL DESIGN

In this section, we consider the following two-sequence, three-period design, which consists of the dual sequences TRR and RTT.

The Two-Sequence Dual Design		Period	
Sequence	I	II	III
1	T	R	R
2	R	T	T

This design is an extra-period design, which is also known as the two-sequence dual design. The two-sequence dual design is balanced in the sense that each formulation follows other formulations, including itself, the same number of times. One advantage of this design is that it allows us to estimate the intra-subject variabilities because each subject receives either the test formulation or the reference formulation twice.

Table 9.3.1 provides the expected values of the sequence-by-period means. The analysis of variance table in terms of degrees of freedom is given in Table

Table 9.3.1 Expected Values of the Sequence-by-Period Means for the Two-Sequence Dual Design

		Period	
Sequence	I	II	III
1	$\mu + G_1 + P_1 + F_T$	$\mu + G_1 + P_2 + F_R + C_T$	$\mu + G_1 + P_3 + F_R + C_R$
2	$\mu + G_2 + P_1 + F_R$	$\mu + G_2 + P_2 + F_T + C_R$	$\mu + G_2 + P_3 + F_T + C_T$

Table 9.3.2   Analysis of Variance Table
for the Two-Sequence Dual Design

Source of variation	Degrees of freedom
Inter-subject	$n_1 + n_2 - 1$
Sequence	1
Residual	$n_1 + n_2 - 2$
Intra-subject	$2(n_1 + n_2)$
Period	2
Formulation	1
Carry-over	1
Residual	$2(n_1 + n_2 - 2)$
Total	$3(n_1 + n_2) - 1$

9.3.2. There are a total of six sequence-by-period means with five degrees of freedom, which can be decomposed as follows:

Source	d.f.
Sequence effect	1
Formulation effect	1
Period effect	2
Carry-over effect	1
Total	5

Therefore, this design allows us to estimate the direct formulation effect in the presence of the carry-over effect. In addition, statistical inference for the carry-over effect can be obtained based upon estimates of intra-subject variabilities. Under normality assumptions of (8.1.2) for S_{ik} and e_{ijk}, the two-sequence dual design design is the optimal design for estimation of the direct formulation effect and the carry-over effect among all two-sequence and three-period designs (see e.g., Cheng and Wu, 1980; Kershner and Federer, 1981; Laska, Meisner and Kushner, 1983; Jones and Kenward, 1989; Lasserre, 1991).

9.3.1. Analysis of Average Bioavailability

Under the two-sequence dual design, like the standard two-sequence, two-period design, the classical confidence interval for the direct formulation effect F and

Schuirmann's two one-sided tests procedure for assessing bioequivalence can be derived similarly.

Let $Y_{ik} = (Y_{i1k}, Y_{i2k}, Y_{i3k})'$ be the vector of the three responses observed on subject i in sequence k. The assumption of compound symmetry requires that the covariance matrix of Y_{ik} has the following structure:

$$\text{Var } (Y_{ik}) = \begin{bmatrix} \sigma_e^2 + \sigma_s^2 & \sigma_s^2 & \sigma_s^2 \\ \sigma_s^2 & \sigma_e^2 + \sigma_s^2 & \sigma_s^2 \\ \sigma_s^2 & \sigma_s^2 & \sigma_e^2 + \sigma_s^2 \end{bmatrix}. \tag{9.3.1}$$

Then, under the normality assumptions stated in (8.1.2), the best linear unbiased estimator for the direct formulation effect, $F = F_T - F_R$, is the following linear contrast of sequence-by-period means (Kershner and Federer, 1981; Jones and Kenward, 1989):

$$\hat{F} = \tfrac{1}{4} [(2\overline{Y}_{\cdot 11} - \overline{Y}_{\cdot 21} - \overline{Y}_{\cdot 31}) - (2\overline{Y}_{\cdot 12} - \overline{Y}_{\cdot 22} - \overline{Y}_{\cdot 32})]. \tag{9.3.2}$$

The expected value and variance of $\hat{F}$ are given by

$$E(\hat{F}) = F_T - F_R, \tag{9.3.3}$$

and

$$\text{Var}(\hat{F}) = \frac{3}{8} \sigma_e^2 \left(\frac{1}{n_1} + \frac{1}{n_2} \right), \tag{9.3.4}$$

Furthermore, an unbiased estimator for the overall mean μ can be obtained as

$$\overline{Y}\ldots = \frac{1}{6} \sum_{k=1}^{2} \sum_{j=1}^{3} \overline{Y}_{\cdot jk}. \tag{9.3.5}$$

Since $F = F_T - F_R$ and $F_T + F_R = 0$, we have

$$F_T = \tfrac{1}{2} F, \quad \text{and} \quad F_R = -\tfrac{1}{2} F.$$

Therefore, unbiased estimates for $\mu_T = \mu + F_T$ and $\mu_R = \mu + F_R$ are given by

$$\hat{\mu}_T = \overline{Y}\ldots + \tfrac{1}{2} \hat{F},$$

and

$$\hat{\mu}_R = \overline{Y}\ldots - \tfrac{1}{2} \hat{F}. \tag{9.3.6}$$

The classical $(1 - 2\alpha) \times 100\%$ confidence interval for F, denoted by (L,U), are given by

$$(L, U) = \hat{F} \pm t[\alpha, 2(n_1 + n_2 - 2)] S \sqrt{\left(\frac{3}{8}\right)\left(\frac{1}{n_1} + \frac{1}{n_2}\right)}. \tag{9.3.7}$$

As a result, we conclude the two formulations are bioequivalent if the interval (L,U) is within the bioequivalent limits, (θ_L, θ_U).

Similarly, Schuirmann's two one-sided tests procedure can be obtained based upon the following two t statistics. We reject interval hypotheses (4.3.1) and conclude bioequivalence at the α level of significance if

$$T_L = \frac{\hat{F} - \theta_L}{S\sqrt{\left(\frac{3}{8}\right)\left(\frac{1}{n_1} + \frac{1}{n_2}\right)}} > t[\alpha, 2(n_1 + n_2 - 2)],$$

and

$$T_U = \frac{\hat{F} - \theta_U}{S\sqrt{\left(\frac{3}{8}\right)\left(\frac{1}{n_1} + \frac{1}{n_2}\right)}} < -t[\alpha, 2(n_1 + n_2 - 2)]. \qquad (9.3.8)$$

Note that, the p-value of Anderson and Hauck's procedure for interval hypotheses (4.3.1) can be obtained by simply plugging T_L and T_U into either (4.3.12) or 4.3.13).

9.3.2. Inference for the Carry-Over Effect

Jones and Kenward (1989) indicated that, under normality assumptions of (8.1.2), the best linear unbiased estimator for the carry-over effect $C = C_T - C_R$ can be obtained by the following linear contrast of sequence-by-period means (also see Kershner and Federer, 1981):

$$\hat{C} = \tfrac{1}{2} [(\overline{Y}_{.21} - \overline{Y}_{.31}) - (\overline{Y}_{.22} - \overline{Y}_{.32})]. \qquad (9.3.9)$$

Under the two-sequence dual design, it can be verified that the cross products of the coefficients in linear contrasts between $\hat{F}$ in (9.3.3) and $\hat{C}$ in (9.3.9) is 0. This indicates that $\hat{C}$ is independent of $\hat{F}$. Therefore, $\hat{F}$ remains the same regardless of presence or absence of unequal carry-over effects. The variance of $\hat{C}$ is given by

$$\text{Var}(\hat{C}) = \tfrac{1}{2} \sigma_e^2 \left(\frac{1}{n_1} + \frac{1}{n_2}\right). \qquad (9.3.10)$$

Based upon (9.3.9) and (9.3.10), we reject the null hypothesis (3.2.4) of equal carry-over effects at the α level of significance if

$$|T_C| > t[\alpha/2, 2(n_1 + n_2 - 2)], \qquad (9.3.11)$$

where

$$T_C = \frac{\hat{C}}{S\sqrt{\frac{1}{2}\left(\frac{1}{n_1} + \frac{1}{n_2}\right)}}.$$

9.3.3. Intra-Subject Contrasts

When the covariance of Y_{ik} does not follow the structure of compound symmetry defined in (9.3.1), we can still assess bioequivalence of average bioavailability based upon a two-sample t test provided that $Var(Y_{i1}) = Var(Y_{i2})$. When the assumption of compound symmetry of $Var(Y_{ik})$ is violated, the model in (9.1.1) becomes

$$Y_{ijk} = \mu + G_k + P_j + F_{(j,k)} + C_{(j-1,k)} + e^*_{ijk}, \tag{9.3.12}$$

where $e^*_{ik} = (e^*_{i1k}, e^*_{i2k}, e^*_{i3k})'$ is distributed with mean O and covariance matrix Σ for $i = 1,2, \ldots, n_k; k = 1,2$.

Define the following intra-subject contrasts

$$d_{ik} = \tfrac{1}{4}[2Y_{i1k} - Y_{i2k} - Y_{i3k}], \qquad i = 1,2, \ldots ,n_k; \quad k = 1,2. \tag{9.3.13}$$

The expected values and variance of d_{ik} are given by

$$E(d_{ik}) = \begin{cases} \tfrac{1}{2}(F_T - F_R) + \tfrac{3}{4}P_1 & k = 1 \\ -\tfrac{1}{2}(F_T - F_R) + \tfrac{3}{4}P_1 & k = 2, \end{cases} \tag{9.3.14}$$

$$Var(d_{ik}) = \sigma_d^2 = c'\Sigma\, c, \tag{9.3.15}$$

where $c' = (0.5, -0.25, -0.25)$ and Σ is the common covariance matrix of Y_{ik}.

Because $\{d_{i1}\}$ and $\{d_{i2}\}$ are two independent samples with common variance σ_d^2, the two one-sided tests procedure and the nonparametric methods discussed in Chapter 4 can be directly applied to assess bioequivalence of average bioavailability. Let $\bar{d}_{\cdot k}$ and SSD be the sample means and pooled sum of squares of d_{ik}, where

$$\bar{d}_{\cdot k} = \frac{1}{n_k} \sum_{i=1}^{n_k} d_{ik},$$

and

$$SSD = \sum_{k=1}^{2} \sum_{i=1}^{n_k} (d_{ik} - \bar{d}_{\cdot k})^2. \tag{9.3.16}$$

Then, unbiased estimates for F and σ_d^2 can be obtained as

$$\hat{F} = \bar{d}._1 - \bar{d}._2,$$

and

$$\hat{\sigma}_d^2 = \frac{SSD}{(n_1 + n_2 - 2)}. \tag{9.3.17}$$

Note that, although $\hat{F}$ obtained from the difference in sample means of intra-subject contrasts between sequences 1 and 2 is the same as that of (9.3.2), $\hat{\sigma}_d^2$ is not the same as S^2 obtained from the analysis of variance table. Moreover, the associated degrees of freedom for $\hat{\sigma}_d^2$ is $(n_1 + n_2 - 2)$ rather than $2(n_1 + n_2 - 2)$. The loss of degrees of freedom is the cost for relaxation of the assumption of compound symmetry for covariance matrix of $\mathbf{Y}_{ik}$.

Based upon the intra-subject contrasts, d_{ik}, $i = 1,2, \ldots ,n_k$; $k = 1,2$, a $(1 - 2\alpha) \times 100\%$ confidence interval for F, denoted by (L_d, U_d), is given by

$$(L_d, U_d) = \bar{d} \pm t(\alpha, n_1 + n_2 - 2)\, \hat{\sigma}_d \sqrt{\frac{1}{n_1} + \frac{1}{n_2}}. \tag{9.3.18}$$

Therefore, we conclude bioequivalence if (L_d, U_d) is within the bioequivalent limits, (θ_L, θ_U). For Schuirmann's two one-sided tests procedure, similarly, we would reject the interval hypotheses of (4.3.1) at the α level if

$$T_L = \frac{\hat{F} - \theta_L}{\hat{\sigma}_d \sqrt{\frac{1}{n_1} + \frac{1}{n_2}}} > t(\alpha, n_1 + n_2 - 2),$$

and

$$T_U = \frac{\hat{F} - \theta_U}{\hat{\sigma}_d \sqrt{\frac{1}{n_1} + \frac{1}{n_2}}} < -t(\alpha, n_1 + n_2 - 2). \tag{9.3.19}$$

Similarly, the p-value of Anderson and Hauck's procedure can also be obtained by plugging T_L and T_U into either (4.3.12) or (4.3.13).

When the normality assumptions of (8.1.2) are seriously in doubt, nonparametric methods described in Section 4.5.1 can be applied directly based upon the intra-subject contrasts, d_{ik}, $i = 1,2, \ldots ,n_k$; $k = 1,2$. For example, for the Wilcoxon-Mann-Whitney two one-sided tests procedure, we simply plug the intra-subject contrasts into b_{hik} in (4.5.2), where $i = 1,2, \ldots ,n_k$; $k = 1,2$; $h = L, U$. To construct a distribution-free confidence interval based upon Hodges-Lehmann's estimator discussed in Section 4.5.2, we simply plug d_{ik} into the computation of all pairwise differences of $D_{i,i'}$, which are defined in Section

4.5.2. However, it should be noted that the distribution-free confidence interval approach is equivalent to the Wilcoxon-Mann-Whitney two one-sided tests procedure.

9.3.4. The Assessment of Intra-Subject Variabilities

As mentioned earlier, in practice, the intra-subject variabilities may differ from formulation to formulation. In this case, we have

$$
\text{Var}(e_{ijk}) = \begin{cases} \sigma_T^2 & \text{if } j = 1 \text{ and } k = 1, \text{ or } j = 2,3 \text{ and } k = 2 \\ \sigma_R^2 & \text{if } j = 1 \text{ and } k = 2, \text{ or } j = 2,3 \text{ and } k = 1. \end{cases}
$$

(9.3.20)

If S_{ik} are i.i.d. with mean 0 and variance σ_s^2, then covariance matrices of Y_{ik} for sequence 1 and sequence 2 are given by

$$
\text{Var}(Y_{i1}) = \begin{bmatrix} \sigma_T^2 + \sigma_s^2 & \sigma_s^2 & \sigma_s^2 \\ \sigma_s^2 & \sigma_R^2 + \sigma_s^2 & \sigma_s^2 \\ \sigma_s^2 & \sigma_s^2 & \sigma_R^2 + \sigma_s^2 \end{bmatrix},
$$

and

$$
\text{Var}(Y_{i2}) = \begin{bmatrix} \sigma_R^2 + \sigma_s^2 & \sigma_s^2 & \sigma_s^2 \\ \sigma_s^2 & \sigma_T^2 + \sigma_s^2 & \sigma_s^2 \\ \sigma_s^2 & \sigma_s^2 & \sigma_T^2 + \sigma_s^2 \end{bmatrix}.
$$

(9.3.21)

Hence,

$$
\text{Var}(d_{ik}) = \begin{cases} \frac{1}{8}(2\sigma_T^2 + \sigma_R^2) & \text{if } k = 1 \\ \frac{1}{8}(\sigma_T^2 + 2\sigma_R^2) & \text{if } k = 2. \end{cases}
$$

(9.3.22)

As a result, if $\sigma_T^2 \neq \sigma_R^2$, the variances of d_{i1} and d_{i2} are not the same. Consequently, the procedures discussed in the previous section based upon either the two-sample t test or Wilcoxon-Mann-Whitney rank sum test are no longer valid. Therefore, unlike the standard two-sequence, two-period crossover design, the two-sequence dual design requires a much stronger assumption of $\sigma_T^2 = \sigma_R^2 = \sigma_e^2$. This assumption is not generally true. However, it can be examined by testing the following hypotheses:

$$
H_0: \sigma_T^2 = \sigma_R^2 \quad \text{vs} \quad H_a: \sigma_T^2 \neq \sigma_R^2
$$

(9.3.23)

Define the following intra-subject contrasts

$$
g_{ik} = (Y_{i2k} - Y_{i3k}), \quad i = 1,2, \ldots ,n_k; \quad k = 1,2.
$$

(9.3.24)

Then, under covariance structure of (9.3.21), unbiased estimates for σ_T^2 and σ_R^2 can be obtained as follows:

$$\hat{\sigma}_T^2 = \frac{1}{2(n_1 - 1)} \sum_{i=1}^{n_1} (g_{i1} - \bar{g}_{\cdot 1})^2,$$

and

$$\hat{\sigma}_R^2 = \frac{1}{2(n_2 - 1)} \sum_{i=1}^{n_2} (g_{i2} - \bar{g}_{\cdot 2})^2, \qquad (9.3.25)$$

where

$$\bar{g}_{\cdot k} = \frac{1}{n_k} \sum_{i=1}^{n_k} g_{ik}.$$

Let $\hat{\sigma}_1^2 = \max(\hat{\sigma}_T^2, \hat{\sigma}_R^2)$, $\hat{\sigma}_2^2 = \min(\hat{\sigma}_T^2, \hat{\sigma}_R^2)$ and ν_1 and ν_2 be the corresponding degrees of freedoms. We then reject H_0 of (9.3.23) at the α level if

$$F_v = \hat{\sigma}_1^2/\hat{\sigma}_2^2 > F(\alpha/2, \nu_1, \nu_2). \qquad (9.3.26)$$

A $(1 - \alpha) \times 100\%$ confidence interval for σ_T^2/σ_R^2, denoted by, (L_F, U_F), can then be obtained, where

$$L_F = \frac{\hat{\sigma}_T^2/\hat{\sigma}_R^2}{F(\alpha/2, n_1 - 1, n_2 - 1)},$$

and

$$U_F = (\hat{\sigma}_T^2/\hat{\sigma}_R^2) \, F(\alpha/2, n_2 - 1, n_1 - 1). \qquad (9.3.27)$$

Note that, under normality assumptions, the intra-subject contrasts d_{ik} are independent of g_{ik}. It follows that $\hat{\sigma}_T^2$ and $\hat{\sigma}_R^2$ are not only independent of each other but also independent of $\hat{F}$ and S^2 given in (9.3.8). Consequently, test statistics T_L and T_U in (9.3.8) are independent of F_v.

As indicated earlier, for the standard two-sequence, two-period crossover design, Schuirmann's two one-sided tests procedure is still valid when the intra-subject variabilities differ from formulation to formulation. This, however, is not true for the two-sequence dual design. Therefore, it is recommended that a preliminary test of hypotheses (9.3.23) for equality of intra-subject variabilities be carried out before the assessment of bioequivalence is performed.

Example 9.3.1   To illustrate the methods discussed in this section, consider the example of a two-sequence dual crossover experiment which was conducted with 18 subjects to compare two formulations of a drug product. In the design, nine subjects were randomly assigned to sequence 1 (TRR) and nine other subjects were randomly assigned to sequence 2 (RTT). Table 9.3.3 lists AUC data for each subject. The sequence-by-period means are given in Table 9.3.4. The intra-

Table 9.3.3 AUC Data and Intra-Subject Contrasts

Sequence	Subject	Period I	II	III	d_{ik}	$\frac{1}{2}g_{ik}$
1: TRR	2	32.14	42.55	36.08	− 3.5875	3.235
	5	51.85	47.37	56.50	− 0.0425	− 4.565
	6	34.28	30.70	33.60	1.0650	− 1.450
	8	27.48	27.87	33.54	− 1.6125	− 2.835
	10	17.32	18.93	19.05	− 0.8350	− 0.060
	11	31.61	24.93	28.21	2.5200	− 1.640
	13	39.26	37.36	45.09	− 0.9825	− 3.865
	17	47.55	40.00	36.00	4.7750	2.000
	18	36.15	35.20	33.26	0.9600	0.970
2: RTT	1	26.79	27.23	31.60	− 1.3125	− 2.185
	3	26.71	30.81	26.39	− 0.9450	2.210
	4	28.70	40.74	30.49	− 3.4575	5.125
	7	35.01	45.85	43.86	− 4.9225	0.995
	9	30.98	34.57	30.18	− 0.6975	2.195
	12	57.70	54.39	44.35	4.1650	5.020
	14	27.87	30.02	27.00	− 0.3200	1.510
	15	31.03	25.77	29.65	1.6600	− 1.940
	16	27.67	21.61	25.64	2.0225	− 2.015

subject mean squared error, $S^2 = 16.67$ which is obtained by fitting model (9.1.1) using PROC GLM of SAS. From (9.3.2) and (9.3.5), we have

$$\hat{F} = 0.674 \quad \text{and} \quad \hat{\mu} = \overline{Y}... = 34.009$$

Hence, estimates for the test and reference formulation means are given by

$$\hat{\mu}_T = \hat{\mu} + \tfrac{1}{2}\hat{F} = 34.346,$$

and

$$\hat{\mu}_R = \hat{\mu} - \tfrac{1}{2}\hat{F} = 33.672.$$

The intra-subject contrasts d_{ik} and $\frac{1}{2}g_{ik}$ are given in Table 9.3.3. It can be verified that $\hat{\sigma}_d^2$, computed from the intra-subject contrasts d_{ik}, is 6.86. Table 9.3.5 summarizes the results of Schuirmann's two one-sided tests procedures as well as the confidence interval based upon the ± 20 rule with μ_R assumed to be $\hat{\mu}_R = 33.672$. The results indicate that all procedures reach the same conclusion regarding bioequivalence of average bioavailability.

Table 9.3.4 Sequence-by-Period Means for
AUC Data in Table 9.3.3

Sequence	Period		
	I	II	III
1	35.29	33.88	35.70
2	32.50	34.55	32.13

$S^2 = 16.67$.

For testing the equality of intra-subject variabilities, estimates of σ_T^2 and σ_R^2 are obtained based upon the intra-subject contrasts g_{ik} as follows:

$$\hat{\sigma}_T = 14.13 \quad \text{and} \quad \hat{\sigma}_R^2 = 15.94.$$

Hence, $F_v = 1.13$ with a p-value of 0.87. Therefore, we fail to reject the null hypothesis of equal intra-subject variability at the 5% level of significance. Since $F(0.025,9,9) = 4.03$, the lower and upper limits of the 95% confidence interval for σ_T^2/σ_R^2 are given by

$$L_F = \frac{14.13/15.94}{4.03} = 0.22, \quad \text{and}$$

$$U_F = (14.13/15.94)(4.03) = 3.57.$$

Table 9.3.5 Summary of Results for AUC Data in Table 9.3.2

Method and assumption	Test statistic[a]		P-Value	90% C.I.
Two one-sided tests	$T_L =$	6.29	<0.0001	$(-1.32, 2.67)$
Compound symmetry	$T_U =$	-5.14	<0.0001	$(96.07\%, 107.93\%)$[b]
Two one-sided tests	$T_L =$	6.00	<0.0001	$(-1.48, 2.83)$
Var $(Y_{ik}) = \Sigma$	$T_U =$	-4.91	<0.0001	$(95.60\%, 108.40\%)$[b]
Wilcoxon-Mann-Whitney				
Two one-sided tests	$W_L =$	80	<0.05	$(-1.65, 2.84)$
Var $(Y_{ik}) = \Sigma$	$W_U =$	3	<0.05	$(95.12\%, 108.43\%)$[b]

[a]The ± 20 rule was applied by assuming $\mu_R = \hat{\mu}_R = 33.67$.
[b]C.I. = confidence interval which was expressed in terms of $\hat{\mu}_R = \hat{\mu} - \frac{1}{2}\hat{F} = 33.67$.

9.4. OPTIMAL FOUR-PERIOD DESIGNS

In this section, we will focus on estimation of the direct formulation effect for two four-period designs, namely the two-sequence, four-period design and the four-sequence, four-period design. We will derive statistical methods for assessment of bioequivalence of average bioavailability under the general model (9.1.1) with normality assumptions of (8.1.2) for each of these four-period designs.

9.4.1. The Two-Sequence, Four-Period Design

The two-sequence, four-period design, which is made of sequences TRRT and RTTR, is summarized as follows:

The Two-Sequence, Four-Period Design			
	Period		
Sequence	I	II	III	IV
1	T	R	R	T
2	R	T	T	R

It can be seen that if the last period is omitted, the two-sequence, four-period design reduces to the two-sequence dual design.

Table 9.4.1 lists expected values of the sequence-by-period means. The analysis of variance table in terms of degrees of freedoms is given in Table 9.4.2. Like the two-sequence dual design, the two-sequence, four-period design can be used to assess bioequivalence of average bioavailability in the presence of the carry-over effect. Moreover, it can also provide estimates of intra-subject variabilities.

Under the assumption of compound symmetry of $\text{Var}(\mathbf{Y}_{ik})$ and normality

Table 9.4.1 Expected Values of the Sequence-by-Period Means for the Two-Sequence, Four-Period Design

Sequence	Period			
	I	II	III	IV
1	$\mu + G_1 + P_1$ $+ F_T$	$\mu + G_1 + P_2$ $+ F_R + C_T$	$\mu + G_1 + P_3$ $+ F_R + C_R$	$\mu + G_1 + P_4$ $+ F_T + C_R$
2	$\mu + G_2 + P_1$ $+ F_R$	$\mu + G_2 + P_2$ $+ F_T + C_R$	$\mu + G_2 + P_3$ $+ F_T + C_T$	$\mu + G_2 + P_4$ $+ F_R + C_T$

Table 9.4.2   Analysis of Variance Table
for the Two-Sequence, Four-Period Design

Source of variation	Degrees of freedom
Inter-subject	$n_1 + n_2 - 1$
Sequence	1
Residual	$n_1 + n_2 - 2$
Intra-subject	$3(n_1 + n_2)$
Period	3
Formulation	1
Carry-over	1
Residual	$3(n_1 + n_2) - 5$
Total	$4(n_1 + n_2) - 1$

assumptions of (8.1.2), the best linear unbiased estimator for the formulation
effect after adjustment for the carry-over effect is given by

$$\hat{F}|C = \tfrac{1}{20}\,[(6\overline{Y}._{11} - 3\overline{Y}._{21} - 7\overline{Y}._{31} + 4\overline{Y}._{41}) \tag{9.4.1}$$
$$- (6\overline{Y}._{12} - 3\overline{Y}._{22} - 7\overline{Y}._{32} + 4\overline{Y}._{42})].$$

Based upon (9.4.1), it can be verified that the variance of $\hat{F}|C$ is given by

$$\text{Var}\,(\hat{F}|C) = \frac{11}{40}\,\sigma_e^2 \left(\frac{1}{n_1} + \frac{1}{n_2}\right). \tag{9.4.2}$$

Since $\overline{Y}...$ is an unbiased estimator of the overall mean μ, unbiased estimates
for μ_T and μ_R can be obtained as follows:

$$\hat{\mu}_T = \overline{Y}... + \tfrac{1}{2}\,(\hat{F}|C),$$

and

$$\hat{\mu}_R = \overline{Y}... - \tfrac{1}{2}\,(F|C).$$

The classical $(1 - 2\alpha) \times 100\%$ confidence interval is then given by

$$(L,U) = \hat{F}|C \pm t[\alpha, 3(n_1 + n_2) - 5]\, S \sqrt{\frac{11}{40} \left(\frac{1}{n_1} + \frac{1}{n_2}\right)}. \tag{9.4.3}$$

Therefore, we conclude bioequivalence if (L,U) is within the bioequivalent limits
(θ_L, θ_U).

Similarly, a two one-sided tests procedure can be obtained from the two t statistics below. We reject the interval hypotheses (4.3.1) and conclude bioequivalence if

$$T_L = \frac{\hat{F}|C - \theta_L}{S\sqrt{\frac{11}{40}\left(\frac{1}{n_1} + \frac{1}{n_2}\right)}} > t[\alpha, 3(n_1 + n_2) - 5],$$

and

$$T_U = \frac{\hat{F}|C - \theta_U}{S\sqrt{\frac{11}{40}\left(\frac{1}{n_1} + \frac{1}{n_2}\right)}} < -t[\alpha, 3(n_1 + n_2) - 5]. \tag{9.4.4}$$

Note that if the covariance matrix of the four responses observed on a subject does not follows the structure of compound symmetry but is the same for all subjects in both sequences, then the method derived for the two-sequence dual design in the previous section can also be applied. Define the following intra-subject contrasts:

$$d_{ik} = \tfrac{1}{20}[6Y_{i1k} - 3Y_{i2k} - 7Y_{i3k} + 4Y_{i4k}],$$
$$i = 1, 2, \ldots, n_k; \quad k = 1, 2. \tag{9.4.5}$$

Then, $\bar{d} = \bar{d}_{.1} - \bar{d}_{.2}$ is an unbiased estimator of F. Therefore, the confidence interval in (9.4.3) and the two one-sided tests procedure in (9.4.4) can be easily carried out by substituting $\hat{F}|C$, S, and $t(\alpha, 3(n_1 + n_2) - 5)$ with $\bar{d}$, $\hat{\sigma}_d$, and $t(\alpha, n_1 + n_2 - 2)$, where

$$\hat{\sigma}_d^2 = \frac{1}{n_1 + n_2 - 2} \sum_{k=1}^{2} \sum_{i=1}^{n_k} (d_{ik} - \bar{d}_{.k})^2$$

with $n_1 + n_2 - 2$ degrees of freedom.

When the normality assumptions in (8.1.2) are questionable, nonparametric methods based upon d_{ik} can be directly applied to obtain the Wilcoxon-Mann-Whitney two one-sided test statistics and a distribution-free confidence interval as described in Section 9.3.3.

It should be noted that if the intra-subject variability differs from formulation to formulation, statistical inference for F may not be valid. This can be seen from the fact that

$$\text{Var}(d_{ik}) = \begin{cases} \frac{1}{400}[52\sigma_T^2 + 58\sigma_R^2], & \text{if } k = 1 \\ \frac{1}{400}[52\sigma_R^2 + 58\sigma_T^2], & \text{if } k = 2. \end{cases} \tag{9.4.6}$$

In other words, random samples $\{d_{1k}\}$ and $\{d_{2k}\}$ do not have the same variance; though, the difference in variances is rather small.

As indicated earlier, the two-sequence, four-period design allows us to assess the carry-over effect. The best linear unbiased estimator for the carry-over effect can be obtained by the following linear contrasts of sequence-by-period means

$$\hat{C}|F = \tfrac{1}{5} [(\overline{Y}_{\cdot 11} + 2\overline{Y}_{\cdot 21} - 2\overline{Y}_{\cdot 31} - \overline{Y}_{\cdot 41})$$
$$- (\overline{Y}_{\cdot 12} + 2\overline{Y}_{\cdot 22} - 2\overline{Y}_{\cdot 32} - \overline{Y}_{\cdot 42})] \qquad (9.4.7)$$

The variance of $\hat{C}|F$ is then given by

$$\text{Var}(\hat{C}|F) = \frac{2}{5} \sigma_e^2 \left(\frac{1}{n_1} + \frac{1}{n_2} \right). \qquad (9.4.8)$$

Therefore, the null hypothesis of equal carry-over effects in (3.2.4) is rejected at the α level of significance if

$$|T_C| > t[\alpha/2, 3(n_1 + n_2) - 5], \qquad (9.4.9)$$

where

$$T_C = \frac{\hat{C}|F}{S\sqrt{\dfrac{2}{5} \left(\dfrac{1}{n_1} + \dfrac{1}{n_2} \right)}}. \qquad (9.4.10)$$

If we fail to reject the null hypothesis H_0: $C_T = C_R$, we might drop the carry-over effect from the model. In this case, the best linear unbiased estimator for F can be expressed by the following linear contrasts of sequence-by-period means

$$\hat{F} = \tfrac{1}{4} [(\overline{Y}_{\cdot 11} - \overline{Y}_{\cdot 21} - \overline{Y}_{\cdot 31} + \overline{Y}_{\cdot 41})$$
$$- (\overline{Y}_{\cdot 12} - \overline{Y}_{\cdot 22} - \overline{Y}_{\cdot 32} + \overline{Y}_{\cdot 42})]. \qquad (9.4.11)$$

The variance of $\hat{F}$ under the assumption of equal carry-over effects is given by

$$\text{Var}(\hat{F}) = \frac{1}{4} \sigma_e^2 \left(\frac{1}{n_1} + \frac{1}{n_2} \right). \qquad (9.4.12)$$

Note that the precision of $\hat{F}$ improves only by a very small fraction compared to that of $\hat{F}|C$. The associated degrees of freedom using $\hat{F}$ for evaluation of bioequivalence based upon either the confidence interval or the two one-sided tests procedure is $3(n_1 + n_2) - 4$.

In absence of unequal carry-over effects, the confidence interval and the two one-sided tests procedure are still valid even when $\sigma_T^2 \neq \sigma_R^2$. To demonstrate this, we note that $\hat{F}$ is the difference of the sample means of the following intra-subject contrasts between sequences 1 and 2

$$d_{ik} = \tfrac{1}{4} [Y_{i1k} - Y_{i2k} - Y_{i3k} + Y_{i4k}], \qquad i = 1, 2, \dots, n_k; \; k = 1, 2. \qquad (9.4.13)$$

It can be verified that the variance of d_{ik} is given by

$$\text{Var}(d_{ik}) = \tfrac{1}{8} (\sigma_T^2 + \sigma_R^2),\qquad(9.4.14)$$

which is the same for both sequences.

In addition, since each subject receives the test formulation and the reference formulation twice, estimates of σ_T^2 and σ_R^2 are available. Therefore, hypotheses (7.2.3) for equivalence in variabilities can be tested using these estimates. Define the following intra-subject contrasts:

$$D_{1ik} = (Y_{i1k} - Y_{i4k}),$$

and

$$D_{2ik} = (Y_{i2k} - Y_{i3k}),\qquad(9.4.15)$$

where $i = 1, 2, \ldots, n_k$; $k = 1, 2$.

Then, under normality assumptions of (8.1.2), the vector $D = (D_{1ik}, D_{2ik})'$ follows a bivariate normal distribution with mean vector and covariance matrix given as follows, respectively.

$$\mu_D = \begin{cases} (P_1 - P_4 - C_R, & P_2 - P_3 + C_T - C_R)', & \text{if } k = 1 \\ (P_1 - P_4 - C_T, & P_2 - P_3 + C_R - C_T)', & \text{if } k = 2 \end{cases}$$

$$\Sigma_D = \begin{cases} \begin{bmatrix} 2\sigma_T^2 & 0 \\ 0 & 2\sigma_R^2 \end{bmatrix}, & \text{if } k = 1 \\[2em] \begin{bmatrix} 2\sigma_R^2 & 0 \\ 0 & 2\sigma_T^2 \end{bmatrix}, & \text{if } k = 2. \end{cases}$$

Since the covariance between D_{1ik} and D_{2ik} is 0, it follows that D_{1ik} and D_{2ik} are independent. Therefore, based upon D_{1ik} and D_{2ik}, independent estimates for σ_T^2 and σ_R^2 can be derived as follows.

Theorem 9.4.1 Let

$$\hat{\sigma}_T^2 = \frac{1}{2(n_1 + n_2 - 2)} \left[\sum_{i=1}^{n_1} (D_{1i1} - \overline{D}_{1\cdot 1})^2 + \sum_{i=1}^{n_2} (D_{2i2} - \overline{D}_{2\cdot 2})^2 \right],$$

$$(9.4.16)$$

and

$$\hat{\sigma}_R^2 = \frac{1}{2(n_1 + n_2 - 2)} \left[\sum_{i=1}^{n_1} (D_{2i1} - \overline{D}_{2\cdot 1})^2 + \sum_{i=1}^{n_2} (D_{1i2} - \overline{D}_{1\cdot 2})^2 \right],$$

where

$$\overline{D}_{h \cdot k} = \frac{1}{n_k} \sum_{i=1}^{n_k} D_{hik}, \qquad h = 1,2; \quad k = 1,2.$$

Then, $\hat{\sigma}_T^2$ and $\hat{\sigma}_R^2$, which are independent, are unbiased estimators of σ_T^2 and σ_R^2, respectively.

Proof (i) Independence

Let $\mathbf{Y}_{ik} = (Y_{i1k}, \ldots, Y_{i4k})'$, $\mathbf{Y}_k = (\mathbf{Y}'_{1k}, \ldots, \mathbf{Y}'_{n_k k})'$, and $\mathbf{D}_{hk} = (D_{h1k}, \ldots, D_{hn_k k})$, $i = 1,2, \ldots, n_k$; $h = 1,2; k = 1,2$. The covariance matrices of $\mathbf{Y}_k$ are given by

$$\text{Cov}(\mathbf{Y}_k) = I_{n_k} \otimes (\Gamma_h + \sigma_s^2 J_{n_k}), \qquad h = 1,2; \quad k = 1,2,$$

where

$$\Gamma_h = \begin{cases} \text{diag}(\sigma_T^2, \sigma_R^2, \sigma_R^2, \sigma_T^2), & \text{if } h = 1 \\ \text{diag}(\sigma_R^2, \sigma_T^2, \sigma_T^2, \sigma_R^2), & \text{if } h = 2, \end{cases}$$

and where $\text{diag}(a_1, a_2, a_3, a_4)$ is a 4×4 diagonal matrix with diagonal elements a_1, a_2, a_3, and a_4. Then,

$$\text{SSD}_{hk} = \sum_{k=1}^{n_k} (D_{hik} - \overline{D}_{h \cdot k})^2, \qquad h = 1,2; \quad k = 1,2,$$

can be expressed in terms of $\mathbf{Y}_k$ as

$$\text{SSD}_{hk} = \mathbf{Y}'_k [I_{n_k} \otimes C_h] \left[I_{n_k} - \frac{1}{n_k} J_{n_k} \right] [I_{n_k} \otimes C'_h] \mathbf{Y}_k,$$

where I_{n_k} is the $n_k \times n_k$ identity matrix J_{n_k} is the $n_k \times n_k$ matrix of 1, and

$$C_h = \begin{cases} (1, 0, 0, -1)', & \text{if } h = 1 \\ (0, 1, -1, 0)', & \text{if } h = 2. \end{cases}$$

It follows that SSD_{hk}, $h = 1,2; k = 1,2$ are mutually independent due to the facts that SSD_{h1} and SSD_{h2} are independent and

$$[I_{n_k} \otimes C_1] \left[I_{n_k} - \frac{1}{n_k} J_{n_k} \right] [I_{n_k} \otimes C'_1] [I_{n_k} \otimes (\Gamma_h + \sigma_s^2 J_{n_k})]$$

$$[I_{n_k} \otimes C_2] \left[I_{n_k} - \frac{1}{n_k} J_{n_k} \right] [I_{n_k} \otimes C'_2] = \mathbf{0}.$$

Since

$$\hat{\sigma}_T^2 = \frac{1}{2(n_1 + n_2 - 2)} [\text{SSD}_{11} + \text{SSD}_{22}]$$

and

$$\hat{\sigma}_R^2 = \frac{1}{2(n_1 + n_2 - 2)} [SSD_{12} + SSD_{21}],$$

the result follows.

(ii) Unbiasedness

It is sufficient to show that $\hat{\sigma}_T^2$ is an unbiased estimator for σ_T^2. The result follows from the facts that

$$SSD_{11} \sim 2\sigma_T^2 \chi^2(n_1 - 1),$$

$$SSD_{22} \sim 2\sigma_T^2 \chi^2(n_2 - 1),$$

and SSD_{11} is independent of SSD_{22}. ∎

Based upon $\hat{\sigma}_T^2$ and $\hat{\sigma}_R^2$, the classical $(1 - 2\alpha) \times 100\%$ confidence interval for σ_T^2/σ_T^2 and the two one-sided tests procedure for equivalence in variabilities can then be obtained by simply plugging σ_T^2 and σ_R^2 and $\nu_1 = \nu_2 = n_1 + n_2 - 2$ into (9.2.16) and (9.2.17), respectively.

Note that although the above procedure is still valid in the presence of unequal carry-over effects; $\hat{\sigma}_T^2$ and $\hat{\sigma}_R^2$ are independent of each other and they are independent of $\hat{F}$ given in (9.4.11), they are not independent of $\hat{F}|C$ in (9.4.1). Therefore, in the presence of unequal carry-over effects, statistical procedures for average bioavailability are not independent of those for variability of bioavailability.

Example 9.4.1 To illustrate statistical methods for assessment of bioequivalence of average bioavailability for the two-sequence, four-period design, let us consider the data set recently published by Ryde, Huitfeldt, and Pettersson (1991). The two-sequence, four-period crossover experiment was conducted to compare the tablets (test) and the capsules (reference) formulations of a prodrug of olsalazine (OLZ) with regard to local bioavailability of N-acetyl-5-aminosalicyclic acid (ac-5-ASA) in the colon.

In essence, this study consists of two standard two-sequence, two-period designs. This study was originally started with a standard two-sequence and two-period crossover design in 10 healthy volunteers (study A). There was a one month washout between periods. However, due to a very large variability in study A, it is decided to repeat the study using the same subjects six months after completion of study A. The same randomization codes and drug batches were used but the order of formulations was reversed (study B). It should be noted that (i) subject 5 did not participate in study B; (ii) one subject, who is not identified in the paper, by mistake, received the formulations in the same order as study A and (iii) no assignment of sequence was given in the paper. For the purpose of illustration, we assign subjects 1 through 5 to sequence 1

Table 9.4.3   AUC Data of ac-5-ASA for Ryde, Huitfeldt, and Pettersson (1991)

| | | Study A | | Study B | |
Sequence	Subject	Period I	Period II	Period III	Period IV
TRRT[a]	1	106.3	36.4	94.7	58.9
	2	149.2	107.1	104.6	119.4
	3	134.8	155.1	132.5	122.0
	4	108.1	84.9	33.2	24.8
	5	92.3	98.5	—	—
RTTR	6	85.0	92.8	81.9	59.5
	7	64.1	112.8	70.4	55.2
	8	15.3	30.1	22.3	17.5
	9	77.4	67.6	72.9	48.9
	10	102.0	106.1	67.9	70.4

[a]T = tablets and R = capsules.
Source: Ryde, Huitfeldt, and Pettersson (1991).

and subjects 6 through 10 to sequence 2, where sequences 1 and 2 are TRRT and RTTR, respectively. Subject 5 is not included in all analyses. We assume that all subjects followed their sequence of of formulations. The AUC data of ac-5-ASA are given in Table 9.4.3. Table 9.4.4 summarizes test results for average bioavailability by the methods discussed in this section as well as those by study A and study B alone.

Table 9.4.4   Summary of Test Results for AUC Data in Table 9.4.3

Data set	Carry-over	Method	μ_T	μ_R	F	90% C.I.[a]
Combined[b]	Yes	ANOVA	87.62	76.63	10.99	(98.94,129.73)
		t	87.62	76.63	10.99	(101.86,126.82)
		Wilcoxon	—	—	10.61	(99.26,127.60)
	No	ANOVA	87.71	76.55	11.16	(100.24,128.93)
		t	87.71	76.55	11.16	(102.04,127.13)
		Wilcoxon	—	—	11.16	(96.34,128.45)
Study A	No	ANOVA	103.24	82.34	20.73	(102.42,148.41)
		Wilcoxon	—	—	21.05	(96.70,155.15)
Study B	No	ANOVA	72.18	70.78	1.40	(87.56,116.40)
		Wilcoxon	—	—	2.88	(85.45,121.19)

[a]In percentage of $\hat{\mu}_R$.
[b]The combination of study A and study B, which results in a two-sequence dual design.

Under the two-sequence, four-period design, test results suggest that the tablets formulation is not equivalent to the capsules formulation in average bioavailability based upon the ± 20 rule. The results are similar regardless of the presence of the carry-over effect. This is because the carry-over effect is not significant ($\hat{C}|F = 0.697$ with a p-value of 0.934). Under study A, the results also indicate that the two formulations are not bioequivalent. However, under study B, the results are in favor of bioequivalence under the assumption of equal carry-over effects. One possible explanation for the inconsistency between test results from study A and study B is that intra-subject variabilities are different between studies. Study B has an estimate of 128.92 for σ_e^2 which is less than one third of that from study A ($\hat{\sigma}_e^2 = 443.71$). Since $\hat{\sigma}_e^2$ for study A is much larger than that of study B, it is doubtful that the assumption of compound symmetry for covariance matrix is satisfied for this data set. We therefore suggest that the nonparametric methods be used. The estimated CVs for study A, study B, and the combined study (A + B) under both the assumption of unequal and equal carry-over effects are 26%, 16%, 25.5% and 25.0%, respectively.

For assessment of intra-subject variabilities, estimates of σ_T^2 and σ_R^2 computed from (9.4.16) are 316.20 and 524.61, respectively. Based upon (9.2.17), the 90% confidence interval for σ_T^2/σ_R^2 is (15.90%, 228.44%). Therefore, if $\delta_L = 80\%$ and $\delta_U = 120\%$, the two formulations are not bioequivalent in variability. Readers can easily verify that σ_T^2 and σ_R^2 are not equivalent in neither study A nor study B by using (7.5.7) and (7.5.8).

9.4.2. The Four-Sequence, Four-Period Design

The following four-sequence, four-period design, which was previously described as design C in Section 2.5, is presented below.

The Four-Sequence, Four-Period Design			
		Period	
Sequence	I	II	III	IV
---	---	---	---	---
1	T	T	R	R
2	R	R	T	T
3	T	R	R	T
4	R	T	T	R

The above four-period, four-sequence design is, in fact, made of two Balaam's designs. The first two periods are exactly the same as those of Balaam's design, while the other two periods are the mirror image of Balaam's design with reversed treatments.

Table 9.4.5 provides the expected values of the sequence-by-period means. The analysis of variance table in terms of degrees of freedom is summarized in Table 9.4.6. Again, the four-sequence, four-period design provides estimates of intra-subject variabilities which can be used to assess the carry-over effects. The direct formulation effect can be estimated in the presence of the carry-over effect. The best linear unbiased estimator for the formulation effect can be found by the following linear contrasts of sequence-by-period means

$$
\begin{aligned}
\hat{F} = \tfrac{1}{8} \, [(&\overline{Y}_{\cdot 11} + \overline{Y}_{\cdot 21} - \overline{Y}_{\cdot 31} - \overline{Y}_{\cdot 41}) \\
- (&\overline{Y}_{\cdot 12} + \overline{Y}_{\cdot 22} - \overline{Y}_{\cdot 32} - \overline{Y}_{\cdot 42}) \\
+ (&\overline{Y}_{\cdot 13} - \overline{Y}_{\cdot 23} - \overline{Y}_{\cdot 33} + \overline{Y}_{\cdot 43}) \\
- (&\overline{Y}_{\cdot 14} - \overline{Y}_{\cdot 24} - \overline{Y}_{\cdot 34} + \overline{Y}_{\cdot 44})]
\end{aligned}
\tag{9.4.17}
$$

It can be verified that, under normality assumptions of (8.1.2), $\hat{F}$ is independent of the estimate of the carry-over effect. Therefore, the estimate of F remains the same regardless of the presence or absence of unequal carry-over effects. The variance of $\hat{F}$ is given by

$$
\mathrm{Var}(\hat{F}) = \frac{1}{16}\,\sigma_e^2 \left(\frac{1}{n_1} + \frac{1}{n_2} + \frac{1}{n_3} + \frac{1}{n_4} \right).
\tag{9.4.18}
$$

In a similar manner, assessment of bioequivalence of average bioavailability can then be carried out with degrees of freedom of $3(n_1 + n_2 + n_3 + n_4) - 5$ and S^2 from the analysis of variance table. It can also be verified that the confidence

Table 9.4.5   Expected Values of the Sequence-by-Period Means for the Four-Sequence, Four-Period Design

Sequence	Period			
	I	II	III	IV
1	$\mu + G_1 + P_1$ $+ F_T$	$\mu + G_1 + P_2$ $+ F_R + C_T$	$\mu + G_1 + P_3$ $+ F_R + C_R$	$\mu + G_1 + P_4$ $+ F_T + C_R$
2	$\mu + G_2 + P_1$ $+ F_R$	$\mu + G_2 + P_2$ $+ F_T + C_R$	$\mu + G_2 + P_3$ $+ F_T + C_T$	$\mu + G_2 + P_4$ $+ F_R + C_T$
3	$\mu + G_3 + P_1$ $+ F_T$	$\mu + G_3 + P_2$ $+ F_R + C_T$	$\mu + G_3 + P_3$ $+ F_R + C_R$	$\mu + G_3 + P_4$ $+ F_T + C_R$
4	$\mu + G_4 + P_1$ $+ F_R$	$\mu + G_4 + P_2$ $+ F_T + C_R$	$\mu + G_4 + P_3$ $+ F_T + C_T$	$\mu + G_4 + P_4$ $+ F_R + C_T$

Table 9.4.6 Analysis of Variance Table for the Four-Sequence, Four-Period Design

Source of variation	Degrees of freedom
Inter-subject	$n_1 + n_2 + n_3 + n_4 - 1$
Sequence	3
Residual	$n_1 + n_2 + n_3 + n_4 - 4$
Intra-subject	$3(n_1 + n_2 + n_3 + n_4)$
Period	3
Formulation	1
Carry-over	1
Residual	$3(n_1 + n_2 + n_3 + n_4) - 5$
Total	$4(n_1 + n_2 + n_3 + n_4) - 1$

interval approach and the two one-sided tests procedure under the four-sequence, four-period design are still valid when $\sigma_T^2 \neq \sigma_R^2$.

For assessment of intra-subject variabilities, define the following intra-subject contrasts

$$D_{1ik} = (Y_{i1k} - Y_{i2k}) \quad \text{if } k = 1,2;$$

$$D_{2ik} = (Y_{i3k} - Y_{i4k}) \quad \text{if } k = 1,2;$$

$$D_{3ik} = (Y_{i1k} - Y_{i4k}) \quad \text{if } k = 3,4;$$

$$D_{4ik} = (Y_{i2k} - Y_{i3k}) \quad \text{if } k = 3,4. \tag{9.4.19}$$

Then, based upon the same argument in the proof of Theorem 9.4.1, it can be verified that $\hat{\sigma}_T^2$ and $\hat{\sigma}_R^2$ are independent and unbiased estimators of σ_T^2 and σ_R^2, respectively, where

$$\hat{\sigma}_T^2 = \frac{1}{2(N-4)} \left[\sum_{i=1}^{n_1} (D_{1i1} - \overline{D}_{1\cdot1})^2 + \sum_{i=1}^{n_2} (D_{2i2} - \overline{D}_{2\cdot2})^2 \right.$$
$$\left. + \sum_{i=1}^{n_3} (D_{3i3} - \overline{D}_{3\cdot3})^2 + \sum_{i=1}^{n_4} (D_{4i4} - \overline{D}_{4\cdot4})^2 \right],$$

and

$$\hat{\sigma}_R^2 = \frac{1}{2(N-4)} \left[\sum_{i=1}^{n_1} (D_{2i1} - \overline{D}_{2\cdot1})^2 + \sum_{i=1}^{n_2} (D_{1i2} - \overline{D}_{1\cdot2})^2 \right.$$
$$\left. + \sum_{i=1}^{n_3} (D_{4i3} - \overline{D}_{4\cdot3})^2 + \sum_{i=1}^{n_4} (D_{3i4} - \overline{D}_{3\cdot4})^2 \right], \tag{9.4.20}$$

and

$$\overline{D}_{h\cdot k} = \frac{1}{n_k} \sum_{i=1}^{n_k} D_{hik}, \qquad h = 1,2,3,4; \quad k = 1,2,3,4.$$

Therefore, the $(1 - 2\alpha) \times 100\%$ confidence interval for σ_T^2/σ_R^2 and the two one-sided tests procedure for equivalence in variabilities can be obtained by plugging $\hat{\sigma}_T^2$ and $\hat{\sigma}_R^2$ into (9.2.17) and (9.2.16) with degrees of freedoms $2(n_1 + n_2 + n_3 + n_4 - 4)$ and $2(n_1 + n_2 + n_3 + n_4 - 4)$, respectively. It should be noted that $\hat{\sigma}_T^2$ and $\hat{\sigma}_R^2$ are independent of $\hat{F}$ given in (9.4.17).

9.5. TRANSFORMATION AND INDIVIDUAL SUBJECT RATIOS

As indicated in Chapter 6, when the responses are skewed, a log-transformation is often considered to remove the skewness and achieve an additive model with relatively homogeneous variance. Let X_{ijk} be the responses and $Y_{ijk} = \log(X_{ijk})$. Suppose Y_{ijk} follows the additive model (9.1.1) with normality assumptions and structure of compound symmetry for covariance matrix. Then, the relative average bioavailability on the original scale for an individual subject can be measured by some individual subject ratios. Therefore, in this section, we will focus on estimation of a ratio of bioavailabilities (i.e., $\delta = \exp(F)$) on the original scale between the two formulations defined in (6.3.1) under Balaam's design, the two-sequence dual design, and the four-sequence, four-period design.

9.5.1. Balaam's Design

Let $r_{ik} = X_{i2k}/X_{i1k}$, $i = 1, 2, \ldots, n_k$; $k = 1,2,3,4$, be the period ratio as defined in (6.4.2). Then, the maximum likelihood estimate of $\delta = \exp(F)$ is given by

$$\hat{\delta}_{ML} = \exp(\hat{F}|C)$$

$$= \frac{\left(\prod_{i=1}^{n_1} r_{i1}\right)^{1/2n_1} \left(\prod_{i=1}^{n_3} r_{i3}\right)^{1/2n_3}}{\left(\prod_{i=1}^{n_2} r_{i2}\right)^{1/2n_2} \left(\prod_{i=1}^{n_4} r_{i4}\right)^{1/2n_4}}. \tag{9.5.1}$$

Therefore, the minimum variance unbiased estimator of δ is given by

$$\hat{\delta}_{MVUE} = \hat{\delta}_{ML} \cdot \Phi_f[-mSS], \tag{9.5.2}$$

where $f = n_1 + n_2 + n_3 + n_4 - 3$, $m = (1/n_1 + 1/n_2 + 1/n_3 + 1/n_4)$, $SS = \frac{1}{2} SSE$, and SSE is the intra-subject residual sum of squares obtained from the analysis of variance table.

9.5.2. The Two-Sequence Dual Design

Under the two-sequence dual design, consider the following ratios

$$r_{ik} = \left[\frac{X_{i1k}^2}{(X_{i2k})(X_{i3k})} \right]^{1/4}, \qquad i = 1, 2, \ldots, n_k; \quad k = 1,2. \qquad (9.5.3)$$

It can be verified that r_{ik} are independently distributed as an univariate lognormal distribution with mean and median given below:

$$E(r_{ik}) = \begin{cases} \exp\{\frac{1}{2} F + \frac{3}{4} P_1 + \frac{3}{16} \sigma_e^2\} & \text{if } k = 1 \\ \exp\{-\frac{1}{2} F + \frac{3}{4} P_1 + \frac{3}{16} \sigma_e^2\} & \text{if } k = 2, \end{cases} \qquad (9.5.4)$$

and

$$Med(r_{ik}) = \begin{cases} \exp\{\frac{1}{2} F + \frac{3}{4} P_1\} & \text{if } k = 1 \\ \exp\{-\frac{1}{2} F + \frac{3}{4} P_1\} & \text{if } k = 2. \end{cases}$$

Hence,

$$\delta = \frac{E(r_{i1})}{E(r_{i2})} = \frac{Med(r_{i1})}{Med(r_{i2})} = \exp(F), \qquad (9.5.5)$$

which is the ratio of bioavailabilities on the original scale between the two formulations defined in (6.3.1). The maximum likelihood estimator of δ is $\exp(\hat{F})$ which always overestimates δ. The minimum variance unbiased estimator of δ is given by

$$\hat{\delta}_{MVUE} = \hat{\delta}_{ML} \cdot \Phi_f[-mSS], \qquad (9.5.6)$$

where $f = 2(n_1 + n_2 - 2)$, $m = (1/n_1 + 1/n_2)$, and $SS = \frac{3}{8} SSE$, where SSE is the sum of squares error obtained from the analysis of variance table.

On the other hand, if, after a log-transformation, Y_{ijk} follows model (9.3.12) with a common error covariance matrix Σ, then the minimum variance unbiased estimator of δ is given by

$$\hat{\delta}_{MVUE} = \hat{\delta}_{ML} \cdot \Phi_f[-mSSD], \qquad (9.5.7)$$

where $f = n_1 + n_2 - 2$ and SSD is as defined in (9.3.16).

9.5.3. The Four-Sequence, Four-Period Design

Under the four-sequence, four-period design, the maximum likelihood estimator of δ is given by

$$\hat{\delta}_{ML} = \exp(\hat{F})$$

$$= \frac{\left[\dfrac{X_{i11} \cdot X_{i21}}{X_{i31} \cdot X_{i41}}\right]^{1/8n_1} \left[\dfrac{X_{i13} \cdot X_{i43}}{X_{i23} \cdot X_{i33}}\right]^{1/8n_3}}{\left[\dfrac{X_{i12} \cdot X_{i22}}{X_{i32} \cdot X_{i42}}\right]^{1/8n_2} \left[\dfrac{X_{i14} \cdot X_{i44}}{X_{i24} \cdot X_{i34}}\right]^{1/8n_4}}. \qquad (9.5.8)$$

Therefore, the minimum variance unbiased estimator of δ is given by

$$\hat{\delta}_{MVUE} = \hat{\delta}_{ML} \cdot \Phi_f[-mSS], \qquad (9.5.9)$$

where $f = 3(n_1 + n_2 + n_3 + n_4) - 5$, $m = (1/n_1 + 1/n_2 + 1/n_3 + 1/n_4)$, $SS = SSE/16$, and SSE is the intra-subject residual sum of squares obtained from the ANOVA table.

9.6. DISCUSSION

As indicated earlier, for assessment of bioequivalence in average bioavailability between two formulations of a drug product, the standard two-sequence, two-period crossover design is often considered. In Chapter 4, under the assumption of no carry-over effects, we have introduced several statistical methods for this assessment. When the carry-over effect is present, the standard two-sequence, two-period crossover design may not be useful because it does not provide an estimate of the formulation effect in the presence of the carry-over effect. In addition, it does not provide independent estimates for intra-subject variabilities, which can be used to establish equivalence in intra-subject variabilities. In this case, as an alternative, a higher-order crossover design is usually preferred because it allows us (i) to estimate the formulation effect when the carry-over effect is present; (ii) to estimate intra-subject variabilities; (iii) to draw inference on the carry-over effect, and (iv) to establish equivalence in variability of bio-availability. Although a higher-order crossover possesses some desirable statistical properties, one should be aware of some practical problems that may be encountered when a high-order crossover design is used. For a higher-order crossover design with more than two periods, the consequences are not only that it may be very time consuming to complete the study, but also that it may increase the number of dropouts. Besides, as indicated in Chapter 2, it may not be desirable to draw too many blood samples from each subject due to medical concerns. On the other hand, if the higher-order crossover design has more than two sequences, it may increase the chance of errors occurring in the randomization schedules.

In this chapter, we have introduced several higher-order crossover designs.

Table 9.6.1 The Distribution of Degrees of Freedom with 24 Subjects

Source	Design				
	2×2^a	4×2^b	2×3^c	2×4^d	4×4^e
Inter-subject	23	23	23	23	23
Sequence	1	3	1	1	3
Residual	22	20	22	22	20
Intra-subject	24	24	48	72	72
Period	1	1	2	3	3
Formulation	1	1	1	1	1
Carry-over	—^{f}	1	1	1	1
Residual	22	21	44	67	67
Total	47	47	71	95	95

$^a 2 \times 2$ = the standard two-sequence, two-period design.
$^b 4 \times 2$ = Balaam's design.
$^c 2 \times 3$ = the two-sequence dual design.
$^d 2 \times 4$ = the two-sequence, four-period design.
$^e 4 \times 4$ = the four-sequence, four-period design.
^{f}The carry-over effect is confounded with the sequence effect.

To compare their relative advantages and disadvantages with the standard two-sequence, two-period crossover design, we summarize the distribution of degrees of freedoms for each study in Table 9.6.1. assuming that there are 24 subjects. From Table 9.6.1, it can be seen that for the standard 2×2 crossover design, the carry-over effect is confounded with the sequence effect which cannot be separated in the analysis. The 4×2 crossover design (i.e., Balaam's design) allows us to estimate the carry-over effect independently. The degrees of freedom for the intra-subject residuals for the 2×2 design and the 4×2 design are 22 and 21, respectively. Therefore, there is not much difference in testing power. For a 2×2 design, if we add one period (or two periods), the design becomes a 2×3 design (or a 2×4 design). It can be seen that the degrees of freedom for the intra-subject residuals increases from 22 to 44 (or 22 to 67). In this case, although the testing power will certainly increase significantly, the overall effects associated with the addition of extra periods should be examined carefully. For the comparison between a 2×4 design and a 4×4 design, there is no difference except for testing of the sequence effect. Based upon the above argument, we suggest that the relative gain and loss due to the increase of additional period or sequence (or subject) be examined before a decision is made.

In the previous section, we briefly introduced the use of log-transformation for a higher-order crossover design. As an alternative, Chow, Peace and Shao (1991) proposed a general approach using individual subject ratios under a multiplicative model. However, this method does not provide statistical inference for the carry-over effect. More discussion regarding a higher-order crossover design can be found in Chow, Liu and Peace (1991).

10

Assessment of Bioequivalence for More than Two Formulations

10.1. INTRODUCTION

In the previous chapter, we have introduced statistical methods for the assessment of bioequivalence under a higher-order crossover design for comparing two formulations of a drug product. The statistical methods depend upon the estimation of the direct formulation effect and its variance. The analysis of a higher-order crossover design for comparing two formulations is quite straightforward because there are only two formulations to be compared regardless of how many sequences or periods in the design. In practice, however, it is often of interest to compare more than two (e.g., three or four) formulations of the same drug in a bioavailability/bioequivalence study. In this case, a standard high-way crossover design is usually considered. For example, for comparing three formulations, we may consider a standard three-sequence, three-period crossover design. The analysis for assessing bioequivalence, however, is much more complicated because there are three pairs of formulation effects to be compared and the variance of these pairs of formulation effects may differ from one another. Moreover, a standard crossover design may not be useful when the carry-over effect is present.

To overcome the disadvantages that a standard crossover design may have, as indicated in Chapter 2, variance-balanced designs are usually recommended because (i) it possesses the property of equal variances for each pairwise differences among formulations; (ii) it provides an estimate for each pairwise difference in the presence of the carry-over effect. A variance-balanced design allows us to estimate each pairwise difference with the same degree of precision and provides analyses for assessing bioequivalence in the presence of carry-over effects. The most common variance-balanced design used for comparing three or four formulations in bioavailability/bioequivalence studies is the so-called Williams design. Williams designs for comparing three or four formulations were briefly described earlier in Chapter 2. In this chapter, statistical methods for assessment of bioequivalence under a Williams design will be discussed.

In many cases, pharmaceutical companies may be interested in comparing a large number of formulations in a bioavailability study. In this case, a complete standard high-way crossover design may not be of practical interest because (i) it is too time consuming to complete the study; (ii) it is not desirable to draw many blood samples from each subject; (iii) a subject is more likely to drop out when he/she is required to return frequently for evaluations. In addition, a complete high-order crossover design may increase the chance of making errors in the randomization schedules which has an impact on valid statistical inference. To accommodate the above concerns, Westlake (1973, 1974) suggested that a balanced incomplete block design be used when comparing a large number of formulations. Several methods for constructing a balanced incomplete block design have been introduced earlier in Chapter 2. It should be noted that, in the absence of carry-over effects, a balanced incomplete block design is also a variance-balanced design. In this chapter, statistical methods for assessment of bioequivalence under a balanced incomplete block design will also be discussed.

In the next section, statistical model and methods, which include the confidence interval and the two one-sided tests procedure, for a general $K \times J$ crossover design are derived. The application of these methods to the two Williams designs and an incomplete balanced block design is given in Sections 3 and 4, respectively. A brief discussion is presented in the last section.

10.2. ASSESSMENT OF AVERAGE BIOAVAILABILITY WITH MORE THAN TWO FORMULATIONS

In this section, statistical methods for assessment of bioequivalence of average bioavailability under a $K \times J$ (i.e., K-sequence and J-period) crossover design for comparing t (t > 2) formulations will be discussed. Again, we will only focus on the confidence interval approach and Schuirmann's two one-sided tests procedure.

10.2.1. Statistical Model and Assumptions

Consider the following statistical model for a K-sequence and J-period crossover design comparing t (t > 2) formulations:

$$Y_{ijk} = \mu + G_k + S_{ik} + P_j + F_{(j,k)} + C_{(j-1,k)} + e_{ijk}, \qquad (10.2.1)$$

where $i = 1,2, \ldots ,n_k$; $j = 1,2, \ldots ,J$; $k = 1,2, \ldots ,K$ and Y_{ijk}, μ, G_k, S_{ik}, P_j, $F_{(j,k)}$, $C_{(j-1,k)}$ and e_{ijk} are as defined in (9.1.1). The assumptions for the fixed effects, compound symmetry, and normality of inter-subject and intra-subject variabilities were outlined in (9.1.1).

For assessment of bioequivalence in average bioavailability under model (10.2.1), the parameters of interest are (i) the unknown formulation means; (ii) pairwise differences in formulation means (or direct formulation effects) and (iii) pairwise first-order carry-over effects. The unknown population mean for the hth formulation is defined as

$$\mu_h = \mu + F_h, \qquad h = 1, 2, \ldots , t. \qquad (10.2.2)$$

Pairwise differences in formulation means (or direct formulation effects) are defined as

$$\theta_{hh'} = F_h - F_{h'}, \qquad 1 \leqslant h \neq h' < t. \qquad (10.2.3)$$

Similarly, pairwise first-order carry-over effects are given by

$$\lambda_{hh'} = C_h - C_{h'}, \qquad 1 \leqslant h \neq h' < t. \qquad (10.2.4)$$

Based upon estimates of μ_h and $\theta_{hh'}$ and the estimates of their variances, the confidence interval approach and Schuirmann's two one-sided tests procedure for assessing bioequivalence among formulations can be derived. Furthermore, statistical inference for the carry-over effects can also be drawn based upon estimates of $\lambda_{hh'}$.

For estimation of $\theta_{hh'}$ and $\lambda_{hh'}$, there exist unbiased estimators, which can be obtained by ordinary least squares (OLS) method based upon model (10.2.1). These unbiased estimators, which can be expressed as a linear contrast of sequence by period means, are given below:

$$\mathscr{L}_a = \sum_{k=1}^{K} \sum_{j=1}^{J} C_{ajk}\overline{Y}_{\cdot jk}, \qquad a = 1,2, \ldots , \frac{t!}{2!(t-2)!}. \qquad (10.2.5)$$

Since

$$\sum_{j=1}^{J} C_{ajk} = \sum_{k=1}^{K} C_{ajk} = 0,$$

under assumptions of normality and $\sigma_h^2 = \sigma_e^2$ for $h = 1,2, \ldots ,t$, the variance of $\mathscr{L}_a$ is given by

$$\text{Var}(\mathscr{L}_a) = \sigma_e^2 \sum_{k=1}^{K} \left[\frac{1}{n_k}\right] \sum_{j=1}^{J} C_{ajk}^2. \qquad (10.2.6)$$

Note that when $n_k = n$ for $k = 1,2, \ldots ,K$, the above variance reduces to

$$\mathrm{Var}(\mathscr{L}_a) = \frac{\sigma_e^2}{n} \sum_{k=1}^{K} \sum_{j=1}^{J} C_{ajk}^2. \qquad (10.2.7)$$

Based upon (10.2.5) and (10.2.6), the $(1 - 2\alpha) \times 100\%$ confidence interval and Schuirmann's two one-sided tests procedure can be obtained to assess bio-equivalence between the hth formulation and the h'th formulation.

10.2.2. Confidence Interval and Two One-Sided Tests Procedure

An unbiased estimator of the overall mean is the arithmetic average of sequence-by-period means, i.e.,

$$\hat{\mu} = \frac{1}{JK} \sum_{j=1}^{J} \sum_{k=1}^{K} \overline{Y}_{\cdot jk}.$$

Let $\hat{F}_h$ be the OLS unbiased estimate of F_h. Then, the population means μ_h, $h = 1,2, \ldots ,t$, can be estimated unbiasedly by

$$\hat{\mu}_h = \hat{\mu} + \hat{F}_h.$$

The OLS unbiased estimates for $\theta_{hh'}$ are then given by

$$\begin{aligned}
\hat{\theta}_{hh'} &= \hat{\mu}_h - \hat{\mu}_{h'} \\
&= \hat{F}_h - \hat{F}_{h'}, \qquad 1 \leqslant h \neq h' < t. \qquad (10.2.8)
\end{aligned}$$

Therefore, the $(1 - 2\alpha) \times 100\%$ confidence interval for $\theta_{hh'}$ is given by

$$[L(\theta_{hh'}),U(\theta_{hh'})] = \hat{\theta}_{hh'} \pm t(\alpha,\nu) \left[S^2 \sum_{k=1}^{K} \frac{1}{n_k} \sum_{j=1}^{J} C_{ajk}^2 \right]^{1/2}, \qquad (10.2.9)$$

where S^2 is the intra-subject mean squared error, with ν degrees of freedom, obtained from the analysis of variance table. Under model (10.2.1), S^2 is an unbiased estimator of σ_e^2. Hence, we conclude that the hth formulation and the h'th formulation are bioequivalent in average bioavailability if $[L(\theta_{hh'}),U(\theta_{hh'})]$ is within bioequivalent limits, (θ_L,θ_U).

Similarly, for the two one-sided tests procedure, we would reject the interval hypotheses (4.3.1) and conclude bioequivalent at the α level of significance if

$$T_L = \frac{\hat{\theta}_{hh'} - \theta_L}{\left[S^2 \sum_{k=1}^{K} \frac{1}{n_k} \sum_{j=1}^{J} C_{ajk}^2 \right]^{1/2}} > t(\alpha,\nu),$$

and

$$T_U = \frac{\hat{\theta}_{hh'} - \theta_U}{\left[S^2 \sum_{k=1}^{K} \frac{1}{n_k} \sum_{j=1}^{J} C_{ajk}^2 \right]^{1/2}} < -t(\alpha, \nu). \tag{10.2.10}$$

Note that the p-values of Anderson and Hauck's procedure for interval hypotheses (4.3.1) can also be obtained by simply plugging T_L and T_U into either (4.3.12) or (4.3.13).

Although the primary hypothesis of interest is to evaluate the equality of carry-over effects, confidence interval for the carry-over effect can also be obtained in a similar manner.

10.2.3. Log-Transformation

When the responses are skewed, a log-transformation may be considered to remove the skewness and achieve model (10.2.1) with relatively homogeneous variance. In this case, a parameter of interest is given by

$$\delta_{hh'} = \exp(\theta_{hh'}) = \exp(F_h - F_{h'})$$

$$= \frac{\exp(\mu + F_h)}{\exp(\mu + F_{h'})}, \qquad 1 \leq h \neq h' \leq t.$$

Under normality assumptions and $\sigma_h^2 = \sigma_e^2$ for $h = 1, 2, \ldots, t$, the maximum likelihood estimate of $\delta_{hh'}$ can be obtained by simply replacing the maximum likelihood estimate of $\theta_{hh'}$ with $\theta_{hh'}$ in the above expression, i.e.,

$$\hat{\delta}_{hh'} = \exp(\hat{\theta}_{hh'}).$$

The minimum variance unbiased estimator of $\delta_{hh'}$ is then given by

$$\hat{\delta}_{hh'} = \exp(\hat{\theta}_{hh'}) \cdot \Phi_f[-mSSE], \tag{10.2.11}$$

where $\theta_{hh'}$ is the OLS unbiased estimate of $\theta_{hh'}$ on the log scale, SSE is the intra-subject sum of squares from the analysis of variance table with degrees of freedom $f = \nu$, and

$$m = \sum_{k=1}^{K} \frac{1}{n_k} \sum_{j=1}^{J} C_{ajk}^2.$$

10.2.4. Variance-Balanced Designs

For assessment of bioequivalence of average bioavailability, as indicated earlier, the confidence interval approach and the two one-sided tests procedure depend upon the estimation of the formulation effects (or difference in formulation means) and their variances. For comparing more than two formulations, there are several possible pairs of differences in formulation means. For example, there are three possible pairwise differences among formulation means (i.e., formulation 1 vs formulation 2; formulation 2 vs formulation 3; formulation 1 vs formulation 3). In this case, it is desirable to estimate these pairwise formulation effects with the same degrees of precision. In other words, it is desirable that there is a common variance for each pair of formulation effect. A design with this property is known as a variance-balanced design. Therefore, an ideal crossover design for comparing more than two formulations is a design which can minimize this common variance. Such a design can lead to the best precision within the class of variance-balanced designs. An example of a variance-balanced design is the so-called Williams design which will be further discussed in the next section.

For a variance-balanced design, the quantity

$$\nu_a = \sum_{k=1}^{K} \sum_{j=1}^{J} C_{ajk}^2$$

is the same for all pairwise formulation effects. In the case where $n_k = n$ for $k = 1, 2, \ldots, t$, the $(1 - 2\alpha) \times 100\%$ confidence interval for $\theta_{hh'}$ given in (10.2.9) reduces to

$$[L(\theta_{hh'}), U(\theta_{hh'})] = \hat{\theta}_{hh'} \pm t(\alpha, \nu) \left[S^2 \frac{\nu_a}{n} \right]^{1/2}. \tag{10.2.12}$$

The two one-sided tests procedure given in (10.2.10) become

$$T_L = \frac{\hat{\theta}_{hh'} - \theta_L}{\left[S^2 \dfrac{\nu_a}{n} \right]^{1/2}} > t(\alpha, \nu),$$

and

$$T_U = \frac{\hat{\theta}_{hh'} - \theta_U}{\left[S^2 \dfrac{\nu_a}{n} \right]^{1/2}} < -t(\alpha, \nu). \tag{10.2.13}$$

Note that the p-values of Anderson and Hauck's procedure for interval hypotheses (4.3.1) can also be obtained by simply plugging T_L and T_U into either (4.3.12) or (4.3.13).

Similarly, the minimum variance unbiased estimator of $\delta_{hh'}$ under a multiplicative model is given by

$$\hat{\delta}_{hh'} = \exp(\hat{\theta}_{hh'}) \cdot \Phi_f \left[-\frac{\nu_a}{n} \, SSE \right]. \tag{10.2.14}$$

When $n_k = n$ for all k, under the structure of a Williams design, ν_a for the ordinary least squares unbiased estimate of $\lambda_{hh'}$ are the same for all $1 \leq h \neq h' \leq t$. Therefore, Williams designs are not only variance-balanced designs for the direct formulation effects but also for the carry-over effects.

10.3. ANALYSES FOR WILLIAMS DESIGNS

In the following two sections, we will focus on statistical analyses for assessment of bioequivalence under Williams designs for comparing three and four formulations.

10.3.1. Williams Designs with Three Formulations

A Williams design for comparing three formulations is a variance-balanced design, which consists of six sequences and three periods (Table 2.5.7). For the sake of convenience, we summarize a Williams design with three formulations as follows:

Williams Design with Three Formulations

Sequence	Period		
	I	II	III
1	R	T_2	T_1
2	T_1	R	T_2
3	T_2	T_1	R
4	T_1	T_2	R
5	T_2	R	T_1
6	R	T_1	T_2

Under model (10.2.1), Table 10.3.1 and Table 10.3.2 provide the expected values of sequence-by-period means and analysis of variance table in terms of degrees of freedom, respectively. As can be seen from Table 10.3.2, there are $2(N - 3)$ degrees of freedom for the intra-subject mean squared error, where $N = \Sigma_{k=1}^{6} n_k$. For estimation of the three formulation effects, F_h, h = 1,2,3, we need to find C_{ajk}, the coefficient of linear contrasts of sequence-by-period means. From Table 10.3.1, the C_{ajk} for each formulation effect can be obtained. Table 10.3.3 lists C_{ajk} for the unbiased ordinary least squares estimators of F_h, h = 1,2,3. Therefore, the three population formulation means μ_h, h = 1,2,3,

Table 10.3.1   Expected Values of the Sequence-by-Period Means for the Williams Design with Three Formulations

Sequence	Period		
	I	II	III
1	$\mu + G_1 + P_1 + F_R$	$\mu + G_1 + P_2$ $+ F_2 + C_R$	$\mu + G_1 + P_3$ $+ F_1 + C_2$
2	$\mu + G_2 + P_1 + F_1$	$\mu + G_2 + P_2$ $+ F_R + C_1$	$\mu + G_2 + P_3$ $+ F_2 + C_R$
3	$\mu + G_3 + P_1 + F_2$	$\mu + G_3 + P_2$ $+ F_1 + C_2$	$\mu + G_3 + P_3$ $+ F_R + C_1$
4	$\mu + G_4 + P_1 + F_1$	$\mu + G_4 + P_2$ $+ F_2 + C_1$	$\mu + G_4 + P_3$ $+ F_R + C_2$
5	$\mu + G_5 + P_1 + F_2$	$\mu + G_5 + P_2$ $+ F_R + C_2$	$\mu + G_5 + P_3$ $+ F_1 + C_R$
6	$\mu + G_6 + P_1 + F_R$	$\mu + G_6 + P_2$ $+ F_1 + C_R$	$\mu + G_6 + P_3$ $+ F_2 + C_1$

Table 10.3.2   Analysis of Variance Table for the Williams Design with Three Formulations

Source of variation	Degrees of freedom[a]
Inter-subject	$N^b - 1$
Sequence	5
Residual	$N - 6$
Intra-subject	$2N$
Period	2
Formulation	2
Carry-over	2
Residual	$2(N - 3)$
Total	$3N - 1$

[a]Degrees of freedom for the intra-subject residual is $2(N - 2)$ if carry-over effects are not included in the model.

[b]$N = n_1 + n_2 + n_3 + n_4 + n_5 + n_6$.

Table 10.3.3 Coefficients[a] for Estimates of Formulations F_R, F_1, and F_2 in the Williams Design with Three Formulations (Adjusted for Carry-Over Effects)

	F_R Period			F_1 Period			F_2 Period		
Sequence	I	II	III	I	II	III	I	II	III
1	3	0	−3	−1	−2	3	−2	2	0
2	−2	2	0	3	0	−3	−1	−2	3
3	−1	−2	3	−2	2	0	3	0	−3
4	−1	−2	3	3	0	−3	−2	2	0
5	−2	2	0	−1	−2	3	3	0	−3
6	3	0	−3	−2	2	0	−1	−2	3

[a]The coefficients are multiplied by 24.

can be estimated unbiasedly by the sum of $\overline{Y}...$ and the linear contrasts of sequence by period means with the coefficients given in Table 10.3.3.

To establish bioequivalence of average bioavailability between formulations h and h', $1 \le h \ne h' \le t$, the confidence interval given in (10.2.9) and Schuirmann's two one-sided tests procedure given in (10.2.10) can be used. For this purpose, Tables 10.3.4 and 10.3.5 lists the coefficients of linear contrasts

Table 10.3.4 Coefficients[a] for Estimates of Pairwise Formulation Effects in the Williams Design with Three Formulations (Adjusted for Carry-Over Effects)

	$\theta_{1R} = F_1 - F_R$ Period				$\theta_{2R} = F_2 - F_R$ Period				$\theta_{21} = F_2 - F_1$ Period			
Seq.	1	2	3	$\Sigma_j C_{ajk}^2$	1	2	3	$\Sigma_j C_{ajk}^2$	1	2	3	$\Sigma_j C_{ajk}^2$
1	−4	−2	6	$56/(24)^2$	−5	2	3	$38/(24)^2$	−1	4	−3	$26/(24)^2$
2	5	−2	−3	$38/(24)^2$	1	−4	3	$26/(24)^2$	−4	−2	6	$56/(24)^2$
3	−1	4	−3	$26/(24)^2$	4	2	−6	$56/(24)^2$	5	−2	−3	$38/(24)^2$
4	4	2	−6	$56/(24)^2$	−1	4	−3	$26/(24)^2$	−5	2	3	$38/(24)^2$
5	1	−4	3	$26/(24)^2$	5	−2	−3	$38/(24)^2$	4	2	−6	$56/(24)^2$
6	−5	2	3	$38/(24)^2$	−4	−2	6	$56/(24)^2$	1	−4	3	$26/(24)^2$
Variance[b]		$\dfrac{5}{12n} \sigma_e^2$				$\dfrac{5}{12n} \sigma_e^2$				$\dfrac{5}{12n} \sigma_e^2$		

[a]Coefficients are multiplied by 24.

[b]Variance when $n = n_k$ for $1 \le k \le 6$.

Table 10.3.5 Coefficients[a] for Estimates of Pairwise Formulation Effects in the Williams Design with Three Formulations (In Absence of Unequal Carry-Over Effects)

| | $\theta_{1R} = F_1 - F_R$ | | | | $\theta_{2R} = F_2 - F_R$ | | | | $\theta_{21} = F_2 - F_1$ | | | |
| | Period | | | | Period | | | | Period | | | |
Seq.	1	2	3	$\Sigma_j C_{ajk}^2$	1	2	3	$\Sigma_j C_{ajk}^2$	1	2	3	$\Sigma_j C_{ajk}^2$
1	-1	0	1	1/18	-1	1	0	1/18	0	1	-1	1/18
2	1	-1	0	1/18	0	-1	1	1/18	-1	0	1	1/18
3	0	1	-1	1/18	1	0	-1	1/18	1	-1	0	1/18
4	1	0	-1	1/18	0	1	-1	1/18	-1	1	0	1/18
5	0	-1	1	1/18	1	-1	0	1/18	1	0	-1	1/18
6	-1	1	0	1/18	-1	0	1	1/18	0	-1	1	1/18
Variance[b]			$\dfrac{1}{3n}\sigma_e^2$				$\dfrac{1}{3n}\sigma_e^2$				$\dfrac{1}{3n}\sigma_e^2$	

[a]Coefficients are multiplied by 6.
[b]Variance when $n = n_k$ for $1 \leqslant k \leqslant 6$.

Table 10.3.6 Coefficients[a] for Estimates of Carry-Over Effects in the Williams Design with Three Formulations (Adjusted for Formulation Effects)

| | $\lambda_{1R} = C_1 - C_R$ | | | | $\lambda_{2R} = C_2 - C_R$ | | | | $\lambda_{21} = C_2 - C_1$ | | | |
| | Period | | | | Period | | | | Period | | | |
Seq.	1	2	3	$\Sigma_j C_{ajk}^2$	1	2	3	$\Sigma_j C_{ajk}^2$	1	2	3	$\Sigma_j C_{ajk}^2$
1	0	-2	2	4/32	-1	-2	3	7/32	-1	0	1	1/32
2	1	2	-3	7/32	1	0	-1	1/32	0	-2	2	4/32
3	-1	0	1	1/32	0	2	-2	4/32	1	2	-3	7/32
4	0	2	-2	4/32	-1	0	1	1/32	-1	-2	3	7/32
5	1	0	-1	1/32	1	2	-3	7/32	0	2	-2	4/32
6	-1	-2	3	7/32	0	-2	2	4/32	1	0	-1	1/32
Variance[b]			$\dfrac{3}{4n}\sigma_e^2$				$\dfrac{3}{4n}\sigma_e^2$				$\dfrac{3}{4n}\sigma_e^2$	

[a]Coefficients are multiplied by 8.
[b]Variance when $n = n_k$ for $1 \leqslant k \leqslant 6$.

for the unbiased ordinary least squares estimates of $\theta_{hh'}$ in the presence and absence of unequal carry-over effects, respectively. Based upon these coefficients, the confidence interval and the two one-sided tests can be easily carried out.

To draw statistical inference on the carry-over effect, the coefficients for estimates of pairwise carry-over effects, $\lambda_{hh'}$, $1 \leq h \neq h' \leq t$, given in Table 10.3.6 are useful.

10.3.2. Williams Design with Four Formulations

A Williams design for comparing four formulations, which consists of four sequences and four periods (Table 2.5.7), is summarized below:

	\multicolumn{4}{c}{Williams Design with Four Formulations}

Williams Design with Four Formulations
Period

Sequence	I	II	III	IV
1	R	T_3	T_1	T_2
2	T_1	R	T_2	T_3
3	T_2	T_1	T_3	R
4	T_3	T_2	R	T_1

Table 10.3.7 and Table 10.3.8 give the expected values of the sequence-by-period means and the analysis of variance table for the degrees of freedoms.

Table 10.3.7 Expected Values of the Sequence-by-Period Means for the Williams Design with Four Formulations

Sequence	\multicolumn{4}{c}{Period}			
	I	II	III	IV
1	$\mu + G_1 + P_1 + F_R$	$\mu + G_1 + P_2 + F_3 + C_R$	$\mu + G_1 + P_3 + F_1 + C_3$	$\mu + G_1 + P_4 + F_2 + C_1$
2	$\mu + G_2 + P_1 + F_1$	$\mu + G_2 + P_2 + F_R + C_1$	$\mu + G_2 + P_3 + F_2 + C_R$	$\mu + G_2 + P_4 + F_3 + C_2$
3	$\mu + G_3 + P_1 + F_2$	$\mu + G_3 + P_2 + F_1 + C_2$	$\mu + G_3 + P_3 + F_3 + C_1$	$\mu + G_3 + P_4 + F_R + C_3$
4	$\mu + G_4 + P_1 + F_3$	$\mu + G_4 + P_2 + F_2 + C_3$	$\mu + G_4 + P_3 + F_R + C_2$	$\mu + G_4 + P_4 + F_1 + C_R$

Table 10.3.8 Analysis of Variance
Table for the Williams Design with Four
Formulations

Source of variation	Degrees of freedom[a]
Inter-subject	$N^b - 1$
Sequence	3
Residual	$N - 4$
Intra-subject	$3N$
Period	3
Formulation	3
Carry-over	3
Residual	$3(N - 3)$
Total	$4N - 1$

[a]Degrees of freedom for the intra-subject residual is
$3(N - 2)$ if carry-over effects are not included in
the model.
[b]$N = n_1 + n_2 + n_3 + n_4$.

Coefficients of the linear contrasts for the unbiased ordinary least squares esti-
mates of F_h, $\theta_{hh'}$, and $\lambda_{hh'}$, $1 \leqslant h \neq h' \leqslant t$, are given in Tables 10.3.9 through
10.3.12. Note that a Williams design with three formulations compares three
pairs of formulations, while there are a total of six pairs of formulations to be
compared for a Williams design with four formulations. The degrees of freedom
for the intra-subject mean squared error is $3(N - 3)$, where $N = \Sigma_{k=1}^4 \, n_k$.
Based upon the coefficients given in Tables 10.3.9 through 10.3.12, bioequi-

Table 10.3.9 Coefficients[a] for Estimates of Formulations F_R, F_1, F_2, and F_3 in the
Williams Design with Four Formulations (Adjusted for Carry-Over Effects)

	F_R Period				F_1 Period				F_2 Period				F_3 Period			
Seq.	1	2	3	4	1	2	3	4	1	2	3	4	1	2	3	4
1	8	0	-4	-4	-3	-4	7	0	-2	-3	-3	8	-3	7	0	4
2	-3	7	0	-4	8	0	-4	-4	-3	-4	7	0	-2	-3	-3	8
3	-2	-3	-3	8	-3	7	0	-4	8	0	-4	-4	-3	-4	7	0
4	-3	-4	7	0	-2	-3	-3	8	-3	7	0	-4	8	0	-4	-4

[a]The coefficients are multiplied by 40.

Table 10.3.10 Coefficients[a] for Estimates of Pairwise Formulation Effects in the Williams Design with Four Formulations (Adjusted for Carry-Over Effects)

| | $\theta_{1R} = F_1 - F_R$ | | | | | $\theta_{2R} = F_2 - F_R$ | | | | |
| | Period | | | | | Period | | | | |
Sequence	1	2	3	4	$\Sigma_j C^2_{ajk}$	1	2	3	4	$\Sigma_j C^2_{ajk}$
1	-11	-4	11	4	$274/(40)^2$	-10	-3	1	12	$254/(40)^2$
2	11	-7	-4	0	$186/(40)^2$	0	-11	7	4	$186/(40)^2$
3	-1	10	3	-12	$254/(40)^2$	10	3	-1	-12	$254/(40)^2$
4	1	1	-10	8	$166/(40)^2$	0	11	-7	-4	$186/(40)^2$
Variance[b]				$\dfrac{11}{20n}\sigma_e^2$					$\dfrac{11}{20n}\sigma_e^2$	

| | $\theta_{3R} = F_3 - F_R$ | | | | | $\theta_{21} = F_2 - F_1$ | | | | |
| | Period | | | | | Period | | | | |
Sequence	1	2	3	4	$\Sigma_j C^2_{ajk}$	1	2	3	4	$\Sigma_j C^2_{ajk}$
1	-11	7	4	0	$186/(40)^2$	1	1	-10	8	$166/(40)^2$
2	1	-10	-3	12	$254/(40)^2$	-11	-4	11	4	$274/(40)^2$
3	-1	-1	10	-8	$166/(40)^2$	11	-7	-4	0	$186/(40)^2$
4	11	4	-11	-4	$274/(40)^2$	-1	10	3	-12	$254/(40)^2$
Variance[b]				$\dfrac{11}{20n}\sigma_e^2$					$\dfrac{11}{20n}\sigma_e^2$	

| | $\theta_{31} = F_3 - F_1$ | | | | | $\theta_{32} = F_3 - F_2$ | | | | |
| | Period | | | | | Period | | | | |
Sequence	1	2	3	4	$\Sigma_j C^2_{ajk}$	1	2	3	4	$\Sigma_j C^2_{ajk}$
1	0	11	-7	-4	$186/(40)^2$	-1	10	3	-12	$254/(40)^2$
2	-10	-3	1	12	$254/(40)^2$	1	1	-10	8	$166/(40)^2$
3	0	-11	7	4	$186/(40)^2$	-11	-4	11	4	$274/(40)^2$
4	10	3	-1	-12	$254/(40)^2$	11	-7	-4	0	$186/(40)^2$
Variance[b]				$\dfrac{11}{20n}\sigma_e^2$					$\dfrac{11}{20n}\sigma_e^2$	

[a]Coefficients are multiplied by 40.

[b]Variance when $n = n_k$ for $1 \leqslant k \leqslant 4$.

Table 10.3.11 Coefficients[a] for Estimates of Carry-Over Effects in the Williams Design with Four Formulations (Adjusted for Formulation Effects)

| | $\lambda_{1R} = C_1 - C_R$ | | | | | $\lambda_{2R} = C_2 - C_R$ | | | | |
| | Period | | | | | Period | | | | |
Sequence	1	2	3	4	$\Sigma_j C^2_{ajk}$	1	2	3	4	$\Sigma_j C^2_{ajk}$
1	-1	-4	1	4	$34/(10)^2$	0	-3	1	2	$14/(10)^2$
2	1	3	-4	0	$26/(10)^2$	0	-1	-3	4	$26/(10)^2$
3	-1	0	3	-2	$14/(10)^2$	0	3	-1	-2	$14/(10)^2$
4	1	1	0	-2	$6/(10)^2$	0	1	3	-4	$26/(10)^2$
Variance[b]					$\dfrac{4}{5n}\sigma^2_e$					$\dfrac{4}{5n}\sigma^2_e$

| | $\lambda_{3R} = C_3 - C_R$ | | | | | $\lambda_{21} = C_2 - C_1$ | | | | |
| | Period | | | | | Period | | | | |
Sequence	1	2	3	4	$\Sigma_j C^2_{ajk}$	1	2	3	4	$\Sigma_j C^2_{ajk}$
1	-1	-3	4	0	$26/(10)^2$	1	1	0	-2	$6/(10)^2$
2	1	0	-3	2	$14/(10)^2$	-1	-4	1	4	$34/(10)^2$
3	-1	-1	0	2	$6/(10)^2$	1	3	-4	0	$26/(10)^2$
4	1	4	-1	-4	$34/(10)^2$	-1	0	3	-2	$14/(10)^2$
Variance[b]					$\dfrac{4}{5n}\sigma^2_e$					$\dfrac{4}{5n}\sigma^2_e$

| | $\lambda_{31} = C_3 - C_1$ | | | | | $\lambda_{32} = C_3 - C_2$ | | | | |
| | Period | | | | | Period | | | | |
Sequence	1	2	3	4	$\Sigma_j C^2_{ajk}$	1	2	3	4	$\Sigma_j C^2_{ajk}$
1	0	1	3	-4	$26/(10)^2$	-1	0	3	-2	$14/(10)^2$
2	0	-3	1	2	$14/(10)^2$	1	1	0	-2	$6/(10)^2$
3	0	-1	-3	4	$26/(10)^2$	-1	-4	1	4	$34/(10)^2$
4	0	3	1	-2	$14/(10)^2$	1	3	-4	0	$26/(10)^2$
Variance[b]					$\dfrac{4}{5n}\sigma^2_e$					$\dfrac{4}{5n}\sigma^2_e$

[a]Coefficients are multiplied by 10.
[b]Variance when $n = n_k$ for $1 \leq k \leq 4$.

Table 10.3.12 Coefficients[a] for Estimates of Pairwise Formulation Effects in the Williams Design with Four Formulations (In Absence of Unequal Carry-Over Effects)

| | $\theta_{1R} = F_1 - F_R$ | | | | | $\theta_{2R} = F_2 - F_R$ | | | | |
| | Period | | | | | Period | | | | |
Sequence	1	2	3	4	$\Sigma_j C^2_{ajk}$	1	2	3	4	$\Sigma_j C^2_{ajk}$
1	-1	0	1	0	1/8	-1	0	0	1	1/8
2	1	-1	0	0	1/8	0	-1	1	0	1/8
3	0	1	0	-1	1/8	1	0	0	-1	1/8
4	0	0	-1	1	1/8	0	1	-1	0	1/8
Variance[b]					$\dfrac{1}{2n}\sigma_e^2$					$\dfrac{1}{2n}\sigma_e^2$

| | $\theta_{3R} = F_3 - F_R$ | | | | | $\theta_{21} = F_2 - F_1$ | | | | |
| | Period | | | | | Period | | | | |
Sequence	1	2	3	4	$\Sigma_j C^2_{ajk}$	1	2	3	4	$\Sigma_j C^2_{ajk}$
1	-1	1	0	0	1/8	0	0	-1	1	1/8
2	0	-1	0	1	1/8	-1	0	1	0	1/8
3	0	0	1	-1	1/8	1	-1	0	0	1/8
4	1	0	-1	0	1/8	0	1	0	-1	1/8
Variance[b]					$\dfrac{1}{2n}\sigma_e^2$					$\dfrac{1}{2n}\sigma_e^2$

| | $\theta_{31} = F_3 - F_1$ | | | | | $\theta_{32} = F_3 - F_2$ | | | | |
| | Period | | | | | Period | | | | |
Sequence	1	2	3	4	$\Sigma_j C^2_{ajk}$	1	2	3	4	$\Sigma_j C^2_{ajk}$
1	0	1	-1	0	1/8	0	1	0	-1	1/8
2	-1	0	0	1	1/8	0	0	-1	1	1/8
3	0	-1	1	0	1/8	-1	0	1	0	1/8
4	1	0	0	-1	1/8	1	-1	0	0	1/8
Variance[b]					$\dfrac{1}{2n}\sigma_e^2$					$\dfrac{1}{2n}\sigma_e^2$

[a]Coefficients are multiplied by 4.

[b]Variance when $n = n_k$ for $1 \leqslant k \leqslant 4$.

Table 10.3.13   AUC Data from a Study in Purich (1980)

Sequence[a]	Subject	Period I	Period II	Period III
(R,T_2,T_1)	1	5.68	4.21	6.83
	2	3.60	5.01	5.78
(T_1,R,T_2)	3	3.55	5.07	4.49
	4	7.31	7.42	7.86
(T_2,T_1,R)	5	6.59	7.72	7.26
	6	9.68	8.91	9.04
(T_1,T_2,R)	7	4.63	7.23	5.06
	8	8.75	7.59	4.82
(T_2,R,T_1)	9	7.25	7.88	9.02
	10	5.00	7.84	7.79
(R,T_1,T_2)	11	4.63	6.77	5.72
	12	3.87	7.62	6.74

[a]R = solution, T_1 = domestic tablet, T_2 = European tablet.
Source: Purich (1980).

valence between formulation h and h′ can be evaluated using either the confidence interval (10.2.9) or the two one-sided tests (10.2.10).

Example 10.3.1   To illustrate the methods discussed in previous sections, we use the AUC data from a study discussed in Purich (1980). This study was conducted with 12 normal subjects to establish bioequivalence among two formulations of a 100 mg beta blocker tablets (domestic and European) and a solution

Table 10.3.14   Sequence by Period Means of Purich's Data

Sequence	Period I	Period II	Period III
1	4.640	4.610	6.305
2	5.430	6.245	6.175
3	8.135	8.315	8.150
4	6.690	7.410	4.940
5	6.125	7.860	8.405
6	4.250	7.195	6.230

Table 10.3.15 Estimates of Formulation Effects for Purich's Data

Carry-over effect	Solution	Domestic	European
Yes	5.97	7.24	6.30
No	6.01	7.06	6.45

(reference). There was a 7-day washout between treatment periods. No assignment of sequence and period was given in Purich's paper. Thus, for the purpose of illustration, we assign subjects 1 and 2 to sequence 1; 3 and 4 to sequence 2; 5 and 6 to sequence 3; 7 and 8 to sequence 4; 9 and 10 to sequence 5; 11 and 12 to sequence 6. Table 10.3.13 gives the AUC data after rearrangement of reference and period according to the Williams design for comparing three formulations. The sequence by period means are given in Table 10.3.14. Table 10.3.15 provides estimates of three formulation means both in the presence and absence of carry-over effects. Test results (without adjustment for significance level) for assessing bioequivalence of average bioavailability are summarized in Table 10.3.16.

The carry-over effects are not statistically significant from 0 (p-value = 0.32). Both the confidence interval approach and the two one-sided tests procedure reach the same conclusion on bioequivalence regardless of the presence of the

Table 10.3.16 Summary of Test Results for Purich's Data

Comparison	Carry-over	Tests[a]	P-Value	90% C.I.
T_1 vs R[b]	Yes	$T_L = 5.08$	<0.001	(0.43, 2.11)
		$T_U = 0.16$	0.56	(107.20%, 135.41%)
	No	$T_L = 5.12$	<0.001	(0.29, 1.80)
		$T_U = -0.37$	0.36	(104.75%, 129.92%)
T_2 vs R	Yes	$T_L = 3.14$	0.003	(−0.51, 1.18)
		$T_U = -1.77$	0.047	(91.47%, 119.68%)
	No	$T_L = 3.73$	0.001	(−0.32, 1.19)
		$T_U = -1.75$	0.047	(94.62%, 119.79%)
T_1 vs T_2	Yes	$T_L = 4.39$	<0.001	(0.097, 1.78)
		$T_U = -0.53$	0.30	(101.63%, 129.83%)
	No	$T_L = 4.13$	<0.001	(−0.15, 1.37)
		$T_U = -1.35$	0.096	(97.54%, 122.72%)

[a]Calculations were based upon ± 20 rule and estimate of solution formation mean which is 5.97 in the presence of carry-over effects and is 6.01 in the absence of carry-over effects.
[b]R = solution, T_1 = domestic tablet, and T_2 = European tablet.

carry-over effect. According to the ± 20 rule, the European 100 mg tablet is equivalent to the solution on average bioavailability; but the domestic tablet is not equivalent to the European tablet or to the solution on average bioavailability.

10.3.3. Heterogeneity of Intra-Subject Variabilities

For assessment of bioequivalence in Williams designs with three formulations or four formulations, we assume that $\sigma_h^2 = \sigma_e^2$ for all h. In practice, this is not always true. However, when the intra-subject variabilities differ from formulation to formulation, bioequivalence of average bioavailability can still be established using a Williams design provided that there are no carry-over effects, which is a reasonable assumption when there is a sufficient length of washout between periods.

In the following, statistical methods for the comparison of test formulation 1 with the reference formulation in a Williams design with four formulations will be derived when the heterogeneity of intra-subject variabilities is present. The method can be easily applied to compare any other pair of formulations in any Williams design.

When the intra-subject variabilities are different, the covariance matrix of the vector of four responses

$$\mathbf{Y}_{ik} = (Y_{i1k}, Y_{i2k}, Y_{i3k}, Y_{i4k})',$$

which are observed on subject i in sequence k, is given by

$$\mathrm{Var}(\mathbf{Y}_{ik}) = \Gamma_k + \sigma_s^2 J_4, \qquad i = 1,2,\ldots,n_k; \quad k = 1,2,3,4,$$

where J_4 is a 4×4 matrix of 1 and

$$\Gamma_k = \begin{cases} \mathrm{diag}(\sigma_R^2, \sigma_3^2, \sigma_1^2, \sigma_2^2), & \text{if } k = 1, \\ \mathrm{diag}(\sigma_1^2, \sigma_R^2, \sigma_2^2, \sigma_3^2), & \text{if } k = 2, \\ \mathrm{diag}(\sigma_2^2, \sigma_1^2, \sigma_3^2, \sigma_R^2), & \text{if } k = 3, \\ \mathrm{diag}(\sigma_3^2, \sigma_2^2, \sigma_R^2, \sigma_1^2), & \text{if } k = 4, \end{cases} \tag{10.3.1}$$

where $\mathrm{diag}(a_1,a_2,a_3,a_4)$ is a 4×4 diagonal matrix with diagonal elements a_1, a_2, a_3, and a_4. Define the period differences as the following intra-subject contrasts:

$$d_{ik} = \begin{cases} \frac{1}{4}[Y_{i31} - Y_{i11}], & \text{if } k = 1, \\ \frac{1}{4}[Y_{i12} - Y_{i22}], & \text{if } k = 2, \\ \frac{1}{4}[Y_{i23} - Y_{i43}], & \text{if } k = 3, \\ \frac{1}{4}[Y_{i44} - Y_{i34}], & \text{if } k = 4. \end{cases} \tag{10.3.2}$$

Then, under normality assumptions, d_{ik} are independently normally distributed with the common variance

$$\mathrm{Var}(d_{ik}) = \tfrac{1}{16}(\sigma_1^2 + \sigma_R^2), \tag{10.3.3}$$

and the means

$$
E(d_{ik}) = \begin{cases} \frac{1}{4}[(F_1 - F_R) + (P_3 - P_1)], & \text{if } k = 1, \\ \frac{1}{4}[(F_1 - F_R) + (P_1 - P_2)], & \text{if } k = 2, \\ \frac{1}{4}[(F_1 - F_R) + (P_2 - P_4)], & \text{if } k = 3, \\ \frac{1}{4}[(F_1 - F_R) + (P_4 - P_3)], & \text{if } k = 4. \end{cases} \tag{10.3.4}
$$

Let $\bar{d}_{.k}$ be the mean of d_{ik} for sequence i and $\bar{d}$ be the sum of $\bar{d}_{.k}$, i.e.,

$$
\bar{d} = \sum_{k=1}^{4} \bar{d}_{.k} = \sum_{k=1}^{4} \frac{1}{n_k} \sum_{i=1}^{n_k} d_{ik}. \tag{10.3.5}
$$

Then $\bar{d}$ is normally distributed with mean $\theta_{1R} = F_1 - F_R$ and variance

$$
\text{Var}(\bar{d}) = \frac{1}{16}(\sigma_1^2 + \sigma_R^2) \sum_{k=1}^{4} \frac{1}{n_k}.
$$

Therefore, $\bar{d}$ is an unbiased estimator of θ_{1R}, and an unbiased estimator of $\frac{1}{16}(\sigma_1^2 + \sigma_R^2)$ is given by

$$
S_d^2 = \frac{1}{N-4} \sum_{k=1}^{4} \sum_{i=1}^{n_k} (d_{ik} - \bar{d}_{.k})^2, \tag{10.3.6}
$$

which is distributed as $\frac{1}{N-4}\chi^2(N - 4)$. Since S_d^2 is independent of $\bar{d}$, the statistic

$$
T_d = \frac{\bar{d}}{\left[S_d^2 \sum_{k=1}^{4} \frac{1}{n_k} \right]^{1/2}}
$$

has a central t distribution with $N - 4$ degrees of freedom. Based upon $\bar{d}$ and S_d^2, assessment of bioequivalence of average bioavailability can be established using either (10.2.8) for the confidence interval approach or (10.2.9) for the two one-sided tests procedure. From the above method, it can be seen that the heterogeneity of intra-subject variabilities causes a decrease in degrees of freedom from $3(N - 3)$ to $(N - 4)$ or a loss in precision. This is because we only use the information from the periods where test formulation 1 and the reference formulation are administered.

For assessment of bioequivalence of intra-subject variabilities between a test formulation and the reference formulation in a Williams design, the test procedure for hypotheses (7.5.2) described in Section 7.5 can be directly applied. The procedure is outlined below.

Let R_{iRk} and R_{iT_1k} be the respective residuals from the sequence by period means corresponding to the periods of sequence k where test formulation 1 and the reference formulation are administered. We first calculate

$$V_{ik} = R_{iT_1k} - R_{iRk},$$
$$U_{Lik} = R_{iT_1k} + \delta_L R_{iRk},$$
$$U_{Uik} = R_{iT_1k} + \delta_U R_{iRk}, \tag{10.3.7}$$

where $i = 1, 2, \ldots, n_k$; $k = 1, 2, 3, 4$.

Then, compute Pearson's (Spearman's) correlation coefficients $\hat{\rho}_L$ ($\tilde{\rho}_L$) between V_{ik} and U_{Lik} and $\hat{\rho}_U$ ($\tilde{\rho}_U$) between V_{ik} and U_{Uik} as described in Chapter 7. If normality assumptions are satisfied, test formulation 1 and the reference formulation are considered to be bioequivalent in intra-subject variabilities at the α level of significance if

$$t_L = \frac{\hat{\rho}_L}{\left[\dfrac{1 - \hat{\rho}_L^2}{N - 5}\right]^{1/2}} > t(\alpha, N - 5),$$

and

$$t_U = \frac{\hat{\rho}_U}{\left[\dfrac{1 - \hat{\rho}_U^2}{N - 5}\right]^{1/2}} < -t(\alpha, N - 5). \tag{10.3.8}$$

On the other hand, if the normality assumptions are in doubt, nonparametric test procedure such as Spearman's rank correlation coefficient can be used. We conclude bioequivalence if

$$\tilde{\rho}_L > r_S(\alpha, N - 4),$$

and

$$\tilde{\rho}_U < -r_S(\alpha, N - 4), \tag{10.3.9}$$

where $r_S(\alpha, N - 4)$ is the αth upper quantile of distribution of Spearman's rank correlation coefficient based upon the $N - 4$ observations.

To illustrate the use of the above methods, we use Purich's data which was described in Example 10.3.1. Test results for assessment of intra-subject variabilities are summarized in Table 10.3.17. The results indicate that all three formulations are not equivalent to each other in intra-subject variability.

10.4. ANALYSIS FOR BALANCED INCOMPLETE BLOCK DESIGN

In Chapter 2, several balanced incomplete block designs were constructed for comparing four and five formulations. These designs include four formulations

Table 10.3.17 Summary of Intra-Subject Variabilities Test Results for Purich's Data

Comparison[a]	Hypotheses[b]	Correlation	
		Pearson (P-Value)	Spearman (P-Value)
T_1 vs R	$\rho_L > 0$	0.55 (0.98)	0.61 (0.05)
	$\rho_U < 0$	0.47 (0.85)	0.55 (>0.05)
T_2 vs R	$\rho_L > 0$	0.45 (0.15)	0.38 (>0.05)
	$\rho_U < 0$	0.36 (0.79)	0.38 (>0.05)
T_1 vs T_2	$\rho_L > 0$	−0.12 (0.60)	−0.03 (>0.05)
	$\rho_U < 0$	−0.21 (0.33)	−0.17 (>0.05)

[a]R = solution, T_1 = domestic tablet, and T_2 = European tablet.
[b]Based upon hypotheses (7.2.3) with equivalent limits ±20%.

with two or three periods and five formulations with two, three, or four periods. For the analyses of these balance incomplete block designs, statistical methods introduced in Section 10.2 can be directly applied. Therefore, in this section, for the purpose of illustration, we will focus on a balanced incomplete block design for comparing four formulations with three periods. Other balanced incomplete block designs can be treated similarly. The balanced incomplete block design for comparing four formulations with three periods is summarized below:

Balanced Incomplete Block Design for Three Formulations with Three Periods			
	Period		
Sequence	I	II	III
1	T_1	T_2	T_3
2	T_2	T_3	R
3	T_3	R	T_1
4	R	T_1	T_2

From the above design, one can see that only subjects in sequences 2, 3, and 4 receive the reference formulation. In other words, comparison within each

Table 10.4.1 Expected Values of the Sequence-by-Period Means for BIBD with Four Formulations and Three Periods

Sequence	Period		
	I	II	III
1	$\mu + G_1 + P_1 + F_1$	$\mu + G_1 + P_2 + F_2 + C_1$	$\mu + G_1 + P_3 + F_3 + C_2$
2	$\mu + G_2 + P_1 + F_2$	$\mu + G_2 + P_2 + F_3 + C_2$	$\mu + G_2 + P_3 + F_R + C_3$
3	$\mu + G_3 + P_1 + F_3$	$\mu + G_3 + P_2 + F_R + C_3$	$\mu + G_3 + P_3 + F_1 + C_R$
4	$\mu + G_4 + P_1 + F_R$	$\mu + G_4 + P_2 + F_1 + C_R$	$\mu + G_4 + P_3 + F_2 + C_1$

subject between a test formulation and the reference formulation can only be made by using data from these sequences. There are a total of six pairs of formulations to be compared.

Table 10.4.1 and Table 10.4.2 give the expected values of the sequence by period means and the analysis of variance table for the distribution of the degrees of freedoms. Coefficients of the linear contrasts for the unbiased ordinary least squares estimates of F_h, $\theta_{hh'}$ and $\lambda_{hh'}$, $1 \leq h \neq h' \leq 4$, are given in Tables

Table 10.4.2 Analysis of Variance Table for the BIBD with Four Formulations and Three Periods

Source of variation	Degrees of freedom[a]
Inter-subject	$N^b - 1$
Sequence	3
Residual	$N - 4$
Intra-subject	$2N$
Period	2
Formulation	3
Carry-over	3
Residual	$2(N - 4)$
Total	$3N - 1$

[a]Degrees of freedom for the intra-subject residual is $2N - 5$ if carry-over effects are not included in the model.

[b]$N = n_1 + n_2 + n_3 + n_4$.

Table 10.4.3 Coefficients[a] for Estimates of Formulations F_R, F_1, F_2, and F_3 in the BIBD with Four Formulations and Three Periods (Adjusted for Carry-Over Effects)

	F_R Period			F_1 Period			F_2 Period			F_3 Period		
Sequence	1	2	3	1	2	3	1	2	3	1	2	3
1	-2	3	-1	6	-3	-3	-2	-1	3	-2	1	1
2	-2	1	1	-2	3	-1	6	-3	-3	-2	-1	3
3	-2	-1	3	-2	1	1	-2	3	-1	6	-3	-3
4	6	-3	-3	-2	-1	3	-2	1	1	-2	3	1

[a]The coefficients are multiplied by 8.

10.4.3 through 10.4.6. The degrees of freedom for the intra-subject mean squared error is $2(N - 4)$, where $N = n_1 + n_1 + n_3 + n_4$. Based upon the coefficients given in Tables 10.4.3 through 10.4.6, bioequivalence of average bioavailability between formulation h and h′ can be evaluated using either the confidence interval (10.2.9) or the two one-sided tests (10.2.10).

From Table 10.4.5, it can be seen that when there are no carry-over effects, the balance incomplete block design is a variance-balanced design. However, in the presence of carry-over effect, this is not true as indicated in Table 10.4.4. This fact suggests that, in the interest of having the property of variance-balanced, a sufficient length of washout should be given between dosing periods to wear off the possible carry-over effects.

Note that the p-values of Anderson and Hauck's procedure for interval hypotheses (4.3.1) can also be obtained by simply plugging T_L and T_U of (10.2.10) into either (4.3.12) or (4.3.13).

10.5. DISCUSSION

In this chapter, bioequivalence in average bioavailability was assessed based upon statistical inference on the difference in formulation means under an additive model (10.2.1). When comparing more than two formulations, Locke (1990) provided a procedure for assessment of bioequivalence based upon the ratio of formulation means under the same additive model which requires a rather weak assumption for covariance matrix. This procedure is basically a generalization of the exact confidence interval based upon Fieller's theorem discussed in Section 4.2.3.

When comparing more than two formulations in a study, multiple comparisons are usually made. For example, for comparing three formulations, there are a total of three comparisons, namely, formulations 1 vs 2; formulations 1 vs 3;

Table 10.4.4 Coefficients[a] for Estimates of Pairwise Formulation Effects in BIBD with Four Formulations and Three Periods (Adjusted for Carry-Over Effects)

	$\theta_{1R} = F_1 - F_R$				$\theta_{2R} = F_2 - F_R$			
	Period				Period			
Sequence	1	2	3	$\Sigma_j C_{ajk}^2$	1	2	3	$\Sigma_j C_{ajk}^2$
1	8	−6	−2	$104/(8)^2$	0	−4	4	$32/(8)^2$
2	0	2	−2	$8/(8)^2$	8	−4	−4	$96/(8)^2$
3	0	2	−2	$8/(8)^2$	0	4	−4	$32/(8)^2$
4	−8	2	6	$104/(8)^2$	−8	4	4	$96/(8)^2$
Variance[b]				$\dfrac{7}{2n}\sigma_e^2$				$\dfrac{4}{n}\sigma_e^2$

	$\theta_{3R} = F_3 - F_R$				$\theta_{21} = F_2 - F_1$			
	Period				Period			
Sequence	1	2	3	$\Sigma_j C_{ajk}^2$	1	2	3	$\Sigma_j C_{ajk}^2$
1	0	−2	2	$8/(8)^2$	−8	2	6	$104/(8)^2$
2	0	−2	2	$8/(8)^2$	8	−6	−2	$104/(8)^2$
3	8	−2	−6	$104/(8)^2$	0	2	−2	$8/(8)^2$
4	−8	6	2	$104/(8)^2$	0	2	−2	$8/(8)^2$
Variance[b]				$\dfrac{7}{2n}\sigma_e^2$				$\dfrac{7}{2n}\sigma_e^2$

	$\theta_{31} = F_3 - F_1$				$\theta_{32} = F_3 - F_2$			
	Period				Period			
Sequence	1	2	3	$\Sigma_j C_{ajk}^2$	1	2	3	$\Sigma_j C_{ajk}^2$
1	−8	4	4	$96/(8)^2$	0	2	−2	$8/(8)^2$
2	0	−4	4	$32/(8)^2$	−8	2	6	$104/(8)^2$
3	8	−4	−4	$96/(8)^2$	8	−6	−2	$104/(8)^2$
4	0	4	−4	$32/(8)^2$	0	2	−2	$8/(8)^2$
Variance[b]				$\dfrac{4}{n}\sigma_e^2$				$\dfrac{7}{2n}\sigma_e^2$

[a]Coefficients are multiplied by 8.
[b]Variance when $n = n_k$ for $1 \leqslant k \leqslant 4$.

Table 10.4.5 Coefficients[a] for Estimates of Pairwise Formulation Effects in BIBD with Four Formulations and Three Periods (In Absence of Unequal Carry-Over Effects)

	$\theta_{1R} = F_1 - F_R$				$\theta_{2R} = F_2 - F_R$			
	Period				Period			
Sequence	1	2	3	$\Sigma_j C_{ajk}^2$	1	2	3	$\Sigma_j C_{ajk}^2$
1	2	-1	-1	$6/(8)^2$	-1	2	-1	$6/(8)^2$
2	1	1	-2	$6/(8)^2$	3	0	-3	$18/(8)^2$
3	0	-3	3	$18/(8)^2$	1	-2	1	$6/(8)^2$
4	-3	3	0	$18/(8)^2$	-3	0	3	$18/(8)^2$
Variance[b]				$\dfrac{3}{4n}\sigma_e^2$				$\dfrac{3}{4n}\sigma_e^2$

	$\theta_{3R} = F_3 - F_R$				$\theta_{21} = F_2 - F_1$			
	Period				Period			
Sequence	1	2	3	$\Sigma_j C_{ajk}^2$	1	2	3	$\Sigma_j C_{ajk}^2$
1	-1	-1	2	$6/(8)^2$	-3	3	0	$18/(8)^2$
2	0	3	-3	$18/(8)^2$	2	-1	-1	$6/(8)^2$
3	3	-3	0	$18/(8)^2$	1	1	-2	$6/(8)^2$
4	-2	1	1	$6/(8)^2$	0	-3	3	$18/(8)^2$
Variance[b]				$\dfrac{3}{4n}\sigma_e^2$				$\dfrac{3}{4n}\sigma_e^2$

	$\theta_{31} = F_3 - F_1$				$\theta_{32} = F_3 - F_2$			
	Period				Period			
Sequence	1	2	3	$\Sigma_j C_{ajk}^2$	1	2	3	$\Sigma_j C_{ajk}^2$
1	-3	0	3	$18/(8)^2$	0	-3	3	$18/(8)^2$
2	-1	2	-1	$6/(8)^2$	-3	3	0	$18/(8)^2$
3	3	0	-3	$18/(8)^2$	2	-1	-1	$6/(8)^2$
4	1	-2	1	$6/(8)^2$	1	1	-2	$6/(8)^2$
Variance[b]				$\dfrac{3}{4n}\sigma_e^2$				$\dfrac{3}{4n}\sigma_e^2$

[a]Coefficients are multiplied by 8.

[b]Variance when $n = n_k$ for $1 \leq k \leq 4$.

Table 10.4.6 Coefficients[a] for Estimates of Carry-over Effects in BIBD with Four Formulations and Three Periods (Adjusted for Formulation Effects)

	$\lambda_{1R} = C_1 - C_R$				$\lambda_{2R} = C_2 - C_R$			
	Period				Period			
Sequence	1	2	3	$\Sigma_j C_{ajk}^2$	1	2	3	$\Sigma_j C_{ajk}^2$
1	4	0	-4	$32/(4)^2$	4	-4	0	$32/(4)^2$
2	-4	4	0	$32/(4)^2$	0	4	-4	$32/(4)^2$
3	0	0	0	$0/(4)^2$	-4	4	0	$32/(4)^2$
4	0	-4	4	$32/(4)^2$	0	-4	4	$32/(4)^2$
Variance[b]				$\dfrac{6}{n}\sigma_e^2$				$\dfrac{8}{n}\sigma_e^2$

	$\lambda_{3R} = C_3 - C_R$				$\lambda_{21} = C_2 - C_1$			
	Period				Period			
Sequence	1	2	3	$\Sigma_j C_{ajk}^2$	1	2	3	$\Sigma_j C_{ajk}^2$
1	4	-4	0	$32/(4)^2$	0	-4	4	$32/(4)^2$
2	0	0	0	$0/(4)^2$	4	0	-4	$32/(4)^2$
3	0	4	-4	$32/(4)^2$	-4	4	0	$32/(4)^2$
4	-4	0	4	$32/(4)^2$	0	0	0	$0/(4)^2$
Variance[b]				$\dfrac{6}{n}\sigma_e^2$				$\dfrac{6}{n}\sigma_e^2$

	$\lambda_{31} = C_3 - C_1$				$\lambda_{32} = C_3 - C_2$			
	Period				Period			
Sequence	1	2	3	$\Sigma_j C_{ajk}^2$	1	2	3	$\Sigma_j C_{ajk}^2$
1	0	-4	4	$32/(4)^2$	0	0	0	$0/(4)^2$
2	4	-4	0	$32/(4)^2$	0	-4	4	$32/(4)^2$
3	0	4	-4	$32/(4)^2$	4	0	-4	$32/(4)^2$
4	-4	4	0	$32/(4)^2$	-4	4	0	$32/(4)^2$
Variance[b]				$\dfrac{8}{n}\sigma_e^2$				$\dfrac{6}{n}\sigma_e^2$

[a]Coefficients are multiplied by 4.

[b]Variance when $n = n_k$ for $1 \leqslant k \leqslant 4$.

and formulations 2 vs 3. In this case, whether the overall type I error rate should be adjusted is a question of great interest. It is our feeling that if the purpose of the study is to establish equivalence between each test formulation and the reference formulation and it is stated in the study protocol, then the adjustment of the significance level is not warranted. However, some comparisons are not specified in advance in the protocol and are carried out after the study, then appropriate adjustment of overall type I error rate is necessary.

Suppose we are interested in comparing k test formulations, T_i, $i = 1,2, \ldots ,k$, and a reference formulation, R. The assessment of bioequivalence basically involves either (i) T_i vs R for all i, or (ii) T_i vs T_j for $i \neq j$, or (iii) pairwise comparison among T_i, $i = 1,2, \ldots ,k$ and R. For the first case, the method described in this chapter can be applied directly. For case (ii), since the reference formulation R is not involved, the conclusion of bioequivalence may depend upon which formulation (i.e., T_i or T_j) is used as the reference formulation. For example, it is possible that T_1 is bioequivalent to T_2, which is treated as the reference formulation, and T_3 is bioequivalent to T_2. However, T_1 may not be bioequivalent to T_3. This may be partially explained by the fact that T_2 is not used as the reference formulation when comparing T_1 and T_3. In this case, it is suggested that the formulation to be used as the reference formulation be specified in the protocol with appropriate adjustment of overall type I error rate if necessary.

For comparing more than two formulations, as indicated earlier, a Williams design or a balance incomplete block design is usually considered. As discussed in previous sections, a Williams design is not only a variance-balanced design for the formulation effect but also a variance-balanced design for the carry-over effect. In other words, it is a variance-balanced design regardless of the presence or absence of the carry-over effect. On the other hand, although the balance incomplete design is a variance-balanced design when there are no carry-over effects, it is not a variance-balanced design in the presence of carry-over effect. Moreover, the degrees of freedom for the mean squared error for the balanced incomplete block design is smaller than that of the Williams design. For example, for comparing four formulations, the degrees of freedom for the balanced incomplete block design is $2(N - 4)$, while the degrees of freedom for the Williams design is $3(N - 3)$. When there are no carry-over effects, the variance for the balanced incomplete block design is much larger than that of the Williams design, i.e.,

$$\frac{3}{4n} \sigma_e^2 \quad \text{vs} \quad \frac{1}{2n} \sigma_e^2.$$

In the presence of the carry-over effect, the variance for the balanced incomplete block design is even larger. Although the Williams design possesses some good properties, it does require the number of periods to at least be equal to the number of formulations being compared, while the balanced incomplete block design does not have this limitation. When there are a large number of formulations to be compared, a balanced incomplete design may be preferred because it can be carried out on a small number of periods. Therefore, to choose between a Williams design and a balanced block design for comparing more than two formulations, the relative gain and loss between the two designs should be taken into consideration.

11

Assessment of Bioequivalence for Drugs with Negligible Plasma Levels

11.1. INTRODUCTION

In previous chapters, bioequivalence between formulations are evaluated based upon pharmacokinetic responses such as AUC, C_{max}, and t_{max}. These responses are usually determined from the blood or plasma concentration-time curve. For some drug products, however, we may have negligible plasma levels because of their intended routes of administration. These drug products include metered dose inhalers (MDI) indicated for the relief of bronchospasm in patients with reversible obstructive airway disease; anti-ulcer agents such as sucralfate; and topical antifungals and vaginal antifungals. Since these products have negligible plasma concentrations, pharmacokinetic responses are no longer adequate for assessment of bioequivalence between drug products. In this case, it is suggested that some other clinical endpoints be used to assess bioequivalence between drug products.

11.2. DESIGN AND CLINICAL ENDPOINTS

In 1989, the FDA issued a guidance for in vivo bioequivalence studies of metaproterenol sulfate and albuterol inhalation aerosols (i.e., MDI). In the FDA guidance, it is suggested that a four-sequence, four-period, double-blind, randomized crossover design be used for comparing drug products with negligible

Table 11.2.1   Design For Assessment of Bioequivalence for MDI Products
Suggested by the FDA 1989 Guidance

	Day			
Sequence	1	2	3	4
1	A	C	B	D
2	C	A	D	B
3	B	D	A	C
4	D	B	C	A
Formulation	Albuterol		Metaproterenol sulfate	
A	Generic 1 puff		Generic 2 puffs	
B	Generic 2 pufs		Generic 3 puffs	
C	Reference 1 puff		Reference 2 puffs	
D	Reference 2 puffs		Reference 3 puffs	
Dose	90 μg/puff		0.65 mg/puff	

plasma levels. The recommended design evaluates clinical endpoints of two products at two different dose levels. This design is summarized in Table 11.2.1. This design is of particular interest because it consists of two levels of crossover factors: the dose and the product. For the dose, we may classify the four sequences into two strata (e.g., sequences 1 and 2 to stratum 1, and sequences 3 and 4 to stratum 2) and the four days into two periods (e.g., days 1 and 2 to period I, and days 3 and 4 to period II). In this case, the design with albuterol MDI can be expressed as follows:

	Time block	
Stratum	I	II
1	90 μg	180 μg
2	180 μg	90 μg

The arrangement based upon the dose is a standard two sequence, two period crossover design. The second level is the product. From the above, it can be seen that each combination of stratum and period (four in total) is also a standard 2×2 crossover design, where the test and reference products are administered at the same dose level.

The advantage of inhalation therapy is that the active ingredient of the drug can be directly delivered to the site of action. In this case, lower dose can be employed and the risk of systematic adverse experiences can be minimized. However, since MDI products produce negligible plasma concentrations, the assessment of bioequivalence will have to based upon other meaningful clinical endpoints. The majority of clinical endpoints recommended by the FDA guidance are derived from the volume of air forced out of lung within one second (FEV_1), which is usually measured at 0, 10, 15, 30, 60, 90, 120, 180, 240, 300, and 360 minutes after dosing. For assessment of bioequivalence between MDI products, the FDA guidance requires that the following information obtained from FEV_1 be provided.

a) onset of therapeutic response;
b) duration of therapeutic response;
c) AUC from the onset of the response to hour 3 based upon FEV_1-time curve;
d) AUC from the onset of the response to the time of termination of the response based upon FEV_1-time curve;
e) FEV_{1max};
f) t_{max};
g) FEV_1 at each time point.

A therapeutic response is defined to be the event such that the postdose FEV_1 measurement exceeds 115% of the baseline value. Onset of the response is the time within 30 minutes postdose when the event of a therapeutic response occurs. Time of onset of the response is calculated by linear interpolation between the first postdose FEV_1 exceeding 115% of the baseline value and the FEV_1 value immediately preceding. The termination of a therapeutic response is defined to be the occurrence of the event of two consecutive FEV_1 measurements falling below 115% of the baseline value before or at hour 6 provided that the event of a therapeutic response has occurred. The time of termination of response is estimated by linear interpolation between the last postdose FEV_1 value exceeding 115% and the first FEV_1 value below 115% after a therapeutic response has occurred. Therefore, the duration of the event of a therapeutic response is the time interval between onset and termination of the response. FEV_{1max} and t_{max} can be computed from the FEV_1-time curve in the same way as C_{max} and t_{max} are computed from the plasma concentration-time curve.

11.3. STATISTICAL CONSIDERATIONS

11.3.1. Continuous Endpoint

For each clinical endpoint, there are two major comparisons, which are (i) to assess bioequivalence between two products at each dose and (ii) to differentiate clinical response affected by a two-fold difference in dose. In other words, statistical comparisons of interest are

a) bioequivalence at 90 μg: treatment A vs treatment C;
b) bioequivalence at 180 μg: treatment B vs treatment D;
c) comparison between 90 μg and 180 μg for generic products: treatment A vs treatment B;
d) comparison between 90 μg and 180 μg for reference products: treatment C vs treatment D.

For continuous variables such as AUC, under the assumption of normality and compound symmetry, analyses can be performed under model (10.2.1). Table 11.3.1 gives the expected values of sequence-by-period means. The coefficients of linear contracts of sequence-by-period means for the unbiased ordinary least squares (OLS) estimates of $\theta_{AC} = F_A - F_C$, $\theta_{BD} = F_B - F_D$, $\theta_{BA} = F_B - F_A$, and $\theta_{DC} = F_D - F_C$ are given in Table 11.3.2 for the situation where the carry-over effect is present. In the absence of carry-over effect, the coefficients are listed in Table 11.3.3. From Table 11.3.2, when there are carry-over effects, it can be seen that the variance of the unbiased OLS estimates for assessment of bioequivalence between two products is $(33\sigma_e^2)/(20n)$, which is three times as large as that for comparison between two doses of the same product. This is because the design is not a variance-balanced design in presence of the carry-over effect. When there are no carry-over effects, it can be seen

Table 11.3.1 Expected Values of Sequence-by-Period Means for Design in Table 11.2.1

Sequence	Period			
	I	II	III	IV
1	$\mu + G_1 +$ $P_1 + F_A$	$\mu + G_1 + P_2$ $+ F_C + C_A$	$\mu + G_1 + P_3$ $+ F_B + C_C$	$\mu + G_1 + P_4$ $+ F_D + C_B$
2	$\mu + G_2 +$ $P_1 + F_C$	$\mu + G_2 + P_2$ $+ F_A + C_C$	$\mu + G_2 + P_3$ $+ F_D + C_A$	$\mu + G_2 + P_4$ $+ F_B + C_D$
3	$\mu + G_3 +$ $P_1 + F_B$	$\mu + G_3 + P_2$ $+ F_D + C_B$	$\mu + G_3 + P_3$ $+ F_A + C_D$	$\mu + G_3 + P_4$ $+ F_C + C_A$
4	$\mu + G_4 +$ $P_1 + F_D$	$\mu + G_4 + P_2$ $+ F_B + C_D$	$\mu + G_4 + P_3$ $+ F_C + C_B$	$\mu + G_4 + P_4$ $+ F_A + C_C$

Table 11.3.2 Coefficients[a] for OLS Estimates of Pairwise Formulation Effects for Design in Table 11.2.1 (Adjusted for Carry-Over Effects)

I. Comparisons between products at the same dose

	90 μg $\theta_{AC} = F_A - F_C$					180 μg $\theta_{BD} = F_B - F_D$				
	Period					Period				
Sequence	1	2	3	4	$\Sigma_j C_{ajk}^2$	1	2	3	4	$\Sigma_j C_{ajk}^2$
1	26	−8	−17	−1	$1030/(40)^2$	14	−2	−3	−9	$290/(40)^2$
2	−26	8	17	1	$1030/(40)^2$	−14	2	3	9	$290/(40)^2$
3	14	−2	−3	−9	$290/(40)^2$	26	−8	−17	−1	$1030/(40)^2$
4	−14	2	3	9	$290/(40)^2$	−26	8	17	1	$1030/(40)^2$
Variance[b]					$\dfrac{33}{20n}\sigma_e^2$					$\dfrac{33}{20n}\sigma_e^2$

II. Comparisons between doses of the same product

	Generic $\theta_{BA} = F_B - F_A$					Reference $\theta_{DC} = F_D - F_C$				
	Period					Period				
Sequence	1	2	3	4	$\Sigma_j C_{ajk}^2$	1	2	3	4	$\Sigma_j C_{ajk}^2$
1	−12	−1	14	−1	$342/(40)^2$	0	−7	0	7	$98/(40)^2$
2	0	−7	0	7	$98/(40)^2$	−12	−1	14	−1	$342/(40)^2$
3	12	1	−14	1	$342/(40)^2$	0	7	0	−7	$98/(40)^2$
4	0	7	0	−7	$98/(40)^2$	12	1	−14	1	$342/(40)^2$
Variance[b]					$\dfrac{11}{20n}\sigma_e^2$					$\dfrac{11}{20n}\sigma_e^2$

[a]Coefficients are multiplied by 40.

[b]Variance when $n = n_k$ for $1 \leq k \leq 4$.

Table 11.3.3 Coefficients[a] for OLS Estimates of Pairwise Formulation Effects for Design in Table 11.2.1 (In Absence of Carry-Over Effects)

I. Comparison between products at the same dose

| | 90 μg $\theta_{AC} = F_A - F_C$ | | | | | 180 μg $\theta_{BD} = F_B - F_D$ | | | | |
| | Period | | | | | Period | | | | |
Sequence	1	2	3	4	$\Sigma_j C_{ajk}^2$	1	2	3	4	$\Sigma_j C_{ajk}^2$
1	4	−4	0	0	$32/(16)^2$	0	0	4	−4	$32/(16)^2$
2	−4	4	0	0	$32/(16)^2$	0	0	−4	4	$32/(16)^2$
3	0	0	4	−4	$32/(16)^2$	4	−4	0	0	$32/(16)^2$
4	0	0	−4	4	$32/(16)^2$	−4	4	0	0	$32/(16)^2$
Variance[b]					$\dfrac{1}{2n}\sigma_e^2$					$\dfrac{1}{2n}\sigma_e^2$

II. Comparisons between doses of the same product

| | Generic $\theta_{BA} = F_B - F_A$ | | | | | Reference $\theta_{DC} = F_D - F_C$ | | | | |
| | Period | | | | | Period | | | | |
Sequence	1	2	3	4	$\Sigma_j C_{ajk}^2$	1	2	3	4	$\Sigma_j C_{ajk}^2$
1	−4	0	4	0	$32/(16)^2$	0	4	0	−4	$32/(16)^2$
2	0	−4	0	4	$32/(16)^2$	−4	0	4	0	$32/(16)^2$
3	4	0	−4	0	$32/(16)^2$	0	4	0	−4	$32/(16)^2$
4	0	4	0	−4	$32/(16)^2$	4	0	−4	0	$32/(16)^2$
Variance[b]					$\dfrac{1}{2n}\sigma_e^2$					$\dfrac{1}{2n}\sigma_e^2$

[a]Coefficients are multiplied by 16.
[b]Variance when $n = n_k$ for $1 \leq k \leq 4$.

from Table 11.3.3 that all OLS estimates have the same variance $\sigma_e^2/(2n)$, which is the same as the variance of the unbiased OLS estimates obtained from the Williams design with four formulations discussed in Chapter 10. Therefore, unless a sufficient length of washout (FDA guidance requires a washout no less than 24 hours between study days) is provided, the Williams design with four formulations is preferred because it provides a smaller (at least the same) variance of the unbiased OLS estimates with the same number of sequences and periods in the presence of carry-over effects.

11.3.2. Binary Endpoint

A patient is considered to be a responder if his/her FEV_1 value exceeds 115% of the baseline value within 30 minutes after dosing. Thus, a therapeutic response is a binary endpoint (i.e., Yes or No). Therefore, for a standard two-sequence (RT and TR), two-period crossover design, we only observe one of the following four possible outcomes on each patient: (N,N), (N,Y), (Y,N), and (Y,Y). Note that (N,Y) denotes that the patient who does not respond to the drug at period I but responds in period II. Table 11.3.4 gives a summary of the response data from a standard 2×2 crossover design in terms of a 2×4 contingency table (observed counts). For example, n_{41} is the number of patients in sequence 1 who responded to the drug in both periods.

The objective of this section is to assess bioequivalence between two products based upon a 90% confidence interval of the ratio of the marginal probability of response for the test product to that for the reference product after adjustment for period effects. In the following sections, we will introduce the following three model-based methods:

1) weighted least squares method (WLS, Koch and Edwards, 1988);
2) log-linear model (Jones and Kenward, 1989);
3) generalized estimating equations (GEEs, Liang and Zeger, 1986).

Table 11.3.4 Summary of Response Data for a Standard 2×2 Crossover Design (Observed Counts by Response)

Sequence	Outcome				Total
	(N,N)	(N,Y)	(Y,N)	(Y,Y)	
1 (RT)	n_{11}	n_{21}	n_{31}	n_{41}	$n_{.1}$
2 (TR)	n_{12}	n_{22}	n_{32}	n_{42}	$n_{.2}$
Total	$n_1.$	$n_2.$	$n_3.$	$n_4.$	$n_{..}$

Source: Liu and Chow (1992b).

Note that the above three methods are derived based upon asymptotic results. To apply these methods, a total sample size of 40 patients as suggested by the FDA guidance might be large enough for MDI products. Let P_T and P_R denote the marginal probabilities of a therapeutic response of the test and reference formulations, respectively. Then, it is suggested that the decision rule for assessment of bioequivalence be based upon a 90% confidence interval of $\delta = P_T/P_R$. Two products are considered to be bioequivalent if the 90% confidence interval of δ is within the prespecified limits (δ_L, δ_U) (Liu and Chow, 1992b).

Similar to continuous data, estimation of direct formulation effect is not valid unless there is no unequal carry-over effects (or direct formulation-by-period interaction, Jones and Kenward, 1989). Therefore, a preliminary test for the presence of unequal carry-over effects should be carried out before assessing bioequivalence. This can be done by simply performing the Fisher's exact test (Armitage and Berry, 1987) for association of the following 2×2 contingency table (Altham, 1971; Hills and Armitage, 1979; Jones and Kenward, 1989).

n_{11}	n_{41}
n_{12}	n_{42}

11.4. WEIGHTED LEAST SQUARES METHOD

Define $\mathbf{P}_k = (P_{1k}, P_{2k}, P_{3k}, P_{4k})'$ as the 4×1 vector of probabilities corresponding to the four possible outcomes in sequence k, where P_{hk}, $h = 1, \ldots, 4$; $k = 1,2$ are summarized in Table 11.4.1. Let $\mathbf{n}_k = (n_{1k}, n_{2k}, n_{3k}, n_{4k})'$ be the 4×1 vector of observed counts of the four possible outcomes in sequence k, $k = 1,2$. Then $\mathbf{n} = (\mathbf{n}_1, \mathbf{n}_2)'$ has a product multinomial distribution with parameter vector $\mathbf{P} = (\mathbf{P}_1, \mathbf{P}_2)'$ and $\mathbf{n}_1$ and $\mathbf{n}_2$ are independent. Let $\mathbf{p} = (\mathbf{p}_1, \mathbf{p}_2)'$ be the 8×1 vector of observed proportions of the four possible outcomes in both sequences, where

$$\mathbf{p}_k = (p_{1k}, p_{2k}, p_{3k}, p_{4k})',$$

$$p_{hk} = \frac{n_{hk}}{n_{\cdot k}}, \quad \text{and}$$

$$n_{\cdot k} = \sum_{k=1}^{4} n_{hk}. \tag{11.4.1}$$

Then, $\mathbf{p}$ is an unbiased estimator of $\mathbf{P}$ whose covariance matrix is given by

$$\text{Var}(\mathbf{p}) = \begin{bmatrix} V_1 & O_{4 \times 4} \\ O_{4 \times 4} & V_2 \end{bmatrix}, \tag{11.4.2}$$

Table 11.4.1 Summary of Response Data for a Standard 2 × 2 Crossover Design (Probabilities by Response)

Sequence	Outcome				Total
	(N,N)	(N,Y)	(Y,N)	(Y,Y)	
1 (RT)	P_{11}	P_{21}	P_{31}	P_{41}	1
2 (TR)	P_{12}	P_{22}	P_{32}	P_{42}	1
Total	$P_{1\cdot}$	$P_{2\cdot}$	$P_{3\cdot}$	$P_{4\cdot}$	

Source: Liu and Chow (1992b).

where

$$V_k = \frac{1}{n_{\cdot k}} [D(\mathbf{P}_k) - \mathbf{P}_k \mathbf{P}_k'], \quad \text{and}$$

$D(\mathbf{P}_k)$ is a 4×4 diagonal matrix with diagonal elements P_{1k}, P_{2k}, P_{3k}, and P_{4k}, $k = 1,2$.

Let $\mathbf{g} = \mathbf{g}(\mathbf{p})$ be a $t \times 1$ vector of linear function $\mathbf{p}$, which can be expressed as

$$\mathbf{g}(\mathbf{p}) = \mathbf{Cp}, \tag{11.4.3}$$

where C is a $t \times 8$ matrix with elements of each row corresponding to the coefficients of linear functions. A consistent estimator of the covariance matrix of g is then given by

$$V_g = \text{Var}(\mathbf{g}) = C \, \hat{\text{V}}\text{ar}(\mathbf{p})C', \tag{11.4.4}$$

where

$$\hat{\text{V}}\text{ar}(\mathbf{p}) = \begin{bmatrix} \frac{1}{n_{\cdot 1}}[D(\mathbf{p}_1) - \mathbf{p}_1 \mathbf{p}_1'] & O_{4 \times 4} \\ O_{4 \times 4} & \frac{1}{n_{\cdot 2}}[D(\mathbf{p}_2) - \mathbf{p}_2 \mathbf{p}_2'] \end{bmatrix}$$

is a consistent estimator of $\text{Var}(\mathbf{p})$ in (11.4.2).

If the total sample size is large enough such that **g** has approximately a multivariate normal distribution, then the formulation effect can be estimated by fitting a linear model using weighted least squares method. Suppose that we can express the expected value of **g** as a linear model given below

$$E(\mathbf{g}) = Z\boldsymbol{\beta}, \tag{11.4.5}$$

where Z is a t × m design matrix of full rank with m ≤ t, and $\boldsymbol{\beta}$ is a m × 1 vector of parameters. The weighted least squares estimator of $\boldsymbol{\beta}$ and its consistent estimate of the covariance matrix are given by

$$\mathbf{b} = (Z'V_g^{-1}Z)^{-1}Z'V_g^{-1}\mathbf{g}, \tag{11.4.6}$$

and

$$V_b = (Z'V_g^{-1}Z)^{-1}, \tag{11.4.7}$$

respectively.

The predicted values of $E(\mathbf{g})$ are given by

$$\hat{\mathbf{g}} = Z\mathbf{b}. \tag{11.4.8}$$

Hence, lack of fit of the model (11.4.5) can be evaluated by the Wald statistic, which has the form of residual chi-square with t − m degrees of freedom, i.e.,

$$\begin{aligned} R_g &= (\mathbf{g} - Z\mathbf{b})'V_g^{-1}(\mathbf{g} - Z\mathbf{b}) \\ &= \mathbf{g}'V_g^{-1}\mathbf{g} - \hat{\mathbf{g}}'V_g^{-1}\hat{\mathbf{g}}. \end{aligned} \tag{11.4.9}$$

It should be noted that $\mathbf{g}$ is not restricted to the linear functions of $\mathbf{p}$. However, for estimation of the formulation effect, which is expressed in terms of observed proportions of responses, we only consider linear functions of $\mathbf{p}$.

As can be seen from Table 11.4.1, the marginal probability of the occurrence of a therapeutic response for test formulation in sequence 1 is the sum of probabilities P_{21} and P_{41}, which is also the marginal probability for observing a therapeutic response during period II in sequence 1. Other marginal probabilities by sequence and formulation can be similarly defined. Table 11.4.2 gives marginal probabilities of a therapeutic response by formulation. In this case, the matrix C in (11.4.3) becomes

$$C = \begin{bmatrix} A & O_{2 \times 4} \\ O_{2 \times 4} & A \end{bmatrix}, \tag{11.4.10}$$

where

$$A = \begin{bmatrix} 0 & 0 & 1 & 1 \\ 0 & 1 & 0 & 1 \end{bmatrix}.$$

Hence, $\mathbf{g} = (p_{1R}, p_{1T}, p_{2T}, p_{2R})'$, where p_{kf} are obtained by replacing p_{hk} in Table 11.4.2 with p_{hk}, h = 1,2,3,4, and k = 1,2. Under condition of absence of unequal carry-over effects, Koch and Edwards (1988) suggested the following linear model:

$$E(\mathbf{g}) = E\begin{bmatrix} p_{1R} \\ p_{1T} \\ p_{2T} \\ p_{2R} \end{bmatrix} = \begin{bmatrix} 1 & 0 & 0 \\ 1 & 1 & 1 \\ 1 & 0 & 1 \\ 1 & 1 & 0 \end{bmatrix} \begin{bmatrix} \beta_1 \\ \beta_2 \\ F \end{bmatrix} = Z\boldsymbol{\beta}, \tag{11.4.11}$$

Table 11.4.2 Marginal Probabilities of a Therapeutic Response by Sequence and Drug Product

Sequence	Marginal probability	
	Reference	Test
1 (RT)	$P_{1R} = P_{31} + P_{41}$	$P_{1T} = P_{21} + P_{41}$
2 (TR)	$P_{2R} = P_{22} + P_{42}$	$P_{2T} = P_{32} + P_{42}$
Total	$P_R = \frac{1}{2}(P_{1R} + P_{2R})$	$P_T = \frac{1}{2}(P_{1T} + P_{2T})$

Source: Liu and Chow (1992b).

where $\boldsymbol{\beta} = (\beta_1, \beta_2, F)'$, β_1 is the probability of a therapeutic response for the reference formulation at period I, β_2 is the period effect, and F is the corresponding formulation effect. Let $\hat{\boldsymbol{\beta}} = (\hat{\beta}_1, \hat{\beta}_2, \hat{F})'$ be the WLS estimates of $\boldsymbol{\beta}$. Then, the predicted marginal probabilities under model (11.4.11) by sequence and formulation are given by

$$\hat{\mathbf{g}} = Z\hat{\boldsymbol{\beta}} = (\hat{P}_{1R}, \hat{P}_{1T}, \hat{P}_{2T}, \hat{P}_{2R})'.$$

Hence, the predicted marginal formulation probabilities of a therapeutic response are given by

$$\hat{\mathbf{P}}_M = \begin{bmatrix} \hat{P}_R \\ \hat{P}_T \end{bmatrix} = \begin{bmatrix} \frac{1}{2} & 0 & 0 & \frac{1}{2} \\ 0 & \frac{1}{2} & \frac{1}{2} & 0 \end{bmatrix} \begin{bmatrix} \hat{P}_{1R} \\ \hat{P}_{1T} \\ \hat{P}_{2T} \\ \hat{P}_{2R} \end{bmatrix}$$

$$= L\hat{\mathbf{g}}. \tag{11.4.12}$$

Thus, a consistent estimator of the covariance matrix of $\hat{\mathbf{P}}_M$ can be obtained as follows

$$\hat{\text{Var}}(\hat{\mathbf{P}}_M) = \begin{bmatrix} V_{RR} & V_{TR} \\ V_{TR} & V_{TT} \end{bmatrix}$$

$$= LZ(Z'V_g^{-1}Z)^{-1}Z'L'. \tag{11.4.13}$$

A consistent estimate for the bioequivalence measure $\delta = P_T/P_R$ is then given by

$$\hat{\delta} = \hat{P}_T/\hat{P}_R.$$

Hence, based upon Fieller's theorem, an approximate $(1 - 2\alpha) \times 100\%$ confidence interval for δ, denoted by (L,U), can be obtained as follows:

$$(L,U) = \frac{-a_2 \pm (a_2^2 - 4a_1a_3)^{1/2}}{2a_1}, \tag{11.4.14}$$

where

$$a_1 = \hat{P}_R^2 - z^2(\alpha)V_{RR},$$

$$a_2 = 2[\hat{P}_T\hat{P}_R - z^2(\alpha)V_{TR}],$$

$$a_3 = \hat{P}_T^2 - z^2(\alpha)V_{TT},$$

and $z(\alpha)$ is the αth upper quantile of a standard normal distribution. Test product is then considered to be bioequivalent to the reference product if (L,U) is within equivalent limits (δ_L,δ_U).

Example 11.4.1   To illustrate the use of the weighted least squares method, we consider the data from Herson (1991) concerning bioequivalence trials for albuterol metered dose inhaler indicated for acute bronchospasm. Table 11.4.3 gives response data and time of onset for the responses, which were calculated according to the FDA guidance. For the purpose of illustration, the response data will be used to assess bioequivalence between the test (generic) and the reference products at the dose of 90 μg. For simplicity, we consider the standard 2 × 2 crossover model. In other words, sequences 2 and 4 are combined as group 1 and sequences 1 and 3 are combined as group 2, while days 1 and 3 are designated as period I and days 2 and 4 are designated as period II. The observed counts for the response data of the four possible outcomes by group are summarized in Table 11.4.4. The carry-over effect can then be evaluated through the association of a 2 × 2 contingency table, which is constructed from the patients who have the same outcomes during both periods, i.e.,

2	10	12
0	13	13
2	23	25

The Fisher's exact test gives a two-tailed p value of 0.220 which indicates no evidence of carry-over effects. Therefore, the weighted least squares method can be applied using PROC CATMOD of SAS. Since no patients in group 2 did not respond in both periods, a small constant (10^{-6}) was added to this cell. Table 11.4.5 gives the elements of vector **g** which are the observed marginal probabilities of a therapeutic response by group and formulation. The estimates of β and their standard errors are given in Table 11.4.6. The results indicate that both period and formulation effects are not significantly different from 0 (p-values > 0.15). Note that the chi-square test statistic for lack of fit is 0.58 with a p-value of 0.45. This indicates that model (11.4.11) is adequate for this data set. The predicted marginal probabilities, $\hat{\mathbf{g}}$, of a therapeutic response by group and product and their estimated covariance matrix are given in Table 11.4.7. Based upon $\hat{\mathbf{g}}$, the predicted marginal formulation probabilities, $\hat{\mathbf{P}}_M$ can be computed according to (11.4.12), which are given by

Table 11.4.3 Response Data and Time of Onset of Response

		Response		Time of onset (min)	
Group	Patient	Period I	Period II	Period I	Period II
1 (RT)	1003	Yes	Yes	3.77	3.13
	1004	Yes	No	6.72	30.00[a]
	1006	Yes	Yes	2.69	3.68
	1007	Yes	No	3.74	30.00[a]
	1008	Yes	Yes	6.38	5.84
	1011	Yes	Yes	7.88	5.56
	1014	No	No	30.00[a]	30.00[a]
	1015	Yes	No	9.36	30.00[a]
	1020	Yes	No	6.02	30.00[a]
	1021	Yes	Yes	3.58	8.44
	1022	Yes	Yes	6.26	11.61
	1024	Yes	No	8.75	30.00[a]
	2003	Yes	Yes	6.75	5.51
	2007	Yes	Yes	1.89	3.29
	2015	Yes	Yes	2.85	3.23
	2016	No	No	30.00[a]	30.00[a]
	2018	Yes	Yes	5.39	6.31
	2021	No	Yes	30.00[a]	3.64
2 (TR)	1002	Yes	Yes	4.05	4.88
	1009	Yes	Yes	4.97	7.47
	1010	No	Yes	30.00[a]	13.98
	1012	No	Yes	30.00[a]	4.86
	1013	Yes	Yes	14.35	3.01
	1016	Yes	Yes	8.41	10.20
	1017	Yes	No	9.46	30.00[a]
	1018	Yes	Yes	2.50	1.69
	1025	Yes	No	13.55	30.00[a]
	2004	Yes	Yes	4.70	4.71
	2005	Yes	Yes	4.02	2.38
	2006	Yes	Yes	3.84	3.36
	2009	Yes	Yes	4.91	4.80
	2010	Yes	Yes	12.22	21.55
	2011	Yes	Yes	21.48	2.14
	2012	Yes	Yes	4.58	5.13
	2014	Yes	Yes	0.99	1.79
	2019	Yes	No	6.69	30.00[a]
	2020	No	Yes	30.00[a]	9.27
	2022	No	Yes	30.00[a]	7.09
	2023	No	Yes	30.00[a]	5.56
	2024	Yes	No	8.25	30.00[a]

[a]Censoring observations.

Source: Herson (1991).

Table 11.4.4 Summary of Observed Counts of a Therapeutic Response for Data Given in Table 11.4.3

		Outcome			
Group	(N,N)	(N,Y)	(Y,N)	(Y,Y)	Total
1(RT)	2	1	5	10	18
2(TR)	0	5	4	13	22
Total	2	6	9	23	40

[a]No = non-response; yes = response.
Source: Liu and Chow (1992b).

$$\hat{\mathbf{P}}_M = \begin{bmatrix} \hat{P}_R \\ \hat{P}_T \end{bmatrix} = \begin{bmatrix} 0.8364 \\ 0.7110 \end{bmatrix}.$$

A consistent estimate of $\mathrm{Var}(\hat{\mathbf{P}}_M)$ is given by

$$\begin{bmatrix} 0.003442 & -0.000179 \\ -0.000179 & 0.004665 \end{bmatrix}.$$

Hence, a consistent estimate and 90% confidence interval for $\delta = P_T/P_R$ are given by

$$\hat{\delta} = \frac{0.7110}{0.8364} = 0.8501,$$

Table 11.4.5 Summary of Observed Marginal Probabilities of a Therapeutic Response by Group and Drug Product for Data Set in Table 11.4.3

	Marginal probability	
Group	Reference	Test
1	0.8333	0.6111
2	0.8182	0.7727
Total	0.8258	0.6919

Source: Liu and Chow (1992b).

Table 11.4.6 Summary of Weighted Least Squares Estimates of β for Herson's Data

Parameter	Estimate	SE[a]	Chi-square	P-Value
β_1	0.8740	0.0697	157.02	<0.0001
β_2	−0.0751	0.0909	0.68	0.4086
F	−0.1254	0.0919	1.86	0.1724
Lack of fit			0.58	0.4467

[a]SE = standard error.

and

$$(L,U) = (0.6984, 1.0344),$$

respectively.

As a result, according to the ±20 rule, the test product is not bioequivalent to the reference product with 90% assurance based upon the response data.

11.5. LOG-LINEAR MODELS

Jones and Kenward (1989) suggested the use of log-linear models for binary response data. In this section, we will apply their results to the predicted marginal

Table 11.4.7 Summary of Predicted Marginal Probabilities by Group and Drug Product

I. Estimates

	$\hat{P}_{1R}$	$\hat{P}_{1T}$	$\hat{P}_{2T}$	$\hat{P}_{2R}$
Estimates	0.8740	0.6735	0.7486	0.7989

II. Covariance matrix

	$\hat{P}_{1R}$	$\hat{P}_{1T}$	$\hat{P}_{2T}$	$\hat{P}_{2R}$
$\hat{P}_{1R}$	0.004859	−0.001808	0.001695	0.001356
$\hat{P}_{1T}$		0.006487	0.002599	0.002079
$\hat{P}_{2T}$			0.006977	−0.002683
$\hat{P}_{2R}$				0.006118

formulation probabilities in order to obtain an approximate $(1 - 2\alpha) \times 100\%$ confidence interval for δ for assessing bioequivalence.

For a standard 2×2 crossover design, a pair of binary responses is observed on each subject. Let (X_{i1k}, X_{i2k}) be the observed paired responses on subject i in sequence k. Jones and Kenward (1989) gave a general representation of the probabilities of possible outcomes as follows:

$$P_{hk} = P(X_{i1k} = x_{i1k}, X_{i2k} = x_{i2k})$$

$$= \exp\{\eta_{0k} + \eta_{1k}x_{i1k} + \eta_{2k}x_{i2k} + \eta_{12k}x_{i1k}.x_{i2k}\}, \qquad (11.5.1)$$

where η_{0k} is the normalizing term, which was chosen so that the sum of probabilities of the four possible outcomes for sequence k is equal to 1; η_{1k} and η_{2k} are the parameters corresponding to the design, and η_{12k} is the intra-subject dependence parameter which is included in the model to assess the dependence of the two binary responses observed on the same subject. Note that, under the above model, the possible values for X_{ijk} can be coded either 1 or -1. $X_{ijk} = 1$ indicates that a therapeutic response has occurred. On the other hand, if $X_{ijk} = -1$, then the patient does not respond. The value of η_{12k} depends upon the correlation between X_{i1k} and X_{i2k}. η_{12k} is positive (negative) if X_{i1k} and X_{i2k} are positively (negatively) correlated. $\eta_{12k} = 0$ indicates that X_{i1k} and X_{i2k} are independent. In this section, we assume that the correlation between two binary responses observed on the same subject is the same from subject to subject in both sequences, i.e., $\eta_{12k} = \eta_{12}$ for all k. Note that the common intra-subject dependence parameter is also known as the average intra-subject dependence parameter.

Let Y_{ijk} be the logit of the probability of a therapeutic response observed during period j on subject i in sequence k, i.e.,

$$Y_{ijk} = \text{logit}[X_{ijk} = "Y"] = \log\left[\frac{P[X_{ijk} = "Y"]}{1 - P[X_{ijk} = "Y"]}\right], \qquad (11.5.2)$$

where $i = 1,2, \ldots ,n_k; j = 1,2; k = 1,2$.

Then, according to Jones and Kenward (1989), the explicit representation of the log-linear model for P_{hk} can be constructed by the following steps:

Step 1: Obtain logit probabilities Y_{ijk} as defined in (11.5.2). Under assumption of independence between two binary responses observed on the same subject, the logits of the probabilities of a therapeutic response by sequence and period have the same representation as the expected values of the sequence-by-period means for continuous variables under a standard 2×2 crossover design. The representations of the logits of probabilities of a therapeutic response by sequence and period are given in Table 11.5.1.

Table 11.5.1   Representation of the Logits of the Therapeutic Response by Sequence and Period

	Period	
Sequence	I	II
1 (RT)	$\mu + P_1 + F_R$	$\mu + P_2 + F_T + C_R$
2 (TR)	$\mu + P_1 + F_T$	$\mu + P_2 + F_R + C_T$

Source: Liu and Chow (1992b).

Step 2: Obtain the joint probabilities, P_{hk}, of four possible outcomes for each sequence under independence by multiplying the probabilities from Step 1.

Step 3: Include the intra-subject dependence parameter $\eta = \eta_{12}$ in model (11.5.1). Table 11.5.2 summarizes the resulting model for $\log(P_{hk})$.

In the log-linear model for a standard 2×2 crossover design given in Table 11.5.2, P_1, P_2, F_T, F_R, C_T, and C_R are the fixed effects for period, direct formulation, and carry-over, μ_1 and μ_2 are normalizing constants, η is the average intra-subject dependence parameter, and τ has no particular interpretation and other effects are defined as deviations from it.

Let $\mathbf{g}(\mathbf{P})$ be a 8×1 vector with elements of $\log(P_{hk})$. The log-linear model discussed above can be expressed as

$$\mathbf{g}(\mathbf{P}) = \mathbf{Z}\boldsymbol{\beta}, \tag{11.5.3}$$

where

$$\boldsymbol{\beta} = (\mu_1, \mu_2, \tau, P_2, F_T, C_T, \eta)',$$

Table 11.5.2   Log-Linear Model for a Standard 2×2 Crossover Design

		Period II	
Sequence	Period I	No	Yes
1	No	$\mu_1 + \eta$	$\mu_1 + \tau + P_2 + F_T + C_R - \eta$
	Yes	$\mu_1 + \tau + P_1 + F_R - \eta$	$\mu_1 + 2\tau + C_R + \eta$
2	No	$\mu_2 + \eta$	$\mu_2 + \tau + P_2 + F_R + C_T - \eta$
	Yes	$\mu_2 + \tau + P_1 + F_T - \eta$	$\mu_2 + 2\tau + C_T + \eta$

and

$$Z = \begin{bmatrix} 1 & 0 & 0 & 0 & 0 & 0 & 1 \\ 1 & 0 & 1 & 1 & 1 & -1 & -1 \\ 1 & 0 & 1 & -1 & -1 & 0 & -1 \\ 1 & 0 & 2 & 0 & 0 & -1 & 1 \\ 0 & 1 & 0 & 0 & 0 & 0 & 1 \\ 0 & 1 & 1 & 1 & -1 & 1 & -1 \\ 0 & 1 & 1 & -1 & 1 & 0 & -1 \\ 0 & 1 & 2 & 0 & 0 & 1 & 1 \end{bmatrix}$$

Note that the estimation of direct formulation effect is valid only in the absence of carry-over effects (Jones and Kenward, 1989). Therefore, the column corresponding to C_T in the design matrix Z can be omitted when there is no evidence of carry-over effects. The carry-over effect can be examined by Fisher's exact test as described in the previous section.

Agresti (1990) pointed out that the correct likelihood ratio test, maximum likelihood estimation of the parameters, and consistent estimates of their covariance matrix can be obtained if the observed counts n_{hk} are independent Poisson variables with no constraints on μ_1 and μ_2. Based upon this result, the current statistical package, GLIM (Generalized Linear Interactive Modeling, Numerical Algorithm Group, 1986) can be directly applied by specifying log link function and Poisson errors.

Similar to the weighted least squares method, the predicted values of $\mathbf{g}(\mathbf{P})$ is obtained as follows.

$$\hat{\mathbf{g}}(\mathbf{P}) = Z\mathbf{b},$$

where $\mathbf{b}$ is the maximum likelihood estimator of $\boldsymbol{\beta}$.
Hence, the maximum likelihood estimator of $\mathbf{P}$ is given by

$$\hat{P}_{hk} = \exp(\hat{g}_{hk}), \qquad h = 1,2,3,4; \quad k = 1,2. \tag{11.5.4}$$

Let $\hat{\mathbf{P}} = (\hat{P}_{11}, \hat{P}_{21}, \ldots, \hat{P}_{42})'$. Then a consistent estimator of the covariance matrix of $\hat{\mathbf{P}}$ is given by

$$\hat{V}ar(\hat{\mathbf{P}}) = Diag(\hat{P}_{hk})\hat{V}ar(\hat{\mathbf{g}})Diag(\hat{P}_{hk}), \tag{11.5.5}$$

where $Diag(\hat{P}_{hk})$ is an 8×8 diagonal matrix with diagonal elements $\hat{P}_{hk}$, and $\hat{V}ar(\hat{\mathbf{g}})$ is a consistent estimator of $Var(\hat{\mathbf{g}})$, which is given by

$$\hat{V}ar(\hat{\mathbf{g}}) = Z \hat{V}ar(\mathbf{b})Z',$$

where $\hat{V}ar(\mathbf{b})$ is a consistent estimator of $Var(\mathbf{b})$.

The predicted marginal formulation probabilities $\hat{\mathbf{P}}_M$ can then be obtained as

$$\hat{\mathbf{P}}_M = (\hat{P}_R, \hat{P}_T)' = LC\hat{\mathbf{P}}, \tag{11.5.6}$$

where L and C are defined in (11.4.12) and (11.4.10), respectively.

A consistent estimator of the covariance matrix of $\hat{\mathbf{p}}_M$ is then given by

$$\hat{V}ar(\hat{\mathbf{P}}_M) = \begin{bmatrix} V_{RR} & V_{TR} \\ V_{TR} & V_{TT} \end{bmatrix}$$

$$= LC \, \hat{V}ar(\hat{\mathbf{P}})C'L'. \tag{11.5.7}$$

Therefore, in a similar manner, the maximum likelihood estimator and an approximate $(1 - 2\alpha) \times 100\%$ confidence interval for δ can be obtained.

11.6. GENERALIZED ESTIMATING EQUATIONS (GEEs)

When the total sample size is large, as an alternative, the technique of generalized estimating equations (GEEs) by Liang and Zeger (1986) may be useful for estimation of period and direct formulation effects in the absence of carry-over effects. However, this method treats X_{i1k} and X_{i2k} as they were independent, though X_{i1k} and X_{i2k} are not independent. The logistic regression utilizes the representation of the logits of a therapeutic response specified in Table 11.5.1 with the deletion of the carry-over effects C_R and C_T. Define

$$\mathbf{X}_{ik} = (X_{i1k}, X_{i2k})',$$

and

$$\mathbf{Z}_{ik} = (\mathbf{z}_{i1k}', \mathbf{z}_{i2k}')',$$

where $\mathbf{z}_{ijk}$ is the row of the design matrix Z corresponding to the jth period of subject i in sequence k, i = 1, 2, . . . , n_k; k = 1, 2. Note that, for the representation of the logits with the carry-over effects deleted, $\mathbf{Z}_{ik}$ is a 2 × 3 matrix. Let $\hat{\Sigma}(\mathbf{b})$ be a "naive" estimator of the covariance matrix of the estimator $\mathbf{b}$ obtained under the assumption that X_{i1k} and X_{i2k} are independent. Then, a consistent estimator of the covariance matrix of $\mathbf{b}$ regardless of the true underlying correlation structure between the two binary responses observed on the same subject is given by (Zeger and Liang, 1986)

$$\hat{V}ar(\mathbf{b}) = \hat{\Sigma}(\mathbf{b})S_R\hat{\Sigma}(\mathbf{b}), \tag{11.6.1}$$

where

$$S_R = \sum_{k=1}^{2} \sum_{i=1}^{n_k} \mathbf{Z}_{ik}'[\mathbf{x}_{ik} - \hat{\mathbf{P}}_{ik}][\mathbf{x}_{ik} - \hat{\mathbf{P}}_{ik}]'\mathbf{Z}_{ik}, \tag{11.6.2}$$

$$\hat{\mathbf{P}}_{ik} = (\hat{P}_{i1k}, \hat{P}_{i2k})',$$

$$\hat{P}_{ijk} = \hat{P}(X_{ijk} = "Y") = [1 + \exp(-\mathbf{z}_{ijk}'\mathbf{b})]^{-1}, \tag{11.6.3}$$

i = 1, 2, . . . , n_k; j = 1, 2; k = 1, 2.

$\hat{V}ar(\mathbf{b})$ is not only a consistent estimator but also a robust estimator of the covariance matrix of $\mathbf{b}$ in the sense that it is still a consistent estimator even though the correlation between X_{i1k} and X_{i2k} is misspecified. $\hat{V}ar(\mathbf{b})$ is referred to as an "information sandwich." Note that there are only four possible values for vector $\mathbf{z}_{ijk}$ if the carry-over effects are excluded from the model. Therefore, only four possible predicted probabilities $\hat{P}_{ijk}$ in (11.6.3) can be obtained. These four probabilities are actually the predicted marginal probabilities by sequence and formulation, i.e., $\hat{P}_{1R}$, $\hat{p}_{1T}$, $\hat{P}_{2T}$, and $\hat{P}_{2R}$. As a result, a consistent estimator of the marginal formulation probabilities $\mathbf{P}_M$ and a robust, consistent, estimator of its covariance matrix are given by, respectively,

$$\hat{\mathbf{P}}_M = L(\hat{P}_{1R}, \hat{P}_{1T}, \hat{P}_{2T}, \hat{P}_{2R})', \tag{11.6.4}$$

and

$$\hat{V}ar(\hat{\mathbf{P}}_M) = LL_T Z \, \hat{V}ar(\mathbf{b}) Z' L_T L', \tag{11.6.5}$$

where L is defined as (11.4.12) and L_T is a 4×4 diagonal matrix with diagonal elements $-\hat{P}_{fk}(1 - \hat{P}_{fk})$, f = R,T; k = 1,2; and

$$Z = \begin{bmatrix} 1 & -1 & -1 \\ 1 & 1 & 1 \\ 1 & -1 & 1 \\ 1 & 1 & -1 \end{bmatrix}.$$

Hence, similarly, based upon the generalized estimating equations, a consistent estimate and an approximate $(1 - 2\alpha) \times 100\%$ confidence interval can be obtained.

Example 11.6.1   Again, we will use the response data given in Table 11.4.3 to illustrate the log-linear model and the generalized estimating equations method. Table 11.6.1 summarizes the results of carry-over effects in the presence of other effects for a log-linear model and the generalized estimating equations method. Since both methods indicate that there is no evidence of unequal carry-

Table 11.6.1   Summary of Test Results of Carry-Over Effects

Estimate	Log-Linear	GEEs[a]
$C = C_T - C_R$	0.6760	0.3332
SE[b]	1.1090	1.0185
P-Value	0.5422	0.7435

[a]GEEs = generalized estimating equations.
[b]SE = standard error.
Source: Liu and Chow (1992b).

Table 11.6.2 Summary of Results from Log-Linear Model and Generalized Estimating Equations Method

Parameter	Log-linear model			GEEs method		
	Estimate	SE	P-Value	Estimate	SE	P-Value
Intercept	0.1306	0.5670	0.8178	1.2153	0.4992	0.0149
μ_1	−0.2680	0.3306	0.4176	—	—	—
τ	1.2210	0.3681	0.0009	—	—	—
$P_2 - P_1$	−0.5020	0.5464	0.3582	−0.5038	0.5938	0.3962
$F_T - F_R$	−0.7586	0.5600	0.1755	−0.7616	0.5738	0.1845
η	−0.0055	0.2318	0.9810	—	—	—

Source: Liu and Chow (1992b).

over effects (p-values > 0.50), the same models were fitted without carry-over effects. The results for both methods are summarized in Table 11.6.2. The deviance for a log-linear model with scale parameter 1 is 3.822 which yields a p-value of 0.15. Therefore, the log-linear model without carry-over effects is an adequate model for this data set. The predicted marginal probabilities by group and drug product are given in Table 11.6.3 for both methods. Note that both methods yield identical estimates for the parameters. This is probably due to the following reasons: (i) the log-linear model in Table 11.5.2 is actually derived from the model for logits in Table 11.5.1 (see also, Agresti, 1990, Ch. 6); (ii) the magnitude of the estimated average intra-subject dependence is too

Table 11.6.3 Summary of Predicted Marginal Probabilities by Group and Drug Product

Method		$\hat{P}_{1R}$	$\hat{P}_{1T}$	$\hat{P}_{2T}$	$\hat{P}_{2R}$
Log-linear	Estimates	0.8639	0.6417	0.7477	0.7932
	$\hat{P}_{1R}$	0.09920	0.06815	−0.00379	−0.00414
	$\hat{P}_{1T}$		0.06277	−0.00229	−0.00256
	$\hat{P}_{2T}$			0.02727	0.01990
	$\hat{P}_{2R}$				0.02871
GEEs		$\hat{P}_{1R}$	$\hat{P}_{1T}$	$\hat{P}_{2T}$	$\hat{P}_{2R}$
	Estimate	0.8639	0.6417	0.7477	0.7932
	$\hat{P}_{1R}$	0.00586	0.00159	0.00546	0.00456
	$\hat{P}_{1T}$		0.02408	0.01086	0.00996
	$\hat{P}_{2T}$			0.01416	0.00305
	$\hat{P}_{2R}$				0.01082

Source: Liu and Chow (1992b).

Table 11.6.4 Summary of Predicted Marginal Formulation Probabilities by Method

Method		$\hat{P}_R$	$\hat{P}_T$	$\hat{\delta} = \hat{P}_T/\hat{P}_R$	90% C.I.
WLS	Estimate	0.8364	0.7110	85.01%	(69.84%, 103.44%)
	$\hat{P}_R$	0.003422	-0.000179		
	$\hat{P}_T$		0.004665		
Log-linear	Estimate	0.8285	0.6947	83.85%	(61.31%, 104.22%)
	$\hat{P}_R$	0.02990	0.02043		
	$\hat{P}_T$		0.02136		
GEEs	Estimate	0.8285	0.6947	83.85%	(67.79%, 106.23%)
	$\hat{P}_R$	0.006449	0.005015		
	$\hat{P}_T$		0.014990		

Source: Liu and Chow (1992b).

small to have any impact on prediction. However, the estimated variances of the predicted marginal probability obtained by the generalized estimating equations method are much smaller than those by the log-linear model.

To compare the three methods discussed in previous sections, the predicted marginal formulation probabilities, consistent estimates of δ, and 90% confidence interval for δ for each method are summarized in Table 11.6.4. According to the ± 20 rule, the three methods reach the same conclusion of bioinequivalence. However, among these methods, the weighted least squares method gives the narrowest confidence interval, while the log-linear model yields a much wider confidence interval as compared to the other two methods.

11.7. ANALYSIS OF TIME TO ONSET OF A THERAPEUTIC RESPONSE

As defined in Section 11.2, a therapeutic response is the event that the FEV_1 measurement evaluated within 30 minutes after dosing exceed 115% of its baseline value. If a patient has the same outcomes of a therapeutic response to both test and reference products, i.e., either his/her FEV_1 measurements for both test and reference products exceed (or do not exceed) 115% of their baseline values within 30 minutes after administration, the time to onset of the response for one product may be longer than that of the other product. Therefore, although the test and reference products may be bioequivalent with respect to occurrence of a therapeutic response, they may not be bioequivalent on the time to onset of the response. Hence, time to onset of a therapeutic response is another clinical endpoint, which can be derived from FEV_1-time curve, to characterize the profile of bioequivalence of MDI products.

For the analysis of time to the onset of a therapeutic response, several methods for paired failure times are available. For example, the paired Prentice-Wilcoxon statistic by O'Brien and Fleming (1987); and the parametric models proposed by Huster, Brookmeyer and Self (1989). However, these methods are either for test of equality or derived under a particular form of distribution. None of these methods takes into account the structure of crossover designs. Holt and Prentice (1974); Kalbfleisch and Prentice (1980), however, modified the proportional hazards model for paired failure times which can be extended to estimate the direct formulation effect and its asymptotic confidence interval under the structure of a standard 2×2 crossover design with the assumption of no carry-over effects (also see France, Lewis, and Kay (1991)). This method is essentially a sign test, which can be derived from a binary logistic model. Hence, bioequivalence for the time to onset of a therapeutic response between two products can be assessed using the $(1 - 2\alpha) \times 100\%$ asymptotic confidence interval for hazard ratio of the test product to the reference product.

Let $\mathbf{X}_{ik} = (X_{iRk}, X_{iTk})'$, where X_{ifk} is the time to onset of a therapeutic response (defined as response time), censored or uncensored, of product f for subject i in sequence k. Also, let $\mathbf{z}_{ik} = (\mathbf{z}'_{iRk}, \mathbf{z}'_{iTk})'$ be the corresponding covariate matrices, $i = 1, 2, \ldots, n_k$; $k = 1, 2$. Then, under a proportional hazards model, the hazard function for f product of subject i in sequence k can be expressed as

$$\lambda_{ifk}(x) = \lambda_{0ik}(x)exp(\mathbf{z}'_{ifk}\boldsymbol{\beta}), \qquad (11.7.1)$$

where $\lambda_{0ik}(x)$ is an underlying baseline hazard function assumed to be unknown but different from subject to subject and $\boldsymbol{\beta} = (P, F)$. In which $P = P_2 - P_1$, and $F = F_T - F_R$ are the period and formulation effects, respectively.

If the time to occurrence is continuous, Kalbfleisch and Prentice (1980) showed that a partial likelihood (or conditional likelihood) with the form of a binary logistic likelihood can be constructed as the product (over all subjects) of the conditional probability of the ranks of the bivariate response times given the small response time of subject i in sequence k, $i = 1, 2, \ldots, n_k$ $k = 1, 2$. This partial likelihood is formed based upon subjects whose smallest response time is not censored. Since we rank the response times within each subject, not across all subjects, there are only two possible ranks for each subject. Without loss of generality, we choose 0 and 1 for the two possible values as follows:

$$y_{ik} = \begin{cases} 0 & \text{if } x_{iRk} > x_{iTk}, \\ 1 & \text{if } x_{iRk} < x_{iTk} \end{cases}, \qquad (11.7.2)$$

where x_{ifk} are the observed response time, $i = 1, 2, \ldots, n_k$; $f = R, T$; $k = 1, 2$.

Table 11.7.1 Values of z'_{ijk} and v'_{ik} for Proportional Hazards Model (In Absence of Carry-Over Effects)

	Formulation		
Sequence	z'_{iRk}	z'_{iTk}	v'_{ik}
1 (RT)	$(-\frac{1}{2}, -\frac{1}{2})$	$(\frac{1}{2}, \frac{1}{2})$	$(1, 1)$
2 (TR)	$(\frac{1}{2}, -\frac{1}{2})$	$(-\frac{1}{2}, \frac{1}{2})$	$(-1, 1)$

Source: Liu and Chow (1992b).

Let $v_{ik} = z_{iTk} - z_{iRk}$, $i = 1, 2, \ldots, n_k$; $k = 1, 2$. Then inference on $\boldsymbol{\beta}$ can be obtained by fitting a binary logistic regression with dependent variable y_{ik} and vector of explanatory covariates v_{ik}. Since the elements of $\boldsymbol{\beta}$ are expressed in terms of the period and formulation effects and $P_1 + P_2 = 0$ and $F_T + F_R = 0$, there are only two possible values of v_{ik} which are given in Table 11.7.1. Let $\hat{P}$ and $\hat{F}$ be the resulting estimators of period and formulation effects obtained from fitting the logistic regression, and S_P^2 and S_F^2 be the corresponding estimated asymptotic variances obtained from the inverse of the observed information matrix. Then an approximate $(1 - 2\alpha) \times 100\%$ confidence interval of the hazard ratio of the test product to the reference product, denoted by (L,U), is given by

$$(L,U) = \exp[\hat{F} \pm z(\alpha)S_F]. \tag{11.7.3}$$

Two products are considered to be bioequivalent on time to onset of a therapeutic response at the α level of significance if (L,U) is within the prespecified equivalent limits (δ_L, δ_U).

It can be easily verified that

$$\hat{F} = \tfrac{1}{2}(\hat{\beta}_1 + \hat{\beta}_2),$$

$$\hat{P} = \tfrac{1}{2}(\hat{\beta}_1 - \hat{\beta}_2),$$

and

$$\hat{S}_F^2 = \hat{S}_P^2 = \tfrac{1}{4}(S_1^2 + S_2^2), \tag{11.7.4}$$

where $\hat{\beta}_k$ is the estimator of the intercept of the logistic regression with dependent variable y_{ik} and intercept term as the only explanatory covariate for sequence k, and S_k^2 is the corresponding estimator for the asymptotic variance, $k = 1, 2$.

Kalbfleisch and Prentice (1980) pointed out that the omission of the subjects whose smallest response times are censored will not introduce systematic bias for estimation of period and formulation effects. However, too many subjects with smallest censored response times will certainly have an impact on the efficiency of these estimates.

Table 11.7.2   Summary of Results of Response Time

Group	No. of patients whose $X_{iRk} > X_{iTk}$	No. of patients whose $X_{iRk} < X_{iTk}$	Formulation effect (SE)	90% C.I.
1	5	11	$-0.789\ (0.539)$	$(-1.676, 0.099)$
2	11	11	$0.000\ (0.426)$	$(-0.701, 0.701)$
Period effect			$-0.394\ (0.344)$	$(-0.960, 0.171)$
Formulation effect			$-0.394\ (0.344)$	$(-0.960, 0.171)$
Hazard ratio[a]			67.42%	$(38.30\%, 118.69\%)$

[a]Hazard ratio of the test product to the reference product.
Source: Liu and Chow (1992b).

Example 11.7.1   The time to onset of a therapeutic response for Herson's data set, listed in Table 11.4.3, is used to illustrate the method suggested in this section. Since a therapeutic response is defined within 30 minutes after dosing, the response time is censored at 30 minutes after administration. Patients 1014 and 2016 of group 1 are not included in the logistic regression from the analysis because both response times of the two patients are censored. The results using the logistic regression are summarized in Table 11.7.2. An estimate of the hazard ratio of the test product to the reference product is 67.42% with the corresponding 90% confidence interval being (38.30%, 118.69%). As a result, according to the ± 20 rule, the test product is not bioequivalent to the reference product based upon the time to onset of therapeutic response.

11.8. DISCUSSION

For the weighted least squares method, since there is an empty cell in the 2×4 contingency table constructed for the data from Table 11.4.3, a small constant (correction factor) was added to the cell in order to fit the linear model. Agresti (1990) pointed out that different small constants may have a strong influence on the results of the weighted least squares methods. To investigate this, a sensitivity analysis was performed to study the influence of adding constants of 4 different sizes on the estimates and 90% confidence interval of δ. The results are summarized in Table 11.8.1. It seems that, at least for the data considered here, the weighted least squares method yields rather robust estimates and 90% confidence interval for δ. However, the weighted least squares method may not be useful in the case where there are continuous covariates such as baseline FEV_1.

Table 11.8.1 Sensitivity Analysis of Weighted Least Squares
for Data Given Table in 11.4.3

Correction constant	$\hat{\delta}$	90% C.I.
1×10^{-6}	85.01%	(69.84%, 103.44%)
1×10^{-4}	85.01%	(69.84%, 103.44%)
1×10^{-2}	85.00%	(69.83%, 103.43%)
0.5	84.56%	(69.36%, 103.03%)

For the log-linear model and the generalized estimating equations method, both methods are maximum likelihood procedures, which depend only upon the marginal totals, not individual cell counts. Therefore, both methods are robust to zero cell counts. However, sometimes they can be computationally intensive. Although we may consider the more complex crossover design by the method described in Jones and Kenward (1989) for binary response data, the generalized estimating equations method seems more appealing. For example, for the design in Table 11.2.1 as recommended by the FDA, we can obtain consistent estimates of various effects simply by fitting a logistic regression with the model specified in Table 11.3.1 as if all binary responses were independent. A consistent estimate of the covariance matrix of various effects can also be obtained by the same procedure as described in Section 11.6. All computations can be performed using PROC LOGISTIC and PROC IML of SAS if the logistic link function is used.

In this chapter, we focused on the assessment of bioequivalence between MDI products using response data and time to onset of a therapeutic response as primary clinical endpoints. However, the FDA guidance suggests that seven clinical endpoints be analyzed. This raises a question of whether one, or some, or all of these endpoints should be used for determination of sample size at the planning stage of trials of this kind. In addition, as mentioned earlier, the methods discussed in this chapter depend upon the asymptotic results. Therefore, further research on the calculation of the corresponding power is needed if binary response data is the primary endpoint.

For other drugs such as topical antifungals or anti-ulcer agents, Huque and Dubey (1990) suggested a three-arm parallel placebo-controlled, randomized design which consists of the test and reference products and a placebo group. Bioequivalence is assessed only after superiority of both the test and reference products over the placebo is established. Huque and Dubey (1990) also discussed issues of interim analysis, choice of equivalence limits, multi-center studies, and assessment of bioequivalence under a generalized mixed effects model for binary responses.

Some Related Problems in Bioavailability Studies

12.1. INTRODUCTION

In addition to the assessment of bioequivalence in average bioavailability in terms of the pharmacokinetic responses such as AUC and C_{max}, it is also important to study the behaviors of those pharmacokinetic responses which may provide useful information for evaluation of drug efficacy and safety. For this purpose, some studies related to bioavailability such as drug interaction studies, dose proportionality studies, steady state analyses, and population pharmacokinetics are often conducted to study the behavior of the plasma concentrations in terms of AUC and other pharmacokinetic responses.

It is a common practice that more than one drug may be given to a patient at the same time. Therefore, it is important to examine the relative bioavailability of the study drug when it is co-administered with food or other medications. For this study purpose, a drug interaction study is usually conducted to study the impact of food or other medications on the study drug. A study of this kind can certainly provide valuable information on the efficacy and/or safety of the study drug.

When a new compound is discovered, it is important to determine an appropriate dose so that the drug can reach its maximum therapeutic effect with minimum toxicity. If there is a linear relationship between AUC and dose, then an optimal dose level can be obtained in order to reach a desired blood AUC for therapeutic effect. Therefore, it is of interest to determine whether there is a dose linearity between AUC and the dose level through a dose proportionality study.

The objective of a steady state analysis is to determine whether the plasma concentrations or AUCs can reach and maintain at an almost constant level after multiple dosing. This analysis provides useful information for evaluation of drug safety.

In the following sections, more details regarding drug interaction studies, dose proportionality studies, steady state analyses, and population pharmaco-kinetics are discussed separately.

12.2. DRUG INTERACTION STUDY

In the pharmaceutical industry, when a new compound is developed, it is often of interest to investigate the impact of concurrent usage of the drug with food or other medications on pharmacokinetic characteristics and therapeutic effects based upon the following reasons:

(i)   The study of pharmacokinetic profile of an active compound is usually conducted with some healthy subjects in such a way that each dosing period is preceded by a fast period and followed by a washout period. However, the pharmacokinetic characteristics of the drug may change when the drug is administered with food.

(ii) In practice, patients may take more than one drug at the same time for medical reasons. For example, in addition to AZT, patients with AIDS need to take antibiotics for opportunistic infection such as PCP; patients with hypertension may have to take other medications to control diabetes, renal failure, or arrhythmia.

To study the impact of concurrent usage of the drug with food or other medications, a crossover design is usually considered. For example, let I be the drug under investigation and A be the drug, which may have impact on I. Then, a drug interaction study may be conducted with a crossover design to compare the differences between treatment I and treatment I + A. Since the main purpose of a drug interaction is to examine the existence of changes or differences between pharmacokinetic responses between I and I + A, crossover designs and statistical methods discussed in previous chapters can be easily applied to study the influence of a co-administered drug.

12.2.1. The Evaluation of Drug Interaction

The hypotheses of primary interest for drug interaction studies are formulated as follows.

$$H_0: \mu_I = \mu_{IA} \quad \text{vs} \quad H_a: \mu_I \neq \mu_{IA}, \tag{12.2.1}$$

where $\mu_I = \mu + F_I$ and $\mu_{IA} = \mu + F_{IA}$ are the population treatment means of I and I + A, respectively. Hypotheses (12.2.1) can be equivalently expressed as follows:

$$H_0: \theta = 0 \quad \text{vs} \quad H_a: \theta \neq 0, \tag{12.2.2}$$

where $\theta = \mu_{IA} - \mu_I = F_{IA} - F_I$.

Let $\hat{F}$ be the best linear unbiased estimator of θ obtained under a particular design discussed in previous chapters and S_F be its corresponding standard error with ν degrees of freedom. Then, a $(1 - \alpha) \times 100\%$ confidence interval for θ is given by

$$(L, U) = \hat{F} \pm S_F \, t(\alpha/2, \nu), \tag{12.2.3}$$

where $t(\alpha/2, \nu)$ is the $\alpha/2$ upper quantile of a central t distribution with ν degrees of freedom. We then reject the null hypothesis of (12.2.1) at the α level of significance and conclude that the pharmacokinetic responses of drug I co-administered with drug A is different from that I given alone if

$$|T| > t(\alpha/2, \nu), \tag{12.2.4}$$

where $T = \hat{F}/S_F$.

Note that hypotheses testing procedure stated in (12.2.2) is equivalent to the method of confidence interval, i.e., (12.2.2) is rejected at the α level if and only if the $(1 - \alpha) \times 100\%$ confidence interval, does not contain zero. The other difference between the test procedure for equality and the two one-sided tests procedure for equivalence is that, for the α significance level, instead of α, drug interaction studies use the $\alpha/2$ upper quantile of a central t-distribution in hypotheses testing and confidence interval.

In some cases, the information regarding the pharmacokinetic properties of the study drug may be available when it is administered alone. However, its characteristics in presence of other drugs may not be known. Therefore, in drug interaction studies, it is important not only that a sufficient length of washout must be given, but also that the presence of the carry-over effect be examined. Although, in practice, the standard two-sequence, two-period crossover design is commonly used in a drug interaction study because of its simplicity, it is recommended that an optimal crossover design for comparing two treatments as discussed in Chapters 2 and 9 be used because it provides the best linear unbiased estimates for both treatment and carry-over effects.

12.2.2. Change in Intra-Subject Variability

In addition to evaluation of the existence of change or difference in pharmacokinetic responses between drug I and drug I + A, it is also important to examine the change in intra-subject variability of drug I in the presence of drug A. This can be examined by testing the following hypotheses:

$$H_0: \sigma_I^2 = \sigma_{IA}^2 \quad \text{vs} \quad H_a: \sigma_I^2 \neq \sigma_{IA}^2. \tag{12.2.5}$$

For a standard two sequence, two period crossover design, Pitman-Morgan's test for equality of intra-subject variabilities, given in (7.4.9) or (7.4.11) can be used. For an optimal design comparing two treatments, since the design provides independent, unbiased estimates of σ_I^2 and σ_{IA}^2, the test statistic given in (9.3.26) can be directly applied by simply replacing $\hat{\sigma}_1^2$ and $\hat{\sigma}_2^2$ with

$$\hat{\sigma}_1^2 = \max(\hat{\sigma}_I^2, \hat{\sigma}_{IA}^2),$$
$$\hat{\sigma}_2^2 = \min(\hat{\sigma}_I^2, \hat{\sigma}_{IA}^2),$$

and the corresponding degrees of freedom with ν_1 and ν_2, respectively.

12.2.3. Interaction with a Marketed Drug

As indicated earlier, the purpose of a drug interaction study is to study the impact a (marketed) drug may have on an investigational drug. In practice, it is also of interest to study the impact of the investigational drug on the marketed drug. In this case, basically, there are three treatments to be compared, namely, I (investigational drug), A (marketed drug), and I + A (co-administration of I and A). The comparisons of primary interest are I vs I + A (the impact of A on I) and A vs I + A (the impact of I on A). In this section, we will discuss statistical methods for assessment of the impact of I on A and the impact of A on I using a Williams design for comparing three treatments. The Williams design for drug interaction study is given in Table 12.2.1. It can be seen that the assessment of the impact of A on I is based upon the data from periods where I and I + A are administered. Similarly, the assessment of the impact of I on A is based upon the data from periods where A and I + A are administered.

Table 12.2.1   The Williams Design for Drug Interaction Studies

	Period		
Sequence	I	II	III
1	I[a]	A[b]	I + A
2	I + A	I	A
3	A	I + A	I
4	I + A	A	I
5	A	I	I + A
6	I	I + A	A

[a] I = investigational drug.

[b] A = marketed drug.

Table 12.2.2 Expected Values of Sequence by Period Means for Comparing I and I + A

Sequence	Period I	Period II	Period III
1	$\mu + G_1 + P_1 + F_I$	—	$\mu + G_1 + P_3 + F_{IA}$
2	$\mu + G_2 + P_1 + F_{IA}$	$\mu + G_2 + P_2 + F_I + C_{IA}$	—
3	—	$\mu + G_3 + P_2 + F_{IA}$	$\mu + G_3 + P_3 + F_I + C_{IA}$
4	$\mu + G_4 + P_1 + F_{IA}$	—	$\mu + G_4 + P_3 + F_I$
5	—	$\mu + G_5 + P_2 + F_I$	$\mu + G_5 + P_3 + F_{IA} + C_I$
6	$\mu + G_6 + P_1 + F_I$	$\mu + G_6 + P_2 + F_{IA} + C_I$	—

Table 12.2.2 provides the expected values of sequence by period means. The analysis of variance table in terms of degrees of freedom is given in Table 12.2.3. The expected values of period differences are given in Table 12.2.4. Let θ_I be the treatment effect between I + A and I (or the impact of A on I). Then,

$$\theta_I = F_{IA} - F_I. \tag{12.2.6}$$

Table 12.2.3 The Analysis of Variance Table for the Williams Design for Drug Interaction Study

Source of variation	Degrees of freedom
Inter-subject	$N^a - 1$
Sequence	5
Residual	$N - 6$
Intra-subject	N
Period	2
Formulation	1
Carry-over	1
Residual	$N - 4$
Total	$2N - 1$

[a] $N = \Sigma_k n_k$.

Table 12.2.4 Expected Values of Period Differences for Comparison between I and I + A

Sequence	Period differences	Expected value	$\Sigma_k C_{jk}^2$
1	$\frac{1}{2}(\overline{Y}_{\cdot 31} - \overline{Y}_{\cdot 11})$	$\frac{1}{2}[(F_{IA} - F_I) + (P_3 - P_1)]$	$\frac{1}{2}$
2	$\frac{1}{2}(\overline{Y}_{\cdot 12} - \overline{Y}_{\cdot 22})$	$\frac{1}{2}[(F_{IA} - F_I) + (P_1 - P_2) - C_{IA}]$	$\frac{1}{2}$
3	$\frac{1}{2}(\overline{Y}_{\cdot 33} - \overline{Y}_{\cdot 23})$	$\frac{1}{2}[(F_I - F_{IA}) + (P_3 - P_2) + C_{IA}]$	$\frac{1}{2}$
4	$\frac{1}{2}(\overline{Y}_{\cdot 14} - \overline{Y}_{\cdot 34})$	$\frac{1}{2}[(F_{IA} - F_I) + (P_1 - P_3)]$	$\frac{1}{2}$
5	$\frac{1}{2}(\overline{Y}_{\cdot 25} - \overline{Y}_{\cdot 35})$	$\frac{1}{2}[(F_I - F_{IA}) + (P_2 - P_3) - C_I]$	$\frac{1}{2}$
6	$\frac{1}{2}(\overline{Y}_{\cdot 26} - \overline{Y}_{\cdot 16})$	$\frac{1}{2}[(F_{IA} - F_I) + (P_2 - P_1) + C_I]$	$\frac{1}{2}$

An unbiased estimate of θ_I can also be expressed as a linear contrast of sequence by period means, for example,

$$\hat{\theta}_I = \frac{1}{2}[(\overline{Y}_{\cdot 31} - \overline{Y}_{\cdot 11}) + (\overline{Y}_{\cdot 14} - \overline{Y}_{\cdot 34})],$$

which is the sum of period differences from sequences 1 and 4. The variance of θ_I is given by

$$\text{Var}(\hat{\theta}_I) = \frac{1}{n}\sigma_e^2, \tag{12.2.7}$$

where $\hat{\sigma}_e^2$ can be estimated unbiasedly by the intra-subject mean squared error from the analysis of variance table with $N - 4$ degrees of freedom. Therefore,

Table 12.2.5 Expected Values of Sequence-by-Period Means for Comparing A and I + A

Sequence	Period I	Period II	Period III
1	—	$\mu + G_1 + P_2 + F_A$	$\mu + G_1 + P_3 + F_{IA} + C_A$
2	$\mu + G_2 + P_1 + F_{IA}$	—	$\mu + G_2 + P_3 + F_A$
3	$\mu + G_3 + P_1 + F_A$	$\mu + G_3 + P_2 + F_{IA} + C_A$	—
4	$\mu + G_4 + P_1 + F_{IA}$	$\mu + G_4 + P_2 + F_A + C_{IA}$	—
5	$\mu + G_5 + P_1 + F_A$	—	$\mu + G_5 + P_3 + F_{IA}$
6	—	$\mu + G_6 + P_2 + F_{IA}$	$\mu + G_6 + P_3 + F_A + C_{IA}$

Table 12.2.6 Expected Values of Period Differences for Comparison between A and I + A

Sequence	Period differences	Expected value	$\Sigma_k C_{jk}^2$
1	$\frac{1}{2}(\overline{Y}_{\cdot 31} - \overline{Y}_{\cdot 21})$	$\frac{1}{2}[(F_{IA} - F_A) + (P_3 - P_2) + C_A]$	$\frac{1}{2}$
2	$\frac{1}{2}(\overline{Y}_{\cdot 12} - \overline{Y}_{\cdot 32})$	$\frac{1}{2}[(F_{IA} - F_A) + (P_1 - P_3)]$	$\frac{1}{2}$
3	$\frac{1}{2}(\overline{Y}_{\cdot 13} - \overline{Y}_{\cdot 23})$	$\frac{1}{2}[(F_A - F_{IA}) + (P_1 - P_2) - C_A]$	$\frac{1}{2}$
4	$\frac{1}{2}(\overline{Y}_{\cdot 24} - \overline{Y}_{\cdot 14})$	$\frac{1}{2}[(F_A - F_{IA}) + (P_2 - P_1) + C_{IA}]$	$\frac{1}{2}$
5	$\frac{1}{2}(\overline{Y}_{\cdot 35} - \overline{Y}_{\cdot 15})$	$\frac{1}{2}[(F_{IA} - F_A) + (P_3 - P_1)]$	$\frac{1}{2}$
6	$\frac{1}{2}(\overline{Y}_{\cdot 26} - \overline{Y}_{\cdot 36})$	$\frac{1}{2}[(F_{IA} - F_A) + (P_2 - P_3) - C_{IA}]$	$\frac{1}{2}$

the treatment effect θ_I can be assessed using either the confidence interval (12.2.3) or hypothesis testing procedure (12.2.4).

For the comparison of A and A + I, Tables 12.2.5 and 12.2.6 summarize the expected values of sequence by period means and the expected values of period differences for each sequence. In a similar manner, a statistical test for the impact of I on A, denoted by θ_A, can also be obtained.

Note that the design and analysis for drug interaction studies discussed in this section can also be directly applied to bioavailability/bioequivalence studies for a combinational drug (Yeh and Chiang, 1991) because I and A represent two drugs and I + A is their combination.

12.3. DOSE PROPORTIONALITY STUDY

How to characterize the pharmacokinetic profile of an active (or potentially active) compound is always of interest to researchers. The dose relationship of a therapeutic compound is important in studying the pharmacokinetic effect of dose range within a given dose range. For this purpose, a crossover experiment at several pre-determined dose levels is usually conducted with some healthy human subjects. Each dosing period is preceded by a fast period and followed by a washout period. Blood samples are collected before dosing and at various time points after dosing. Based upon these collected blood samples, the relationship between the pharmacokinetic parameters such as AUC or C_{max} and the dose levels is then evaluated within the dose range tested.

In many cases, it is of particular interest to determine whether the relationship between the pharmacokinetic parameters and dose levels is linear (or log-linear). If the relationship is linear, the rate of change in pharmacokinetic effect over a given dose range is a constant. In this case, the pharmacokinetic effect of a dose change can be easily predicted and dose can be adjusted accordingly to achieve the desired magnitude of effect. In the following, we will characterize the dose proportionality and introduce the classical methods for assessment of dose linearity.

12.3.1. Dose Linearity Model

Let Y be AUC or C_{max} and X be the dose level. Since, in many cases, the standard deviation of Y increases as the dose increases, the primary assumption of dose proportionality is that the standard deviation of Y is proportional to X, i.e.,

$$Var(Y) = X^2\sigma^2,$$

where σ^2 usually consists of inter-subject and intra-subject variabilities. Under this assumption, the following models are usually considered to evaluate the relationship between the response Y and the dose X.

Model 1: $E(Y) = bX$;
Model 2: $E(Y) = a + bX$ where $a \neq 0$;
Model 3: $E(Y) = a \cdot X^b$ where $a > 0$ and $b \neq 0$;
Model 4: $E(Y) = a + cX^b$ where $a \neq 0$ and/or $b \neq 1$.

Model 1 indicates that the relationship between AUC and the dose is linear. The dose response curve is a straight line, which goes through the origin. This indicates that a double dose will result in a double AUC. Model 1 is usually referred to as dose proportionality. It can be seen that model 1 can be used to evaluate dose proportionality by testing the following hypotheses:

$$H_{01}: b = 0 \quad vs \quad H_{a1}: b \neq 0.$$

This can be done using a weighted linear regression with weights equal to X^{-1} based upon the original data (X,Y). It also can be tested examining the 95% confidence interval for the slope b. Note that failure to reject H_{01} may indicate that the AUC is independent of the dose level.

Model 2 indicates that the relationship between AUC and the dose is a straight line with non-zero intercept, a. Again, it can be tested using a weighted linear regression with weights equal to X^{-1} and with the original data (X,Y). The hypotheses of primary interest are given below:

$$H_{02}: a = 0 \quad vs \quad H_{a2}: a \neq 0.$$

Similarly, model 2 can also be evaluated by examining the 95% confidence interval for the intercept a.

Model 3 can be equivalently expressed in terms of the following logarithmic form:

$$\log(E(Y)) = \log(a) + b \log X.$$

Therefore, like model 2, model 3 can be tested a weighted linear regression with log-transformed data (log X, log Y). The hypotheses of primary interest are

$$H_{03}: b = 0 \quad vs \quad H_{a3}: b \neq 0.$$

Again, it can be tested indirectly by examining the 95% confidence interval for the exponent b.

Model 4 is, in fact, the combination of models 2 and 3 and requires non-linear weighted regression techniques to evaluate. This can be done using PROC NLIN of SAS.

Let $X_1 < X_2 < \ldots < X_k$ be the k dose levels and Y_i, $i = 1, \ldots, k$ be the corresponding responses. Under the assumption of $Var(Y) = X^2\sigma^2$, the ratio Y_i/X_i is approximately equal to the slope b for all doses. In most cases, the ratio Y_i/X_i is commonly used as normalized responses. Draper and Smith (1981), however, indicated that the average of these ratios

$$\hat{R} = \frac{1}{k} \sum_{i=1}^{k} \frac{Y_i}{X_i} \qquad (12.3.1)$$

is the best weighted linear estimate of b. Therefore, in some cases, the ratio $Y_i/\hat{R}$ is sometimes considered as the normalized response of Y_i. Note that, under models 1 and 2 (linear), the normalized response should be approximately equal to the dose X_i. Moreover, the linear regression of the normalized response versus X_i should result approximately in a straight line through the origin with slope equal to 1. This is, however, not true for models 3 and 4 (non-linear). Linear regression of the normalized response X_i will result in a straight line with slope not necessarily equal to 1. It is then of interest to measure the departure from dose linearity under models 3 and 4 in the case where the model is incorrectly assumed to be linear (i.e., models 1 or 2).

12.3.2. Departure from Dose Linearity

In this section, we will compare the relationship between model 2 (linear) and model 3 (non-linear) and the effects of curvature and normalization. Suppose model 3 is the true model for the data. Then there is a relationship between estimates obtained assuming model 3 and estimates obtained assuming (incorrectly) model 2. This relationship can be examined in terms of the difference between $Y = a \cdot X^b$ and a least squares linear approximation (Smith, 1986), which is given by

$$Y_l \approx a_l + b_l X,$$

where

$$a_l = \frac{2(1 - b)}{(b + 1)(b + 2)} a \quad \text{and} \quad b_l = \frac{6b}{(b + 1)(b + 2)} a. \tag{12.3.2}$$

From (12.3.2), we note that $a_l < 0$ if and only if $b > 1$. If model 2 is assumed to be the true model and a weighted linear regression is fitted to the data, what results is the following estimated regression equation

$$Y = \hat{a} + \hat{b}X,$$

then we have

$$E(\hat{a}) \approx a_l \quad \text{and} \quad E(\hat{b}) = b_l.$$

This indicates that the estimated regression equation is an estimator of the linearized form of the true response. We would expect a negative (positive) intercept if the data is convex (concave). To measure the degree of curvature, we first normalize the domain of X so that $X = 1$ corresponds to the highest dose under study. Therefore, the ratio of Y and Y_l at $X = 1$ compares the true response to its linearized response, which is given below

$$\mathcal{H} = Y/Y_l = \frac{(b + 1)(b + 2)}{4b + 2}. \tag{12.3.3}$$

From (12.3.3), it can be seen that $\mathcal{H}$ ranges from 0.98 to 1.05 for $b \in (0.75, 1.25)$. Therefore, the true response is within 5% of the linearized response at the highest dose. This indicates that a lack of dose linearity with b within $(0.75, 1.25)$ has little practical significance.

For the normalized response, under model 3, if we assume that $X = 1$ corresponds to the highest dose, then $\hat{R}$ in (12.3.1) is an estimator of the following parameter

$$\sum_{i=1}^{k} \frac{1}{k} \frac{Y_i}{X_i} = \sum_{i=1}^{k} \frac{1}{k} \frac{a \cdot X_i^b}{X_i} = \int_0^1 a \cdot x^{b-1} dx = \frac{a}{b}.$$

Therefore, the normalized dose response is

$$Y_n = \frac{Y}{(a/b)} = b \cdot X^b.$$

Hence, the normalized linearized response is given by

$$Y_{nl} = \frac{Y_l}{(a/b)} = \frac{2b(1 - b)}{(b + 1)(b + 2)} + \frac{6b^2}{(b + 1)(b + 2)} X.$$

Thus, the normalized linearized response Y_{nl} can also give some indication of departures from dose linearity because

$$\frac{Y_n}{Y_{nl}} = \frac{(b + 1)(b + 2)}{4b + 2} = \mathcal{H}.$$

Based upon these results, Smith (1986) proposed some decision criteria using the 95% confidence interval of the slope to determine the departure from dose linearity. If the confidence interval (L,U) satisfies

$$0.75 < L < 1 < U < 1.25,$$

then we conclude that there is no departure from dose linearity. If the confidence interval satisfy

(i) $1 < L < U < 1.25,$ and

(ii) $0.75 < L < U < 1.0,$

then we conclude that there is a slight departure but no practical significance from does linearity. Finally, if $L > 1.25$ or $U < 0.75$, then reject the hypothesis of dose linearity.

Note that the departure from dose linearity can also be examined by a lack of fit test through stepwise polynomial regression (Draper and Smith, 1981) which will not be further discussed here. However, it is recommended that the mean response vs dose level be plotted before an appropriate model is chosen for analysis.

12.3.3.   Assessment of Dose Proportionality

In the interest of balance for carry-over effects, a dose proportionality study is usually conducted with a Latin squares or a balanced incomplete block design. Let Y_{ijk} be the responses for the ith subject at the jth dose from a $K \times K$ Latin squares with n_k subjects in each sequence. Then, under the assumption that the standard deviation of the response is proportional to the dose, dose proportionality can be assessed using the following statistical models based upon the normalized responses:

Model I: $\dfrac{Y_{ijk}}{X_j} = G_k S_{ik} X_j^{\beta j} C_{(j-1,k)} P_{(j,k)} e_{ijk}$

Model II: $\dfrac{Y_{ijk}}{X_j} = G_k + S_{ik} + \beta_j X_j + C_{(j-1,k)} + P_{(j,k)} + e_{ijk},$

where G_k is the fixed effect of the kth sequence; S_{ik}, $C_{(j-1,k)}$, $P_{(j,k)}$, and e_{ijk} are as defined in (4.1.2), $i = 1,2, \ldots ,n_k$; $j = 1,2, \ldots , K$; $k = 1,2, \ldots ,K$. It is assumed that $\{S_{ik}\}$ are independently and identically distributed with mean 0 and variance σ_s^2 and $\{e_{ij}\}$ are independently identically distributed with mean 0 and variance σ_e^2. $\{S_{ik}\}$ and $\{e_{ijk}\}$ are mutually independent. The analysis of

Table 12.3.1 The Analysis of Variance Table for the K × K Latin Squares

Source of variation	Degrees of freedom
Inter-subject	$N - 1$
Sequence	$K - 1$
Residual	$N - K$
Intra-subject	$(K - 1)N$
Period	$K - 1$
Carry-over	$K - 1$
Dose	$K - 1$
Linear	1
Quadratic	1
Cubic	1
$\vdots$	$\vdots$
Residual	$(K - 1)(N - 3)$
Total	$KN - 1$

variance table in terms of degrees of freedom for a Latin squares is given in Table 12.3.1.

Note that a log transformation on model I (multiplicative) leads to model II (additive). For model II, the variance of Y_{ijk} is given by

$$\mathrm{Var}(Y_{ijk}) = X_j^2(\sigma_s^2 + \sigma_e^2).$$

Therefore, the assumption that the standard deviation of Y_{ijk} is proportional to X_j is met.

Under either model I or model II, we first test whether the model is adequate before the assessment of dose proportionality. This can be done by testing the following hypotheses:

$$H_0: \beta_j = \beta \quad \text{for all } j,$$
$$\text{vs} \quad H_a: \beta_j \neq \beta \quad \text{for at least one } j. \tag{12.3.4}$$

We then conclude that the model is adequate if we fail to reject the null hypothesis. In this case, dose proportionality can be assessed by testing the following hypotheses:

$$H_0: \beta = 0 \quad \text{vs} \quad H_a: \beta \neq 0. \tag{12.3.5}$$

A dose proportionality relationship between Y and X is concluded if we fail to reject the null hypothesis.

For assessment of dose linearity, the following models based upon the original data may be useful:

Model III: $Y_{ijk} = G_k S_{ik} X_j^{\beta j} C_{(j-1,k)} P_{(j,k)} e_{ijk}$

Model IV: $Y_{ijk} = G_k + S_{ik} + \beta_j X_j + C_{(j-1,k)} + P_{(j,k)} + e_{ijk}$,

In a similar manner, we first test whether the model is adequate. If the model is adequate, then test hypotheses (12.3.5) for dose linearity.

Note that tests for hypotheses (12.3.4) and (12.3.5) under each of models I through IV can be easily performed using PROC GLM of SAS.

12.3.4. Discussion

As indicated earlier, a weighted linear regression is often used to evaluate dose linearity. The coefficient of correlation obtained from the model is always mis-used as an indicator of linearity. It should be noted that if the relationship between Y and X is linear, the coefficient of correlation is close to 1. However, a high coefficient of correlation does not necessarily imply a linear relationship between Y and X.

In model 1, under the assumption of $Var(Y) = X^2\sigma^2$, we have $Var(\log(Y)) \approx E(Y)^{-2} Var(Y) = \sigma^2/b^2$, which is the square of the coefficient of variation of Y. This indicates that log-transformation is a variance stabilizing transformation for Y if the standard deviation of Y is proportional to X.

In many applications, each subject may be bled serially. The AUC and C_{max} are obtained for each subject at each dose. In this case, weighted linear regression techniques can be applied either to the individual data $\{(X_i, Y_{ij}), i = 1, 2, \ldots, k; j = 1, \ldots, n_i\}$ or to the mean data $\{(X_i, \overline{Y}_i), i = 1, 2, \ldots, k\}$. Smith (1986) developed a computer program for assessing dose linearity in animal experiments with serial and non-serial collection of blood samples. It should be noted that narrower confidence intervals are often obtained with the individual data than with the mean data because the individual data provides more degrees of freedom for error.

The evaluation of model 4 requires a non-linear weighted regression statistical package such as PROC NLIN of SAS. This procedure sometimes is very sensitive to the selected initial values. To characterize a nonlinear dose response curve, Chow, Hsu, and Tse (1991) proposed a nonparametric test procedure by comparing the mean rate of change (slope) in response among adjacent dose intervals. Although this method does not provide an exact test for the dose response curve, it is more sensible and easier to implement in practice.

In many cases, a dose response curve is assumed to be linear with zero intercept (dose proportionality), i.e., $E(Y|X) = a + bX$ with $a = 0$. However, there exist situations where the dose response curve is nonlinear at the first pass. In practice, it is generally very difficult to verify that a = 0 if the lowest dose included in the study is not small enough even though the response is zero at zero dose. A reasonable approach is to assume that there is a simple linear relationship between Y and X over the range (X_1, X_k) with an arbitrary intercept at dose zero and test its significance against the hypothesis of a nonlinear dose response curve.

12.4. STEADY STATE ANALYSIS

For a multiple dose regimen, the amount of drug in the body is said to have reached a steady state level if the amount or average concentration of the drug in the body remains stable. To determine whether the steady state has been reached, the pharmacokinetic parameter AUC from 0 to infinity is usually considered because it measures the extent of absorption. AUC from 0 to infinity is probably the most reliable pharmacokinetic response for determination of steady state. However, it requires a large amount of blood samples from each subject over a period which investigators feel that steady state might be reached. Therefore, in practice, the use of AUC from 0 to infinity for determination of a steady state of a drug may not be totally feasible.

As an alternative, if the preliminary information of the pharmacokinetic profile of the active ingredient is available, then the trough and peak values of plasma concentrations are usually used to determine whether the steady state has been reached. In a multiple dose regimen, the trough values are defined as the plasma concentrations at the lowest dose level which is usually measured at the time immediately prior to dosing. Peak plasma concentration, which is measured at the estimated t_{max} obtained from previous studies, provides information for determination of whether the plasma concentration exceeds the toxicity level. On the other hand, trough value is an indication of whether the plasma concentration is above the efficacious therapeutic level. The peak to trough ratio is usually used as an indicator of fluctuation of drug efficacy and safety. For example, a large peak to trough value may indicate that either the drug is not effective (trough concentration is too low), or the drug has harmful adverse effects (peak concentration is too high), or both. Therefore, an effective and safe drug should have a relatively small peak to trough ratio.

12.4.1. Univariate Analysis

Let Y_{ij} be the trough value of the ith subject at the jth dose. Then, the following model can be used to evaluate changes in trough values:

$$Y_{ij} = \mu + S_i + F_j + e_{ij}$$
$$= \alpha_j + S_i + e_{ij}, \qquad (12.4.1)$$

where μ is the overall mean, S_i is the random effect of subject i, i = 1,2, . . . ,N, F_j is the fixed effect of the jth dose, j = 1,2, . . . ,d, and e_{ij} is the random error.

Under assumptions of normality and compound symmetry of S_i and e_{ij}, the maximum likelihood estimators of α_j, σ_e^2, and σ_s^2 are given by

$$\hat{\alpha}_j = \overline{Y}_{.j},$$
$$\hat{\sigma}_e^2 = \text{MSE},$$
$$\hat{\sigma}_s^2 = \frac{(\text{MSB} - \text{MSE})}{d},$$

where $\overline{Y}_{.j}$, MSB and MSE are as defined in Section 8.5.

For a total of d doses, the Helmert transformation (Searle, 1971) consists of the following d − 1 row vectors

$$\mathbf{C}_1' = \left(-1, \frac{1}{d-1}, \ldots, \frac{1}{d-1}\right)$$

$$\mathbf{C}_2' = \left(0, -1, \frac{1}{d-2}, \ldots, \frac{1}{d-2}\right)$$

$$\mathbf{C}_{d-1}' = (0, 0, \ldots, -1, 1) \tag{12.4.2}$$

Note that $\mathbf{C}_j'\mathbf{C}_{j'} = 0$. Let

$$\hat{\boldsymbol{\alpha}} = (\hat{\alpha}_1, \hat{\alpha}_2, \ldots, \hat{\alpha}_d),$$
$$= (\overline{Y}_{.1}, \overline{Y}_{.2}, \ldots, \overline{Y}_{.d})',$$

and

$$T_j = \frac{\mathbf{C}_j'\hat{\boldsymbol{\alpha}}}{[\text{MSE } \mathbf{C}_j'\mathbf{C}_j]^{1/2}}. \tag{12.4.3}$$

Then, we conclude that the steady state is reached at the jth dose if

$$|T_j| \leq t[\alpha/2, (N-1)(d-1)], \tag{12.4.4}$$

provided that $|T_1|, \ldots, |T_{j-1}|$ are greater than $t[\alpha/2, (N-1)(d-1)]$.

12.4.2. Multivariate Analysis

If the assumption of compound symmetry is not satisfied, one can either apply a log transformation of the trough plasma concentrations, or use a multivariate analysis based upon the one-sample Hotelling T^2 statistic to determine whether the steady state is reached.

Let $\mathbf{Y}_i = (Y_{i1}, \ldots, Y_{id})'$ be d × 1 vector of the trough plasma concentrations at d different dosing times observed on subject i. Also, let S be the sample covariance matrix computed from $\mathbf{Y}_i$, i.e.,

$$S = \frac{1}{N-1} \sum_{i=1}^{N} [Y_i - \hat{\alpha}][Y_i - \hat{\alpha}]'. \tag{12.4.5}$$

Then, the steady state is considered to be reached at the jth dose if

$$|t_j| \leqslant T(\alpha, d-1, N-d+1), \tag{12.4.6}$$

provided that $|t_1|, \ldots, |t_{j-1}| > T(\alpha, d-1, N-d+1)$,
where

$$t_j^2 = \frac{N(C_j'\hat{\alpha})^2}{C_j'SC_j}, \tag{12.4.7}$$

and

$$
\begin{aligned}
T^2(\alpha, d-1, N-d+1) \\
= \left[\frac{(N-1)(d-1)}{N-d+1} \right] F(\alpha, d-1, N-d+1). \tag{12.4.8}
\end{aligned}
$$

12.4.3. Discussion

The procedure for determination of the steady state, either univariate or multi-variate, is to search the first dose (time) where T_j (t_j) is not rejected at the α significance level. However, although T_j (t_j) is not rejected, $T_{j'}$ ($t_{j'}$) might be rejected for $j' > j$. This indicates that the trough plasma concentrations still fluctuate after the jth dose. It should be noted that although T_j (t_j) and $T_{j'}(t_{j'})$ are independent of each other ($j' \neq j$), in order to have an overall α type I error rate, an adjustment for the test significance level is necessary.

As indicated earlier, peak to trough ratios can also be used to determine whether the steady state is reached. A relatively small peak to trough ratio indicates that the study drug is relatively effective and safe. Recently, analysis of peak to trough values has become very important for assessing efficacy and safety of the drug in anti-hypertensive, anti-angina, and anti-arrhythmia therapies. However, the procedure discussed above cannot be applied directly because the peak to trough ratios may not satisfy normality assumptions. In fact, even when the peak values and the trough values follow normal distributions, the peak to trough ratios,

$$R_i = \frac{Y_{pi}}{Y_{ti}}, \qquad i = 1, 2, \ldots, N,$$

where Y_{pi} and Y_{ti} are the peak and trough values for subject i. R_i may follow a Cauchy distribution which does not have first and any higher moments (e.g., mean and variance). As an alternative, one may consider analyzing the peak to trough ratios using nonparametric summary statistics such as median and inter-quantile range, or apply the Fieller's theorem discussed in Chapter 4 to construct a confidence interval for the true population peak to trough ratio.

12.5. POPULATION PHARMACOKINETICS

The study of individual pharmacokinetic parameters has become very popular because it is a representative sample of a population of such individuals. Individual pharmacokinetic parameters are usually estimated by fitting a pharmacokinetic model. The analysis of pharmacokinetic responses has emerged as a useful approach to ascertaining the nature of drug disposition in patient populations. Although such an analysis permits the evaluation of associations between patient characteristics and variation in drug disposition, it usually involves a nonlinear mixed effects model which presents certain statistical and pharmacokinetic modeling difficulties. Such an analysis is usually performed after a considerably large database of plasma concentrations, doses, patient characteristics is established.

For estimation of population pharmacokinetic parameters, a typical approach is the so-called standard two-stage method (STS). At the first stage, we estimate individual parameters. Then, we treat the estimates as samples to obtain a confidence interval for the population parameters. This method, however, does not account for the variability of the estimates obtained from the first stage. To avoid individual estimates, Sheiner et al. (1972, 1977) proposed an alternative first order method for estimation of the mean and variance of the population parameters by minimizing an extended least squares criterion. This method, which utilizes a nonlinear mixed effect model, is available in the software NONMEM (Beal and Sheiner, 1980).

Recently, several other methods, including parametric and nonparametric, have been developed. These methods include the use of the EM algorithm (Dempster et al., 1977) and the Bayesian approach (Racine-Poon, 1985). For example, Prevost proposed an iterative two-stage method (ITS), which uses a linearization of the model around the estimated parameters, under the assumption of a normal distribution of parameters (Steimer et al., 1984). The ITS method has recently been investigated by Lindstrom and Bates (1990) under different approximations. For nonparametric methods, Mallet (1986) developed a method based upon maximum likelihood of the whole set of individual measurements. Schumitzky (1990) also proposed a nonparametric algorithm to estimate the distribution using an EM estimation approach. For a one-compartment model, these methods are available on the software NPEM (Jellife et al., 1990).

Since the analysis of population pharmacokinetic parameters usually involves a nonlinear mixed effects model, the estimates of inter-subject (or inter-individual), intra-subject (or intra-individual) variability, and measurement error are extremely important. A relatively large inter-subject variability may indicate that more patients are needed in order to have sound statistical inference for the parameters. On the other hand, if the intra-subject variability is much larger than the inter-subject variability, more plasma samples may be needed to characterize the plasma concentration-time curve. An appropriate nonlinear mixed effects model should be able to account for these variations.

For a pharmacokinetic model, it is of interest to estimate the population parameters and the difference in parameters between groups such as treatment group, sex, age and race. The estimation of population parameters can be obtained based upon plasma concentrations over time. Statistical inference for the difference in parameters between groups may be considered for assessing bio-equivalence based upon a pre-selected decision rule. The interpretation of these parameters, however, is important in evaluation of drug performances within and between groups.

References

Agresti, A. (1990). Categorical Data Analysis. John Wiley and Sons, New York, NY.

Altham, P.M.E. (1971). The analysis of matched proportions. Biometrika, 58, 561–576.

Anderson, R.L. (1982). Analysis of Variance Components, Unpublished Lecture Notes. University of Kentucky, Lexington, KY.

Anderson, S. and Hauck, W.W. (1983). A new procedure for testing equivalence in comparative bioavailability and other clinical trials. Communications in Statistics— Theory and Methods, 12, 2663–2692.

Anderson, S. and Hauck, W.W. (1990). Consideration of individual bioequivalence. J. of Pharmacokin. and Biopharm., 18, 259–273.

Armitage, P. and Berry, G. (1987). Statistical Methods in Medical Research. 2nd Edition, Blackwell Scientific Publications, Oxford, England.

Balaam, L.N. (1968). A two-period design with t^2 experimental units. Biometrics, 24, 61–73.

Balant, L.P. (1991). Is there a need for more precise definitions of bioavailability. European J. of Clinical Pharmacology, 40, 123–126.

Beal, S.L. and Sheiner, L.B. (1980). The NONMEM system. Amer. Stat., 34, 118–119.

Beckhofer, R.E., Dunnett, C.W., and Sobel, M. (1954). A two sample multiple decision procedure for linking means of normal populations with a common unknown variance. Biometrika, 41, 170–176.

Bickel, P.J. and Doksum, A.D. (1977). Mathematical Statistics. Holden-Day, San Francisco, CA.

Blackwelder, W.C. (1982). Proving the null hypothesis in clinical trials. Controlled Clinical Trials, 3, 345–353.

361

Boardman, T.J. (1974). Confidence intervals for variance components—a comparative Monte Carlo study. Biometrics, 30, 251–262.

Bose, R.C., Clatworthy, W.H. and Shrikhande (1954). Tables of partially balanced incomplete block design with two associate classes. J. Amer. Stat. Assoc., 47, 151–184.

Box, G.E.P. and Tiao, G.C. (1973). Bayesian Inference in Statistical Analysis. Addison-Wesley, Reading, MA.

Bradu, D. and Mundlak, Y. (1970). Estimation in lognormal linear model. J. of Amer. Stat. Assoc., 65, 198–211.

Brown, B.W. (1980). The crossover experiment for clinical trials. Biometrics, 36, 69–79.

Buonaccorsi, J.P. and Gatsonis, C.A. (1988). Bayesian inference for ratios of coefficients in a linear model. Biometrics, 44, 87–101.

Chen, M.L. and Pelsor, F.R. (1991). Half-life revisited: implication in clinical trials and bioavailability/bioequivalence evaluation. J. Amer. Pharm. Assoc., 80, 406–408.

Cheng, C.S. and Wu, C.F. (1980). Balanced repeated measurements designs. Annals of Statistics, 8, 1272–83. (Corrigendum, 11, p349 (1983).)

Chinchilli, V.M. and Durham, B.S. (1989). Testing for bioequivalence via R-estimation. Presented at the Forty-Fifth Conference on Applied Statistics, Atlantic City, New Jersey.

Chow, S.C. (1985). Resampling procedures for the estimation of non-linear functions of parameters. Unpublished Ph.D. Thesis. University of Wisconsin, Madison, WI.

Chow, S.C. (1989). Some results on bioavailability/bioequivalence studies. Proceedings of the Biopharmaceutical section of the American Statistical Association, 260–268.

Chow, S.C. (1990). Alternative approaches for assessing bioequivalence regarding normality assumptions. Drug Information Journal, 24, No. 4, 753–762. (Corrigendum, 25, p. 161 (1991).)

Chow, S.C., Cheng, J. and Shao, J. (1990). A nonparametric bootstrap approach for assessing bioequivalence. Unpublished manuscript.

Chow, S.C., Hsu, J.P., and Tse, S.K. (1991). Characterizing dose reponse curve in clinical trials. Unpublished manuscript.

Chow, S.C., Liu, J.P., and Peace, K.E. (1991). On assessment of bioequivalence under a higher-order crossover design. Submitted to J. of Biopharmaceutical Statistics.

Chow, S.C., Peace, K.E., and Shao, J. (1991). Assessment of bioequivalence using a multiplicative model. Journal of Biopharmaceutical Statistics, 1, No. 2, 193–203.

Chow, S.C. and Shao, J. (1988). A new procedure for the estimation of variance components. Statistics & Probability Letters, 6, 349–355.

Chow, S.C. and Shao, J. (1989). Tests for batch-to-batch variation in stability analysis. Statistics in Medicine, 8, 883–890.

Chow, S.C. and Shao, J. (1990). An alternative approach for the assessment of bioequivalence between two formulations of a drug. Biometrical Journal, 32, No. 8, 969–976.

Chow, S.C. and Shao, J. (1991). A note on decision rules in bioequivalence studies. Unpublished manuscript.

Chow, S.C. and Tse, S.K. (1988). Intrasubject variability estimation in bioavailability/bioequivalence studies. Proceedings of the Biopharmaceutical section of the American Statistical Association, 142–147.

Chow, S.C. and Tse, S.K. (1990a). Outlier detection in bioavailability/bioequivalence studies. Statistics in Medicine, 9, 549–558.

Chow, S.C. and Tse, S.K. (1990b). A related problem in bioavailability/bioequivalence studies—estimation of intrasubject variability with a common CV. Biometrical Journal, 32, No. 5, 597–607.

Clayton, D. and Leslie, A. (1981). The bioavailability of erythromycin stearate versus enteric-coated erythromycin base when taken immediately before and after food. J. Int. Med. Res., 9, 470–477.

Cochran, W.G. and Cox, G.M. (1957). Experimental Designs. 2nd Ed. John Wiley and Sons, New York, NY.

Colton, T. (1974). Statistics in Medicine. Little, Brown and Company, Boston, MA.

Cornell, R.G. (1980). Evaluation of Bioavailability Data Using Nonparametric Statistics in Drug Absorption and Disposition: Statistical Considerations. ed. by K.S. Albert, American Pharmaceutical Association, Academy of Pharmaceutical Sciences, 51–57. Washington, D.C.

Cornell, R.G. (1990). The evaluation of bioequivalence using nonparametric procedures. Communications in Statistics—Theory and Methods, 19, 4153–4165.

Crow, E.L. and Shimizu, K. (1988). Lognormal Distribution. Marcel Dekker, New York, NY.

Dare, J.G. (1964). Particle size in relation to formulation. Aust. J. Pharm., 45, p. S58.

Dempster, A.P., Laird, N.M. and Rubin, D.B. (1977). Maximum likelihood from incomplete data via EM algorithm. J. Roy. Stat. Soc., B, 39, 1–38.

Diletti, E., Hauschke, D., and Steinijans, V.W. (1991). Sample size determination for bioequivalence assessment by means of confidence interval. International Journal of Clinical Pharmacology, Therapy and Toxicology, 29, 1–8.

Draper, N.R. and Smith, H. (1981). Applied Regression Analysis. 2nd Ed., John Wiley & Sons, New York, NY.

Drug Bioequivalence Study Panel (1974). Drug Bioequivalence. A report from the Drug Bioequivalence Study Panel to the Office of Technology Assessment, Congress of the United States, prepared by Family Health Care, Inc., Washington, D.C. under contract OTA C-1 with the Office of Technology Assessment, Congress of the United States.

Dunnett, C.W. and Gent, M. (1977). Significance testing to establish equivalence between treatments, with special reference to data in the form of 2×2 tables. Biometrics, 33, 593–602.

Edwards, W., Lindman, H., and Savage, L.J. (1963). Bayesian statistical inference for psychological research. Psychological Review, 70, p. 193.

Efron, B. (1982). The Jackknife, Bootstrap and Other Resampling Plans. SIAM, Philadelphia, PA.

Ekbohm, G. and Melander, H. (1989). The subject-by-formulation interaction as a criterion of interchangeability of drugs. Biometrics, 45, 1249–1254.

Esinhart, J.D. and Chinchilli, V.M. (1990). Statistical inference on intra-subject variability in bioequivalence studies. Proceedings of the Biopharmaceutical Section of the American Statistical Association, 37–42.

Esinhart, J.D. and Chinchilli, V.M. (1991). The analysis of bioequivalence studies using generalized estimating equations. Presented at the 1991 ENAR Meeting of the Biometric Society, Houston, Texas.

Fieller, E. (1954). Some problems in interval estimation. J. Roy. Stat. Soc., B, 16, 175–185.

Fisher, R.A. and Yates, F. (1953). Statistical Tables for Biological, Agricultural, and Medical Research. 4th Ed. Oliver and Boyd, Edinburgh, Scotland.

Fluehler, H., Grieve, A.P., Mandallaz, D., Mau, J., and Moser, H.A. (1983). Bayesian approach to bioequivalent assessment: an example. Journal of Pharmaceutical Sciences, 72, 1178–1181.

Fluehler, H., Hirtz, J. and Moser, H.A. (1981). An aid to decision-making in bioequivalence assessment. Journal of Pharmacokin. Biopharm., 9, 235–243.

France, L.A., Lewis, J.A., and Kay, R. (1991). The analysis of failure time data in crossover studies. Statistics in Medicine, 10, 1099–1113.

Frick, H. (1987). On level and power of Anderson and Hauck's procedure for testing equivalence in comparative bioavailability. Communications in Statistics—Theory and Methods, 16, 2771–2778.

Gibaldi, M. and Perrier, D. (1982). Pharmacokinetics. Marcel Dekker, New York, NY.

Graybill, F.A. (1961). An Introduction to Linear Statistical Models. McGraw-Hill, New York, NY.

Graybill, F.A. (1976). Theory and Application of the Linear Model. Duxbury, Boston, MA.

Grieve, A.P. (1985). A Bayesian analysis of the two-period crossover design for clinical trials. Biometrics, 41, 979–990.

Grizzle, J.E. (1965). The two-period changeover design and its use in clinical trials. Biometrics, 21, 467–480.

Grizzle, J.E. and Allen, D.M. (1969). Analysis of growth and dose response curves. Biometrics, 25, 357–381.

Guidelines for Biopharmaceutical Studies in Man, American Pharmaceutical Association, Academy of Pharmaceutical Sciences, Washington, D.C., 1972, Appendix I, p.17.

Hauck, W.W. and Anderson, S. (1984). A new statistical procedure for testing equivalence in two-group comparative bioavailability trials. J. Pharmacokin. Biopharm., 12, 83–91.

Hauschke, D., Steinijans, V.W., and Diletti, E. (1990). A distribution-free procedure for the statistical analyses of bioequivalence studies. Int. J. Clin. Pharmacol. Ther. Toxic., 28, 72–78.

Hawkins, P.M. (1980). Identification of Outliers. Chapman and Hall., New York and London.

Haynes, J.D. (1981). Statistical simulation study of new proposed uniformity requirement for bioequivalency studies. Journal of Pharmaceutical Sciences, 70, 673–675.

Herson, J. (1991). Statistical controversies in design and analysis of bioequivalence trial for pharmaceuticals with negligible blood levels: the metered dose inhaler trial. Presented at the 14th Midwest Biopharmaceutical Statistics Workshop, Muncie, Indiana.

Hills, M. and Armitage, P. (1979). The two-period cross-over clinical trial. British Journal of Clinical Pharmacology, 8, 7–20.

Hinkley, D.V. (1969). On the ratio of two correlated normal variables. Biometrika, 56, 635–639.

Ho, I. and Patel, H. I. (1988). Comparison of variances in a bioequivalence study. Presented at the American College of Clinical Pharmacology 17th Annual Meeting. Orlando, Florida.

Hochberg, Y. and Tamhane, A.C. (1987). Multiple Comparison Procedures. John Wiley & Sons, New York, NY.

Hocking, R.P. (1985). The Analysis of Linear Models. Brooks/Cole, Monterey, CA.

Hodges, J.L. and Lehmann, E.L. (1963). Estimates of location based on rank tests. Ann. Math. Stat., 34, 598–611.

Hollander, M. and Wolfe, D.A. (1973). Nonparametrics Statistical Methods. John Wiley & Sons, New York, NY.

Holt, J.D. and Prentice, R.L. (1974). Survival analysis in twin studies and matched-pair experiments. Biometrika, 61, 17–30.

Hoyle, M.H. (1968). The estimation of variances after using a Gaussinating transformation. Ann. Math. Stat., 39, 1125–1143.

Huitson, A., Poloniecki, J., Hews, R., and Barker, N. (1982). A review of cross-over trials. The Statistician, 31, 71–80.

Huque, M. and Dubey, S.D. (1990). A three arm design and analysis for clinical trials in establishing therapeutic equivalence with clinical endpoints. Proceedings of the Biopharmaceutical Section of the American Statistical Association, 91–98.

Huster, W.J., Brookmeyer, R., and Self, S.G. (1989). Model paired survival data with covariates. Biometrics, 45, 145–156.

Jacques, J.A. (1972). Compartmental Analysis in Biology and Medicine. Elsevier, New York, NY.

Jelliffe, R.W., Gomis, P. and Schumitzky, A. (1990). A population model of gentamicin made with a new nonparametric EM algorithm. Technical Report 90-4, USC.

John, P.W.M. (1971). Statistical Design and Analysis of Experiments. Macmillan, New York, NY.

Johnson, R.A. and Wichern, D.W. (1982). Applied Multivariate Statistical Analysis. Prentice Hall Inc., Englewood Cliffs, New Jersey.

Jones, B. and Kenward, M.G. (1989). Design and Analysis of Crossover Trials. Chapman-Hall, London, England.

Kalbfleisch, J.P. and Prentice, R.L. (1980). The Statistical Analysis of Failure Time Data. John Wiley and Sons, New York, NY.

Kendall, M.G. and Stuart, A. (1961). The Advance Theory of Statistics. Vol. II, Griffen, London, England.

Kershner, R.P. and Federer, W.T. (1981). Two-treatment crossover design for estimating a variety of effects. J. Amer. Stat. Assoc., 76, 612–618.

Kirkwood, T.B.L. and Westlake, W.J. (1981). Bioequivalence testing—A need to re-think. Biometrics, 37, 589–594.

Koch, G.G. (1972). The use of nonparametric methods in the statistical analysis of the two-period change-over design. Biometrics, 28, 577–584.

Koch, G.G. and Edwards, S. (1988). Clinical Efficacy Trials with Categorical Data. Chapter 8 in Biopharmaceutical Statistics for Drug Development ed. by K. Peace, 403–457, Marcel Dekker, New York, NY.

Kunin, C.M. (1966). Absorption, distribution, excretion and fate of Kanamycin. Ann. NY Acad. Scien., 132, p. 811.

Land, C.E. (1988). Hypothesis tests and interval estimates, Chapter 3 in Lognormal Distribution, ed. by E.L. Crow and Shimizu, 87–112, Marcel Dekker, New York, NY.

Laska, E.M. and Meisner, M. (1985). A variational approach to optimal two-treatment crossover designs: applications to carryover effect methods. J. Amer. Stat. Assoc., 80, 704–710.

Laska, E.M., Meisner, M. and Kushner, H.B. (1983). Optimal crossover designs in the presence of carryover effects. Biometrics, 39, 1089–1091.

Lasserre, V. (1991). Determination of optimal designs using linear models in crossover trials. Statistics in Medicine, 10, 909–924.

Lehmann, E.L. (1959). Testing Statistical Hypotheses. John Wiley and Sons, New York, NY.

Lehmann, E.L. (1975). Nonparametrics: Statistical Methods Based on Ranks. Holden-Day, San Francisco, CA.

Levy, N.W. (1986). Bioequivalence of Solid Oral Dosage Forms. A presentation to the U.S. Food and Drug Administration Hearing on Bioequivalence of Solid Oral Dosage Forms, September 29–October 1, Pharmaceutical Manufacturer Association, Section II, 9–11.

Liang, K.Y. and Zeger, S.L. (1986). Longitudinal data analysis using generalized linear models, Biometrika, 73, 13–22.

Lin, J.S., Chow, S.C., and Tse, S.K. (1991). A SAS procedure for outlier detection in bioavailability/bioequivalence studies. Proceedings for the 16th SAS Users Group International Conference, 1433–1437.

Lin, T.L. and Tsong, Y. (1990). Removal of a statistical outlier—Impact on the subsequent statistical test. Presented at Joint Statistical Meetings, Anaheim, California.

Lindley, D.V. (1965). Introduction to Probability and Statistics for a Bayesian Viewpoint, Part II Inference. Cambridge University Press, Cambridge, Great Britain.

Lindstrom, M.J. and Bates, D.M. (1990). Nonlinear mixed effects models for repeated measures data. Biometrics, 46, 673–687.

Liu, J.P. (1991). Bioequivalence and intrasubject variability. Journal of Biopharmaceutical Statistics, 1, No. 2 205–219.

Liu, J.P. and Chow, S.C. (1992a). On power calculation of Schuirmann's two one-sided tests procedure in bioequivalence. J. Pharmacokin. & Biopharm. 20, 101–104.

Liu, J.P. and Chow, S.C. (1992b). On assessment of bioequivalence for drugs with negligible plasma levels. Biometrical Journal, To appear.

Liu, J.P. and Chow, S.C. (1992c). On assessment of bioequivalence in variability of bioavailability. Communications in Statistics—Theory and Methods, To appear.

Liu, J.P. and Weng, C.S. (1991). Detection of outlying data in bioavailability/bioequivalence studies. Statistics in Medicine, 10, No. 9, 1375–1389.

Liu, J.P. and Weng, C.S. (1992). Estimation of direct formulation effect under lognormal distribution in bioavailability/bioequivalence studies. Statistics in Medicine, To appear.

Locke, C.S. (1984). An exact confidence interval for untransformed data for the ratio of two formulation means. J. Pharmacokin. Biopharm., 12, 649–655.

Mallet, A. (1986). A maximum likelihood estimation method for random coefficient regression models. Biometrika, 73, 645–656.

Mandallaz, D. and Mau, J. (1981). Comparison of different methods for decision-making in bioequivalence assessment. Biometrics, 37, 213–222.

Mann, H.B. and Whitney, D.R. (1947). On a test of whether one of two random variables is stochastically larger than the other. Ann. Math. Stat., 18, 50–60.

Mantel, N. (1977). Do we want confidence interval symmetric about the null value? Biometrics, 33, 759–760.

Martinez, M.N. and Jackson, A.J. (1991). Suitability of various noninfinity area under the plasma concentration-time curve (AUC) estimates for use in bioequivalence determination: relationship to AUC from zero to time infinity (AUC0-inf). Pharm. Research, 18, 512–517.

McCulloch, C.E. (1987). Tests for equality of variances with paired data. Communications in Statistics—Theory and Methods, 16, 1377–1391.

McQuarrie, D.A. (1967). Stochastic approach to chemical kinetics. Journal of Applied Probability, 4, 413–478.

Mehran, F. (1973). Variance of the MVUE for the lognormal mean. J. Amer. Stat. Assoc., 68, 726–727.

Metzler, C.M. (1974). Bioavailability: A problem in equivalence. Biometrics, 30, 309–317.

Metzler, C.M. (1988). Statistical methods for deciding bioequivalence of formulations. in Oral Substained Released Formulations: Design and Evaluation. ed. by Yacobi, A., Halperin-Walega, 217–238, E. Pergamon Press, NY.

Metzler, C.M. and Huang, D.C. (1983). Statistical methods for bioavailability and bioequivalence. Clin. Res. Practices & Drug Reg. Affairs, 1, 109–132.

Morgan, W.A. (1939). A test for the significance of the difference between the two variances in a sample from a normal bivariate population. Biometrika, 31, 13–19.

Müller-Cohrs, J. (1990). The power of the Anderson-Hauck test and the double t-test. Biometrical Journal, 32, 259–266.

Neyman, J. and Scott, E.L. (1960). Correction for bias introduced by a transformation of variables. Ann. Math. Stat., 31, 643–655.

O'Brien, P.C. and Fleming, T.R. (1987). A paired Prentice-Wilcoxon test for censored paired data. Biometrics, 43, 169–180.

Oser, B.L., Melnick, D., and Hochberg, M. (1945). Physiological availability of the vitamins—study of methods for determining availability in pharmaceutical products. Ind. Eng. Chem. Anal. Ed., 17, 401–411.

Ott, L. (1984). An Introduction to Statistical Method and Data Analysis. 2nd Edition, Duxbury Press, Boston, MA.

Owen, D.B. (1965). A special case of a noncentral t distribution. Biometrika, 52, 437–446.

Patil, V.H. (1965). Approximation to the Behrens-Fisher distributions. Biometrika, 52, 267–271.

Peace, K.E. (1986). Estimating the degree of equivalence and non-equivalence; An alternative to bioequivalence testing. Proceedings of the Biopharmaceutical section of the American Statistical Association, 63–69.

Peace, K.E. (1990). Statistical Issues in Drug Research and Development. Marcel Dekker, New York, NY.

Phillips, K.F. (1990). Power of the two one-sided tests procedure in bioequivalence. J. Pharmacokin. Biopharm., 18, 137–144.

Pitman. E.J.G. (1939). A note on normal correlation. Biometrika, 31, 9–12.

Pocock, S.J. (1983). Clinical Trials—A Practical Approach. John Wiley & Sons, New York, NY.

Purich, E. (1980). Bioavailability/Bioequivalency Regulations: An FDA Perspective in Drug Absorption and Disposition: Statistical Considerations. ed. by K.S. Albert, American Pharmaceutical Association, Academy of Pharmaceutical Sciences, 115–137, Washington, D.C.

Racine-Poon, A. (1985). A Bayesian approach to nonlinear random effects models. Biometrics, 41, 1015–1023.

Racine-Poon, A., Grieve, A.P., Fluehler, H., and Smith, A.F. (1986). Bayesian methods in practice: experiences in the pharmaceutical industry (with discussion). Applied Statistics, 35, 93–150.

Racine-Poon, A., Grieve, A.P., Fluehler, H., and Smith, A.F. (1987). A two-stage procedure for bioequivalence studies. Biometrics, 43, 847–856.

Randles, R.H. and Wolfe, D.A. (1979). Introduction to the Theory of Nonparametric Statistics. John Wiley and Sons, New York, NY.

Rocke, D.M. (1984). On testing for bioequivalence. Biometrics, 40, 225–230.

Rodda, B.E. (1986). Bioequivalence of Solid Oral Dosage Forms. A presentation to the U.S. Food and Drug Administration Hearing on Bioequivalence of Solid Oral Dosage Forms, September 29–October 1, Pharmaceutical Manufacturers Association, Section III, 12–15.

Rodda, B.E. and Davis, R.L. (1980). Determining the probability of an important difference in bioavailability. Clin. Pharmacol. Ther., 28, 247–252.

Rowland, M. and Tozer, T.N. (1980). Clinical Pharmacokinetics Concepts and Applications. Lea & Febiger, Philadelphia, PA.

Ryde, M., Huitfeldt, B., and Pettersson, R. (1991). Relative bioavailability of Olsalazine from tables and capsules: a drug targeted for local effect in the colon. Biopharm. & Drug Disposition, 12, 233–246.

SAS (1990). Statistical Analysis System. SAS User's Guide: Statistics, Version 6, SAS Institute, Cary, N.C.

Schuirmann, D.J. (1981). On hypothesis testing to determine if the mean of a normal distribution is continued in a known interval. Biometrics, 37, p. 617 (Abstract).

Schuirmann, D.J. (1987). A comparison of the two one-sided tests procedure and the power approach for assessing the equivalence of average bioavailability. J. of Pharmacokin. Biopharm., 15, 657–680.

Schuirmann, D.J. (1989). Confidence intervals for the ratio of two means from a crossover study. Proceedings of the Biopharmaceutical Section of the American Statistical Association, 121–126. Washington, D.C.

Schumitzky, A. (1990). Nonparametric EM algorithms for estimating prior distributions. Technical Reports, 90-2, USC.

Searle, S.R. (1971). Linear Models. John Wiley & Sons, New York, NY.

Selwyn, M.R., Dempster, A.P., and Hall, N.R. (1981). A Bayesian approach to bioequivalence for the 2×2 changeover design. Biometrics, 37, 11–21.

Selwyn, M.R. and Hall, N.R. (1984). On Bayesian methods for bioequivalence. Biometrics, 40, 1103–1108.

Shapiro, S.S. and Wilk, M.B. (1965). An analysis of variance test for normality (complete samples). Biometrika, 52, 591–611.

Sheiner, L.B., Rosenberg, B. and Marathe, V.V. (1977). Estimation of population characteristics of pharmacokinetic parameters from routine clinical data. J. Pharmacokin. Biopharm., 5, 445–479.

Sheiner, L.B., Rosenberg, B. and Melmon, K.L. (1972). Modelling of individual pharmacokinetics for computer-aided drug dosage. Comp. Biomed. Res., 5, 441–459.

Shimizu, K. (1988). Point Estimation, Chapter 2 in Lognormal Distribution, ed. by Crow and Shimizu, 27–86, Marcel Dekker, New York, NY.

Shirley, E. (1976). The use of confidence intervals in biopharmaceutics. J. Pharm. Pharmacol., 28, 312–313.

Smith, S.J. (1988). Evaluating the efficiency of the Δ distribution mean estimator. Biometrics, 44, 485–493.

Smith, T. (1986). Statistical methods—dose proportionality. Technical Report, Ayerst Laboratories, New York, NY.

Snedecor, G.W. and Cochran, W.G. (1980). Statistical Methods. 7th Edition, Iowa State University Press, Ames, Iowa.

Snee, R.D. (1972). On the analysis of response curve data. Technometrics, 14, 47–62.

Srinivasan, R. and Langenberg, P. (1986). A two-stage procedure with controlled error probabilities for testing bioequivalence. Biometrical Journal, 28, 825–833.

Steimer, J.L., Mallet, A., Golmard, J.F., and Boisvieux, J.F. (1984). Alternative approaches to estimation of population pharmacokinetic parameters; comparison with the nonlinear mixed effect model. Drug. Metab. Rev., 15, 265–292.

Steinijans, V.W. and Diletti, E. (1983). Statistical analysis of bioavailability studies: parametric and nonparametric confidence intervals. Europ. J. Clin. Pharmacol., 24, 127–136.

Steinijans, V.W. and Diletti, E. (1985). Generalization of distribution-free confidence intervals for bioavailability ratios. Europ. J. Clin. Pharmacol., 28, 85–88.

Thiyagarajan, B. and Dobbins, T.W. (1987). An assessment of the 75/75 rule in bioequivalence. Proceedings of the Biopharmaceutical section of the American Statistical Association. 143–148.

Tse, S.K. (1990). A comparison of interval estimation procedures in bioavailability/bioequivalence. Preecedings of the Biopharmaceutical section of the American Statistical Association. 43–46.

Tukey, J.W. (1951). Components in regression. Biometrics, 7, 33–69.

Wagner, J.G. (1971). Biopharmaceutics and Relevant Pharmacokinetics. Drug Intelligence Publications, Hamilton, Illinois.

Wagner, J.G. (1975). Fundamentals of Clinical Pharmacokinetics. Drug Intelligence Publications, Hamilton, Illinois.

Walsh, J.E. (1949). Some significance tests for the median which are valid under very general conditions. Ann. Math. Stat., 20, 64–81.

Wang, C.M. (1990). On the lower bound of confidence coefficients for a confidence interval on variance components. Biometrics, 46, 187–192.

Ware, J.H., Mosteller, F., Ingelfinger, J.A. (1986). "P-Values", Chapter 8, in Medical Use of Statistics, ed. by J.C. Bailar and F. Mosteller, NEJM Books, Waltham, MA.

Weiner, D. (1989). Bioavailability. Notes of the training course for new clinical statisticians sponsored by the Biostatistics Subsections of Pharmaceutical Manufacturer Association, March, 1989, Washington, D.C.

Westlake, W.J. (1972). Use of confidence intervals in analysis of comparative bioavailability trials. Journal of Pharmaceutical Sciences, 61, 1340–1341.

Westlake, W.J. (1973). The design and analysis of comparative blood-level trials. Current Concepts in the Pharmaceutical Sciences. ed. by J. Swarbrick, Lea & Febiger, Philadelphia, PA.

Westlake, W.J. (1974). The use of balanced incomplete block designs in comparative bioavailability trials. Biometrics, 30, 319–327.

Westlake, W.J. (1976). Symmetrical confidence intervals for bioequivalence trials. Biometrics, 32, 741–744.

Westlake, W.J. (1979). Statistical aspects of comparative bioavailability trials. Biometrics, 35, 273–280.

Westlake, W.J. (1981). Bioequivalence testing—A need to rethink (Reader reaction response). Biometrics, 37, 591–593.

Westlake, W.J. (1986). Bioavailability and Bioequivalence of Pharmaceutical Formulations. Biopharmaceutical Statistics for Drug Development, ed. by K. Peace, 329–352, Marcel Dekker, New York, NY.

Wijnard, H.P. and Timmer, C.J. (1983). Mini-computer programs for bioequivalence testing of pharmaceutical drug formulations in two-way crossover studies. Computer Programs in Biomedicine, 17, 73–88.

Wilcoxon, F. (1945). Individual comparisons by ranking methods. Biometrics, 1, 80–83.

Williams, E.J. (1949). Experimental designs balanced for the residual effects of treatment. Australian Journal of Scientific Research, 2, 149–168.

Williams, J.S. (1962). A confidence interval for variance components. Biometrika, 49, 278–281.

Yee, K.F. (1986). The calculation of probabilities in rejecting bioequivalence. Biometrics, 42, 961–965.

Yeh, C.M. and Chiang, T. (1991). Bioavailability in a 3×3 crossover study of a combination drug. Presented at 1991 Drug Information Association Statistics Workshop. Hilton Head, South Carolina.

Yeh, K.C. and Kwan, K.C. (1978). A comparison of numerical integrating algorithms by trapezoidal, Lagrange, and spline approximations. J. Pharmacokin. Biopharm., 6, 79–81.

Appendix A: Statistical Tables

Appendix A.1 Areas of Upper Tail of the Standard Normal Distribution

z	0.00	0.01	0.02	0.03	0.04	0.05	0.06	0.07	0.08	0.09
0.0	0.5000	0.4960	0.4920	0.4880	0.4840	0.4801	0.4761	0.4721	0.4681	0.4641
0.1	0.4602	0.4562	0.4522	0.4483	0.4443	0.4404	0.4364	0.4325	0.4286	0.4247
0.2	0.4207	0.4168	0.4129	0.4090	0.4052	0.4013	0.3974	0.3936	0.3897	0.3859
0.3	0.3821	0.3783	0.3745	0.3707	0.3669	0.3632	0.3594	0.3557	0.3520	0.3483
0.4	0.3446	0.3409	0.3372	0.3336	0.3300	0.3264	0.3228	0.3192	0.3156	0.3121
0.5	0.3085	0.3050	0.3015	0.2981	0.2946	0.2912	0.2877	0.2843	0.2810	0.2776
0.6	0.2743	0.2709	0.2676	0.2643	0.2611	0.2578	0.2546	0.2514	0.2483	0.2451
0.7	0.2420	0.2389	0.2358	0.2327	0.2296	0.2266	0.2236	0.2206	0.2177	0.2148
0.8	0.2119	0.2090	0.2061	0.2033	0.2005	0.1977	0.1949	0.1922	0.1894	0.1867
0.9	0.1841	0.1814	0.1788	0.1762	0.1736	0.1711	0.1685	0.1660	0.1635	0.1611
1.0	0.1587	0.1562	0.1539	0.1515	0.1492	0.1469	0.1446	0.1423	0.1401	0.1379
1.1	0.1357	0.1335	0.1314	0.1292	0.1271	0.1251	0.1230	0.1210	0.1190	0.1170
1.2	0.1151	0.1131	0.1112	0.1093	0.1075	0.1056	0.1038	0.1020	0.1003	0.0985
1.3	0.0968	0.0951	0.0934	0.0918	0.0901	0.0885	0.0869	0.0853	0.0838	0.0823
1.4	0.0808	0.0793	0.0778	0.0764	0.0749	0.0735	0.0721	0.0708	0.0694	0.0681
1.5	0.0668	0.0655	0.0643	0.0630	0.0618	0.0606	0.0594	0.0582	0.0571	0.0559
1.6	0.0548	0.0537	0.0526	0.0516	0.0505	0.0495	0.0485	0.0475	0.0465	0.0455
1.7	0.0446	0.0436	0.0427	0.0418	0.0409	0.0401	0.0392	0.0384	0.0375	0.0367
1.8	0.0359	0.0351	0.0344	0.0336	0.0329	0.0322	0.0314	0.0307	0.0301	0.0294
1.9	0.0287	0.0281	0.0274	0.0268	0.0262	0.0256	0.0250	0.0244	0.0239	0.0233
2.0	0.02275	0.02222	0.02169	0.02118	0.02068	0.02018	0.01970	0.01923	0.01876	0.01831
2.1	0.01786	0.01743	0.01700	0.01659	0.01618	0.01578	0.01539	0.01500	0.01463	0.01426
2.2	0.01390	0.01355	0.01321	0.01287	0.01255	0.01222	0.01191	0.01160	0.01130	0.01101
2.3	0.01072	0.01044	0.01017	0.00990	0.00964	0.00939	0.00914	0.00889	0.00866	0.00842
2.4	0.00820	0.00798	0.00776	0.00755	0.00734	0.00714	0.00695	0.00676	0.00657	0.00639
2.5	0.00621	0.00604	0.00587	0.00570	0.00554	0.00539	0.00523	0.00508	0.00494	0.00480
2.6	0.00466	0.00453	0.00440	0.00427	0.00415	0.00402	0.00391	0.00379	0.00368	0.00357
2.7	0.00347	0.00336	0.00326	0.00317	0.00307	0.00298	0.00289	0.00280	0.00272	0.00264
2.8	0.00256	0.00248	0.00240	0.00233	0.00226	0.00219	0.00212	0.00205	0.00199	0.00193
2.9	0.00187	0.00181	0.00175	0.00169	0.00164	0.00159	0.00154	0.00149	0.00144	0.00139

Source: Table 3 of *Statistical Tables for Science, Engineering and Management* by J. Murdock and J. A. Barnes (Macmillian, London, 1968).

Appendix A.2 Upper Quantiles of a χ^2 Distribution

v/α	0.995	0.990	0.975	0.950	0.900	0.100	0.050	0.025	0.010	0.005
1	392704.10^{-10}	157088.10^{-9}	982069.10^{-9}	393214.10^{-8}	0.0157908	2.70554	3.84146	5.02389	6.63490	7.87944
2	0.0100251	0.0201007	0.0506356	0.102587	0.210720	4.60517	5.99147	7.37776	9.21034	10.5966
3	0.0717212	0.114832	0.215795	0.351846	0.584375	6.25139	7.81473	9.34840	11.3449	12.8381
4	0.206990	0.297110	0.484419	0.710721	1.063623	7.77944	9.48773	11.1433	13.2767	14.8602
5	0.411740	0.554300	0.831211	1.145476	1.61031	9.23635	11.0705	12.8325	15.0863	16.7496
6	0.675727	0.872085	1.237347	1.63539	2.20413	10.6446	12.5916	14.4494	16.8119	18.5476
7	0.989265	1.239043	1.68987	2.16735	2.83311	12.0170	14.0671	16.0128	18.4753	20.2777
8	1.344419	1.646482	2.17973	2.73264	3.48954	13.3616	15.5073	17.5346	20.0902	21.9550
9	1.734926	2.087912	2.70039	3.32511	4.16816	14.6837	16.9190	19.0228	21.6660	23.5893
10	2.15585	2.55821	3.24697	3.94030	4.86518	15.9871	18.3070	20.4831	23.2093	25.1882
11	2.60321	3.05347	3.81575	4.57481	5.57779	17.2750	19.6751	21.9200	24.7250	26.7569
12	3.07382	3.57056	4.40379	5.22603	6.30380	18.5494	21.0261	23.3367	26.2170	28.2995
13	3.56503	4.10691	5.00874	5.89186	7.04150	19.8119	22.3621	24.7356	27.6883	29.8194
14	4.07468	4.66043	5.62872	6.57063	7.78953	21.0642	23.6848	26.1190	29.1413	31.3193
15	4.60094	5.22935	6.26214	7.26094	8.54675	22.3072	24.9958	27.4884	30.5779	32.8013
16	5.14224	5.81221	6.90766	7.96164	9.31223	23.5418	26.2962	28.8454	31.9999	34.2672
17	5.69724	6.40776	7.56418	8.67176	10.0852	24.7690	27.5871	30.1910	33.4087	35.7185
18	6.26481	7.01491	8.23075	9.39046	10.8649	25.9894	28.8693	31.5264	34.8053	37.1564
19	6.84398	7.63273	8.90655	10.1170	11.6509	27.2036	30.1435	32.8523	36.1908	38.5822

20	7.43386	8.26040	9.59083	10.8508	12.4426	28.4120	31.4104	34.1696	37.5662	39.9968
21	8.03366	8.89720	10.28293	11.5913	13.2396	29.6151	32.6705	35.4789	38.9321	41.4010
22	8.64272	9.54249	10.9823	12.3380	14.0415	30.8133	33.9244	36.7807	40.2894	42.7956
23	9.26042	10.19567	11.6885	13.0905	14.8479	32.0069	35.1725	38.0757	41.6384	44.1813
24	9.88623	10.8564	12.4011	13.8484	15.6587	33.1963	36.4151	39.3641	42.9798	45.5585
25	10.5197	11.5240	13.1197	14.6114	16.4734	34.3816	37.6525	40.6465	44.3141	46.9278
26	11.1603	12.1981	13.8439	15.3791	17.2919	35.5631	38.8852	41.9232	45.6417	48.2899
27	11.8076	12.8786	14.5733	16.1513	18.1138	36.7412	40.1133	43.1944	46.9630	49.6449
28	12.4613	13.5648	15.3079	16.9279	18.9392	37.9159	41.3372	44.4607	48.2782	50.9933
29	13.1211	14.2565	16.0471	17.7083	19.7677	39.0875	42.5569	45.7222	49.5879	52.3356
30	13.7867	14.9535	16.7908	18.4926	20.5992	40.2560	43.7729	46.9792	50.8922	53.6720
40	20.7065	22.1643	24.4331	26.5093	29.0505	51.8050	55.7585	59.3417	63.6907	66.7659
50	27.9907	29.7067	32.3574	34.7642	37.6886	63.1671	67.5048	71.4202	76.1539	79.4900
60	35.5346	37.4848	40.4817	43.1879	46.4589	74.3970	79.0819	83.2976	88.3794	91.9517
70	43.2752	45.4418	48.7576	51.7393	55.3290	85.5271	90.5312	95.0231	100.425	104.215
80	51.1720	53.5400	57.1532	60.3915	64.2778	96.5782	101.879	106.629	112.329	116.321
90	59.1963	61.7541	65.6466	69.1260	73.2912	107.565	113.145	118.136	124.116	128.299
100	67.3276	70.0648	74.2219	77.9295	82.3581	118.498	224.342	129.561	135.807	140.169

Source: Tables of Percentage Points of the χ^2-Distribution by C. M. Thompson, *Biometrika* (1941), Vol. 32, pp. 188–189.

Appendix A.3   Upper Quantiles of a Central t Distribution

v/α	0.050	0.025	0.010	0.005
1	6.3138	12.706	25.452	63.657
2	2.9200	4.3027	6.2053	9.9248
3	2.3534	3.1825	4.1765	5.8409
4	2.1318	2.7764	3.4954	4.6041
5	2.0150	2.5706	3.1634	4.0321
6	1.9432	2.4469	2.9687	3.7074
7	1.8946	2.3646	2.8412	3.4995
8	1.8595	2.3060	2.7515	3.3554
9	1.8331	2.2622	2.6850	3.2498
10	1.8125	2.2281	2.6338	3.1693
11	1.7959	2.2010	2.5931	3.1058
12	1.7823	2.1788	2.5600	3.0545
13	1.7709	2.1604	2.5326	3.0123
14	1.7613	2.1448	2.5096	2.9768
15	1.7530	2.1315	2.4899	2.9467
16	1.7459	2.1199	2.4729	2.9208
17	1.7396	2.1098	2.4581	2.8982
18	1.7341	2.1009	2.4450	2.8784
19	1.7291	2.0930	2.4334	2.8609
20	1.7247	2.0860	2.4231	2.8453
21	1.7207	2.0796	2.4138	2.8314
22	1.7171	2.0739	2.4055	2.8188
23	1.7139	2.0687	2.3979	2.8073
24	1.7109	2.0639	2.3910	2.7969
25	1.7081	2.0595	2.3846	2.7874
26	1.7056	2.0555	2.3788	2.7787
27	1.7033	2.0518	2.3734	2.7707
28	1.7011	2.0484	2.3685	2.7633
29	1.6991	2.0452	2.3638	2.7564
30	1.6973	2.0423	2.3596	2.7500
40	1.6839	2.0211	2.3289	2.7045
60	1.6707	2.0003	2.2991	2.6603
120	1.6577	1.9799	2.2699	2.6174
∞	1.6449	1.9600	2.2414	2.5758

Source: Tables of Percentage Points of the t-Distribution by M. Merrington, *Biometrika* (1941), Vol. 32, p. 300.

Appendix A.4 Upper Quantiles of an F Distribution. $\alpha = 0.05$.

ν_2 \ ν_1	1	2	3	4	5	6	7	8	9
1	161.45	199.50	215.71	224.58	230.16	233.99	236.77	238.88	240.54
2	18.513	19.000	19.164	19.247	19.296	19.330	19.353	19.371	19.385
3	10.128	9.5521	9.2766	9.1172	9.0135	8.9406	8.8868	8.8452	8.8123
4	7.7086	6.9443	6.5914	6.3883	6.2560	6.1631	6.0942	6.0410	5.9988
5	6.6079	5.7861	5.4095	5.1922	5.0503	4.9503	4.8759	4.8183	4.7725
6	5.9874	5.1433	4.7571	4.5337	4.3874	4.2839	4.2066	4.1468	4.0990
7	5.5914	4.7374	4.3468	4.1203	3.9715	3.8660	3.7870	3.7257	3.6767
8	5.3177	4.4590	4.0662	3.8378	3.6875	3.5806	3.5005	3.4381	3.3881
9	5.1174	4.2565	3.8626	3.6331	3.4817	3.3738	3.2927	3.2296	3.1789
10	4.9646	4.1028	3.7083	3.4780	3.3258	3.2172	3.1355	3.0717	3.0204
11	4.8443	3.9823	3.5874	3.3567	3.2039	3.0946	3.0123	2.9480	2.8962
12	4.7472	3.8853	3.4903	3.2592	3.1059	2.9961	2.9134	2.8486	2.7964
13	4.6672	3.8056	3.4105	3.1791	3.0254	2.9153	2.8321	2.7669	2.7144
14	4.6001	3.7389	3.3439	3.1122	2.9582	2.8477	2.7642	2.6987	2.6458
15	4.5431	3.6823	3.2874	3.0556	2.9013	2.7905	2.7066	2.6408	2.5876
16	4.4940	3.6337	3.2389	3.0069	2.8524	2.7413	2.6572	2.5911	2.5377
17	4.4513	3.5915	3.1968	2.9647	2.8100	2.6987	2.6143	2.5480	2.4943
18	4.4139	3.5546	3.1599	2.9277	2.7729	2.6613	2.5767	2.5102	2.4563
19	4.3808	3.5219	3.1274	2.8951	2.7401	2.6283	2.5435	2.4768	2.4227
20	4.3513	3.4928	3.0984	2.8661	2.7109	2.5990	2.5140	2.4471	2.3928
21	4.3248	3.4668	3.0725	2.8401	2.6848	2.5727	2.4876	2.4205	2.3661
22	4.3009	3.4434	3.0491	2.8167	2.6613	2.5491	2.4638	2.3965	2.3419
23	4.2793	3.4221	3.0280	2.7955	2.6400	2.5277	2.4422	2.3748	2.3201
24	4.2597	3.4028	3.0088	2.7763	2.6207	2.5082	2.4226	2.3551	2.3002
25	4.2417	3.3852	2.9912	2.7587	2.6030	2.4904	2.4047	2.3371	3.2821
26	4.2252	2.3690	2.9751	2.7426	2.5868	2.4741	2.3883	2.3205	2.2655
27	4.2100	3.3541	2.9604	2.7278	2.5719	2.4591	2.3732	2.3053	2.2501
28	4.1960	3.3404	2.9467	2.7141	2.5581	2.4453	2.3593	2.2913	2.2360
29	4.1830	3.3277	2.9340	2.7014	2.5454	2.4324	2.3463	2.2782	2.2229
30	4.1709	3.3158	2.9223	2.6896	2.5336	2.4205	2.3343	2.2662	2.2107
40	4.0848	3.2317	2.8387	2.6060	2.4495	2.3359	2.2490	2.1802	2.1240
60	4.0012	3.1504	2.7581	2.5252	2.3683	2.2540	2.1665	2.0970	2.0401
120	3.9201	3.0718	2.6802	2.4472	2.2900	2.1750	2.0867	2.0164	1.9588
∞	3.8415	2.9957	2.6049	2.3719	2.2141	2.0986	2.0096	1.9384	1.8799

Source: Tables of Percentage Points of the Inverted beta (F)-Distribution by M. Merrington and C. M. Thompson, *Biometrika* (1942), Vol. 33, pp. 73–88.

Appendix A.4   Continued $\alpha = 0.05$.

ν_2 \ ν_1	10	12	15	20	24	30	40	60	12	∞
1	241.88	243.91	245.95	248.01	249.05	250.09	251.14	252.20	253.25	254.32
2	19.396	19.413	19.429	19.446	19.454	19.462	19.471	19.479	19.487	19.496
3	8.7855	8.7446	8.7029	8.6602	8.6385	8.6166	8.5944	8.5720	8.5494	8.5265
4	5.9644	5.9117	5.8578	5.8025	5.7744	5.7459	5.7170	5.6878	5.6581	5.6281
5	4.7351	4.6777	4.6188	4.5581	4.5272	4.4957	4.4638	4.4314	4.3984	4.3650
6	4.0600	3.9999	3.9381	3.8742	3.8415	3.8082	3.7743	3.7398	3.7047	3.6688
7	3.6365	3.5747	3.5108	3.4445	3.4105	3.3758	3.3404	3.3043	3.2674	3.2298
8	3.3472	3.2840	3.2184	3.1503	3.1152	3.0794	3.0428	3.0053	2.9669	2.9276
9	3.1373	3.0729	3.0061	2.9365	2.9005	2.8637	2.8259	2.7872	2.7475	2.7067
10	2.9782	2.9130	2.8450	2.7740	2.7372	2.6996	2.6609	2.6211	2.5801	2.5379
11	2.8536	2.7876	2.7186	2.6464	2.6090	2.5705	2.5309	2.4901	2.4480	2.4045
12	2.7534	2.6866	2.6169	2.5436	2.5055	2.4663	2.4259	2.3842	2.3410	2.2962
13	2.6710	2.6037	2.5331	2.4589	2.4202	2.3803	2.3392	2.2966	2.2524	2.2064
14	2.6021	2.5342	2.4630	2.3879	2.3487	2.3082	2.2664	2.2230	2.1778	2.1307
15	2.5437	2.4753	2.4035	2.3275	2.2878	2.2468	2.2043	2.1601	2.1141	2.0658
16	2.4935	2.4247	2.3522	2.2756	2.2354	2.1938	2.1507	2.1058	2.0589	2.0096
17	2.4499	2.3807	2.3077	2.2304	2.1898	2.1477	2.1040	2.0584	2.0107	1.9604
18	2.4117	2.3421	2.2686	2.1906	2.1497	2.1071	2.0629	2.0166	1.9681	1.9168
19	2.3779	2.3080	2.2341	2.1555	2.1141	2.0712	2.0264	1.9796	1.9302	1.8780
20	2.3479	2.2776	2.2033	2.1242	2.0825	2.0391	1.9938	1.9464	1.8963	1.8432
21	2.3210	2.2504	2.1757	2.0960	2.0540	2.0102	1.9645	1.9165	1.8657	1.8117
22	2.2967	2.2258	2.1508	2.0707	2.0283	1.9842	1.9380	1.8895	1.8380	1.7831
23	2.2747	2.2036	2.1282	2.0476	2.0050	1.9605	1.9139	1.8649	1.8128	1.7570
24	2.2547	2.1834	2.1077	2.0267	1.9838	1.9390	1.8920	1.8424	1.7897	1.7331
25	2.2365	2.1649	2.0889	2.0075	1.9643	1.9192	1.8718	1.8217	1.7684	1.7110
26	2.2197	2.1479	2.0716	1.9898	1.9464	1.9010	1.8533	1.8027	1.7488	1.6906
27	2.2043	2.1323	2.0558	1.9736	1.9299	1.8842	1.8361	1.7851	1.7307	1.6717
28	2.1900	2.1179	2.0411	1.9586	1.9147	1.8687	1.8203	1.7689	1.7138	1.6541
29	2.1768	2.1045	2.0275	1.9446	1.9005	1.8543	1.8055	1.7537	1.6981	1.6377
30	2.1646	2.0921	2.1048	1.9317	1.8874	1.8409	1.7918	1.7396	1.6835	1.6223
40	2.0772	2.0035	1.9245	1.8389	1.7929	1.7444	1.6928	1.6373	1.5766	1.5089
60	1.9926	1.9174	1.8364	1.7480	1.7001	1.6491	1.5943	1.5343	1.4673	1.3893
120	1.9105	1.8337	1.7505	1.6587	1.6084	1.5543	1.4952	1.4290	1.3519	1.2539
∞	1.8307	1.7522	1.6664	1.5705	1.5173	1.4591	1.3940	1.3180	1.2214	1.0000

Appendix A.4 Continued $\alpha = 0.025$.

ν_2 \ ν_1	1	2	3	4	5	6	7	8	9
1	647.79	799.50	864.16	899.58	921.85	937.11	948.22	956.66	963.28
2	38.506	39.000	39.165	39.248	39.298	39.331	39.355	39.373	39.387
3	17.443	16.044	15.439	15.101	14.885	14.735	14.624	14.540	14.473
4	12.218	10.649	9.9792	9.6045	9.3645	9.1973	9.0741	8.9796	8.9047
5	10.007	8.4336	7.7636	7.3879	7.1464	6.9777	6.8531	6.7572	6.6810
6	8.8131	7.2598	6.5988	6.2272	5.9876	5.8197	5.6955	5.5996	5.5234
7	8.0727	6.5415	5.8898	5.5226	5.2852	5.1186	4.9949	4.8994	4.8232
8	7.5709	6.0595	5.4160	5.0526	4.8173	4.6517	4.5286	4.4332	4.3572
9	7.2093	5.7147	5.0781	4.7181	4.4844	4.3197	4.1971	4.1020	4.0260
10	6.9367	5.4564	4.8256	4.4683	4.2361	4.0721	3.9498	3.8549	3.7790
11	6.7241	5.2559	4.6300	4.2751	4.0440	3.8807	3.7586	3.6638	3.5879
12	6.5538	5.0959	4.4742	4.1212	3.8911	3.7283	3.6065	3.5118	3.4358
13	6.4143	4.9653	4.3472	3.9959	3.7667	3.6043	3.4827	3.3880	3.3120
14	6.2979	4.8567	4.2417	3.8919	3.6634	3.5014	3.3799	3.2853	3.2093
15	6.1995	4.7650	4.1528	3.8043	3.5764	3.4147	3.2934	3.1987	3.1227
16	6.1151	4.6867	4.0768	3.7294	3.5021	3.3406	3.2194	3.1248	3.0488
17	6.0420	4.6189	4.0112	3.6648	3.4379	3.2767	3.1556	3.0610	2.9849
18	5.9781	4.5597	3.9539	3.6083	3.3820	3.2209	3.0999	3.0053	2.9291
19	5.9216	4.5075	3.9034	3.5587	3.3327	3.1718	3.0509	2.9563	2.8800
20	5.8715	4.4613	3.8587	3.5147	3.2891	3.1283	3.0074	2.9128	2.8365
21	5.8266	4.4199	3.8188	3.4754	3.2501	3.0895	2.9686	2.8740	2.7977
22	5.7863	4.3828	3.7829	3.4401	3.2151	3.0546	2.9338	2.8392	2.7628
23	5.7498	4.3492	3.7505	3.4083	3.1835	3.0232	2.9024	2.8077	2.7313
24	5.7167	4.3187	3.7211	3.3794	3.1548	2.9946	2.8738	2.7791	2.7027
25	5.6864	4.2909	3.6943	3.3530	3.1287	2.9685	2.8478	2.7531	2.6766
26	5.6586	4.2655	3.6697	3.3289	3.1048	2.9447	2.8240	2.7293	2.6528
27	5.6331	4.2421	3.6472	3.3067	3.0828	2.9228	2.8021	2.7074	2.6309
28	5.6096	4.2205	3.6264	3.2863	3.0625	2.9027	2.7820	2.6872	2.6106
29	5.5878	4.2006	3.6072	3.2674	3.0438	2.8840	2.7633	2.6686	2.5919
30	5.5675	4.1821	3.5894	3.2499	3.0265	2.8667	2.7460	2.6513	2.5746
40	5.4239	4.0510	3.4633	3.1261	2.9037	2.7444	2.6238	2.5289	2.4519
60	5.2857	3.9253	3.3425	3.0077	2.7863	2.6274	2.5068	2.4117	2.3344
120	5.1524	3.8046	3.2270	2.8943	2.6740	2.5154	2.3948	2.2994	2.2217
∞	5.0239	3.6889	3.1161	2.7858	2.5665	2.4082	2.2875	2.1918	2.1136

Appendix A.4   Continued $\alpha = 0.025$.

ν_2 \ ν_1	10	12	15	20	24	30	40	60	120	∞
1	968.63	976.71	984.87	993.10	997.25	1001.4	1005.6	1009.8	1014.0	1018.3
2	39.398	39.415	39.431	39.448	39.456	39.465	39.473	39.481	39.490	39.498
3	14.419	14.337	14.253	14.167	14.124	14.081	14.037	13.992	13.947	13.902
4	8.8439	8.7512	8.6565	8.5599	8.5109	8.4613	8.4111	8.3604	8.3092	8.2573
5	6.6192	6.5246	6.4277	6.3285	6.2780	6.2269	6.1751	6.1225	6.0693	6.0153
6	5.4613	5.3662	5.2687	5.1684	5.1172	5.0652	5.0125	4.9589	4.9045	4.8491
7	4.7611	4.6658	4.5678	4.4667	4.4150	4.3624	4.3089	4.2544	4.1989	4.1423
8	4.2951	4.1997	4.1012	3.9995	3.9472	3.8940	3.8398	3.7844	3.7279	3.6702
9	3.9639	3.8682	3.7694	3.6669	3.6142	3.5604	3.5055	3.4493	3.3918	3.3329
10	3.7168	3.6209	3.5217	3.4186	3.3654	3.3110	3.2554	3.1984	3.1399	3.0798
11	3.5257	3.4296	3.3299	3.2261	3.1725	3.1176	3.0613	3.0035	2.9441	2.8828
12	3.3736	3.2773	3.1772	3.0728	3.0187	2.9633	2.9063	2.8478	2.7874	2.7249
13	3.2497	3.1532	3.0527	2.9477	2.8932	2.8373	2.7797	2.7204	2.6590	2.5955
14	3.1469	3.0501	2.9493	2.8437	2.7888	2.7324	2.6742	2.6142	2.5519	2.4872
15	3.0602	2.9633	2.8621	2.7559	2.7006	2.6437	2.5850	2.5242	2.4611	2.3953
16	2.9862	2.8890	2.7875	2.6808	2.6252	2.5678	2.5085	2.4471	2.3831	2.3163
17	2.9222	2.8249	2.7230	2.6158	2.5598	2.5021	2.4422	2.3801	2.3153	2.2474
18	2.8664	2.7689	2.6667	2.5590	2.5027	2.4445	2.3842	2.3214	2.2558	2.1869
19	2.8173	2.7196	2.6171	2.5089	2.4523	2.3937	2.3329	2.2695	2.2032	2.1333
20	2.7737	2.6758	2.5731	2.4645	2.4076	2.3486	2.2873	2.2234	2.1562	2.0853
21	2.7348	2.6368	2.5338	2.4247	2.3675	2.3082	2.2465	2.1819	2.1141	2.0422
22	2.6998	2.6017	2.4984	2.3890	2.3315	2.2718	2.2097	2.1446	2.0760	2.0032
23	2.6682	2.5699	2.4665	2.3567	2.2989	2.2389	2.1763	2.1107	2.0415	1.9677
24	2.6396	2.5412	2.4374	2.3273	2.2693	2.2090	2.1460	2.0799	2.0099	1.9353
25	2.6135	2.5149	2.4110	2.3005	2.2422	2.1816	2.1183	2.0517	1.9811	1.9055
26	2.5895	2.4909	2.3867	2.2759	2.2174	2.1565	2.0928	2.0257	1.9545	1.8781
27	2.5676	2.4688	2.3644	2.2533	2.1946	2.1334	2.0693	2.0018	1.9299	1.8527
28	2.5473	2.4484	2.3438	2.2324	2.1735	2.1121	2.0477	1.9796	1.9072	1.8291
29	2.5286	2.4295	2.3248	2.2131	2.1540	2.0923	2.0276	1.9591	1.8861	1.8072
30	2.5112	2.4210	2.3072	2.1952	2.1359	2.0739	2.0089	1.9400	1.8664	1.7867
40	2.3882	2.2882	2.1819	2.0677	2.0069	1.9429	1.8752	1.8028	1.7242	1.6371
60	2.2702	2.1692	2.0613	1.9445	1.8817	1.8152	1.7440	1.6668	1.5810	1.4822
120	2.1570	2.0548	1.9450	1.8249	1.7597	1.6899	1.6141	1.5299	1.4327	1.3104
∞	2.0483	1.9447	1.8326	1.7085	1.6402	1.5660	1.4835	1.3883	1.2684	1.0000

Appendix A.4 Continued $\alpha = 0.010$

v_2 \ v_1	1	2	3	4	5	6	7	8	9
1	4052.2	4999.5	5403.3	5624.6	5763.7	5859.0	5928.3	5981.6	6022.5
2	98.503	99.000	99.166	99.249	99.299	99.332	99.356	99.374	99.388
3	34.116	30.817	29.457	28.710	28.237	27.911	27.672	27.489	27.345
4	21.198	18.000	16.694	15.977	15.522	15.207	14.976	14.799	14.659
5	16.258	13.274	12.060	11.392	10.967	10.672	10.456	10.289	10.158
6	13.745	10.925	9.7795	9.1483	8.7459	8.4661	8.2600	8.1016	7.9761
7	12.246	9.5466	8.4513	7.8467	7.4604	7.1914	6.9928	6.8401	6.7188
8	11.259	8.6491	7.5910	7.0060	6.6318	6.3707	6.1776	6.0289	5.9106
9	10.561	8.0215	6.9919	6.4221	6.0569	5.8018	5.6129	5.4671	5.3511
10	10.044	7.5594	6.5523	5.9943	5.6363	5.3858	5.2001	5.0567	4.9424
11	9.6460	7.2057	6.2167	5.6683	5.3160	5.0692	4.8861	4.7445	4.6315
12	9.3302	6.9266	5.9526	5.4119	5.0643	4.8206	4.6395	4.4994	4.3875
13	9.0738	6.7010	5.7394	5.2053	4.8616	4.6204	4.4410	4.3021	4.1911
14	8.8616	6.5149	5.5639	5.0354	4.6950	4.4558	4.2779	4.1399	4.0297
15	8.6831	6.3589	5.4170	4.8932	4.5556	4.3183	4.1415	4.0045	3.8948
16	8.5310	6.2262	5.2922	4.7726	4.4374	4.2016	4.0259	3.8896	3.7804
17	8.3997	6.1121	5.1850	4.6690	4.3359	4.1015	3.9267	3.7910	3.6822
18	8.2854	6.0129	5.0919	4.5790	4.2479	4.0146	3.8406	3.7054	3.5971
19	8.1850	5.9259	5.0103	4.5003	4.1708	3.9386	3.7653	3.6305	3.5225
20	8.0960	5.8489	4.9382	4.4307	4.1027	3.8714	3.6987	3.5644	3.4567
21	8.0166	5.7804	4.8740	4.3688	4.0421	3.8117	3.6396	3.5056	3.3981
22	7.9454	5.7190	4.8166	4.3134	3.9880	3.7583	3.5867	3.4530	3.3458
23	7.8811	5.6637	4.7649	4.2635	3.9392	3.7102	3.5390	3.4057	3.2986
24	7.8229	5.6136	4.7181	4.2184	3.8951	3.6667	3.4959	3.3629	3.2560
25	7.7698	5.5680	4.6755	4.1774	3.8550	3.6272	3.4568	3.3239	3.2172
26	7.7213	5.5263	4.6366	4.1400	3.8183	3.5911	3.4210	3.2884	3.1818
27	7.6767	5.4881	4.6009	4.1056	3.7848	3.5580	3.3882	3.2558	3.1494
28	7.6356	5.4529	4.5681	4.0740	3.7539	3.5276	3.3581	3.2259	3.1195
29	7.5976	5.4205	4.5378	4.0449	3.7254	3.4995	3.3302	3.1982	3.0920
30	7.5625	5.3904	4.5097	4.0179	3.6990	3.4735	3.3045	3.1726	3.0665
40	7.3141	5.1785	4.3126	3.8283	3.5138	3.2910	3.1238	2.9930	2.8876
60	7.0771	4.9774	4.1259	3.6491	3.3389	3.1187	2.9530	2.8233	2.7185
120	6.8510	4.7865	3.9493	3.4796	3.1735	2.9559	2.7918	2.6629	2.5586
∞	6.6349	4.6052	3.7816	3.3192	3.0173	2.8020	2.6393	2.5113	2.4073

Appendix A.4 Continued $\alpha = 0.010$

ν_2 \ ν_1	10	12	15	20	24	30	40	60	120	∞
1	6055.8	6106.3	6157.3	6208.7	6234.6	6260.7	6286.8	6313.0	6339.4	6366.0
2	99.399	99.416	99.432	99.449	99.458	99.466	99.474	99.483	99.491	99.501
3	27.229	27.052	26.872	26.690	26.598	26.505	26.411	26.316	26.221	26.125
4	14.546	14.374	14.198	14.020	13.929	13.838	13.745	13.652	13.558	13.463
5	10.051	9.8883	9.7222	9.5527	9.4665	9.3793	9.2912	9.2020	9.1118	9.0204
6	7.8741	7.7183	7.5590	7.3958	7.3127	7.2285	7.1432	7.0568	6.9690	6.8801
7	6.6201	6.6491	6.3143	6.1554	6.0743	5.9921	5.9084	5.8236	5.7372	5.6495
8	5.8143	5.6668	5.5151	5.3591	5.2793	5.1981	5.1156	5.0316	4.9460	4.8588
9	5.2565	5.1114	4.9621	4.8080	4.7290	4.6486	4.5667	4.4831	4.3978	4.3105
10	4.8492	4.7059	4.5582	4.4054	4.3269	4.2469	4.1653	4.0819	3.9965	3.9090
11	4.5393	4.3974	4.2509	4.0990	4.0209	3.9411	3.8596	3.7761	3.6904	3.6025
12	4.2961	4.1553	4.0096	3.8584	3.7805	3.7008	3.6192	3.5355	3.4494	3.3608
13	4.1003	3.9603	3.8154	3.6646	3.5868	3.5070	3.4253	3.3413	3.2548	3.1654
14	3.9394	3.8001	3.6557	3.5052	3.4274	3.3476	3.2556	3.1813	3.0942	3.0040
15	3.8049	3.6662	3.5222	3.3719	3.2940	3.2141	3.1319	3.0471	2.9595	2.8684
16	3.6909	3.5527	3.4089	3.2588	3.1808	3.1007	3.0182	2.9330	2.8447	2.7528
17	3.5931	3.4552	3.3117	3.1615	3.0835	3.0032	2.9205	2.8348	2.7459	2.6530
18	3.5082	3.3706	3.2273	3.0771	2.9990	2.9185	2.8354	2.7493	2.6597	2.5660
19	3.4338	3.2965	3.1533	3.0031	2.9249	2.8442	2.7608	2.6742	2.5839	2.4893
20	3.3682	3.2311	3.0880	2.9377	2.8594	2.7785	2.6947	2.6077	2.5168	2.4212
21	3.3098	3.1729	3.0299	2.8796	2.8011	2.7200	2.6359	2.5484	2.4568	2.3603
22	3.2576	3.1209	2.9780	2.8274	2.7488	2.6675	2.5831	2.4951	2.4029	2.3055
23	3.2106	3.0740	2.9311	2.7805	2.7017	2.6202	2.5355	2.4471	2.3542	2.2559
24	3.1681	3.0316	2.8887	2.7380	2.6591	2.5773	2.4923	2.4035	2.3099	2.2107
25	3.1294	2.9931	2.8502	2.6993	2.6203	2.5383	2.4530	2.3637	2.2695	2.1694
26	3.0941	2.9579	2.8150	2.6640	2.5848	2.5026	2.4170	2.3273	2.2325	2.1315
27	3.0618	2.9256	2.7827	2.6316	2.5522	2.4699	2.3840	2.2938	2.1984	2.0965
28	3.0320	2.8959	2.7530	2.6017	2.5223	2.4397	2.3535	2.2629	2.1670	2.0642
29	3.0045	2.8685	2.7256	2.5742	2.4946	2.4118	2.3253	2.2344	2.1378	2.0342
30	2.9791	2.8431	2.7002	2.5487	2.4689	2.3860	2.2992	2.2079	2.1107	2.0062
40	2.8005	2.6648	2.5216	2.3689	2.2880	2.2034	2.1142	2.0194	1.9172	1.8047
60	2.6318	2.4961	2.3523	2.1978	2.1154	2.0285	1.9360	1.8363	1.7263	1.6006
120	2.4721	2.3363	2.1915	2.0346	1.9500	1.8600	1.7628	1.6557	1.5330	1.3805
∞	2.3209	2.1848	2.0385	1.8783	1.7908	1.6964	1.5923	1.4730	1.3246	1.0000

Appendix A.4   Continued $\alpha = 0.005$.

ν_2 \ ν_1	1	2	3	4	5	6	7	8	9
1	16211	20000	21615	22500	23056	23437	23715	23925	24091
2	198.50	199.00	199.17	199.25	199.30	199.33	199.36	199.37	199.39
3	55.552	49.799	47.467	46.195	45.392	44.838	44.434	44.126	43.882
4	31.333	26.284	24.259	23.155	22.456	21.975	21.622	21.352	21.139
5	22.785	18.314	16.530	15.556	14.940	14.513	14.200	13.961	13.722
6	18.635	14.544	12.917	12.028	11.464	11.073	10.786	10.566	10.391
7	16.236	12.404	10.882	10.050	9.5221	9.1554	8.8854	8.6781	8.5138
8	14.688	11.042	9.5965	8.8051	8.3018	7.9520	7.6942	7.4960	7.3386
9	13.614	10.107	8.7171	7.9559	7.4711	7.1338	6.8849	6.6933	6.5411
10	12.826	9.4270	8.0807	7.3428	6.8723	6.5446	6.3025	6.1159	5.9676
11	12.226	8.9122	7.6004	6.8809	6.4217	6.1015	5.8648	5.6821	5.5368
12	11.754	8.5096	7.2258	6.5211	6.0711	5.7570	5.5245	5.3451	5.2021
13	11.374	8.1865	6.9257	6.2335	5.7910	5.4819	5.2529	5.0761	4.9351
14	11.060	7.9217	6.6803	5.9984	5.5623	5.2574	5.0313	4.8566	4.7173
15	10.798	7.7008	6.4760	5.8029	5.3721	5.0708	4.8473	4.6743	4.5464
16	10.575	7.5138	6.3034	5.6378	5.2117	4.9134	4.6920	4.5207	4.3838
17	10.384	7.3536	6.1556	5.4967	5.0746	4.7789	4.5594	4.3893	4.2535
18	10.218	7.2148	6.0277	5.3746	4.9560	4.6627	4.4448	4.2759	4.1410
19	10.073	7.0935	5.9161	5.2681	4.8526	4.5614	4.3448	4.1770	4.0428
20	9.9439	6.9865	5.8177	5.1743	4.7616	4.4721	4.2569	4.0900	3.9564
21	9.8295	6.8914	5.7304	5.0911	4.6808	4.3931	4.1789	4.0128	3.8799
22	9.7271	6.8064	5.6524	5.0168	4.6088	4.3225	4.1094	3.9440	3.8116
23	9.6348	6.7300	5.5823	4.9500	4.5441	4.2591	4.0469	3.8822	3.7502
24	9.5513	6.6610	5.5190	4.8898	4.4857	4.2019	3.9905	3.8264	3.6949
25	9.4753	6.5982	5.4615	4.8351	4.4327	4.1500	3.9394	3.7758	3.6447
26	9.4059	6.5409	5.4091	4.7852	4.3844	4.1027	3.8928	3.7297	3.5989
27	9.3423	6.4885	5.3611	4.7396	4.3402	4.0594	3.8501	3.6875	3.5571
28	9.2838	6.4403	5.3170	4.6977	4.2996	4.0197	3.8110	3.6487	3.5186
29	9.2297	6.3958	5.2764	4.6591	4.2622	3.9830	3.7749	3.6130	3.4832
30	9.1797	6.3547	5.2388	4.6233	4.2276	3.9492	3.7416	3.5801	3.4505
40	8.8278	6.0664	4.9759	4.3738	3.9860	3.7129	3.5088	3.3498	3.2220
60	8.4946	5.7950	4.7290	4.1399	3.7600	3.4918	3.2911	3.1344	3.0083
120	8.1790	5.5393	4.4973	3.9207	3.5482	3.2849	3.0874	2.9330	2.8083
∞	7.8794	5.2983	4.2794	3.7151	3.3499	3.0913	2.8968	2.7444	2.6210

Appendix A.4   Continued $\alpha = 0.005$.

ν_2 \ ν_1	10	12	15	20	24	30	40	60	12	∞
1	24224	24426	24630	24836	24940	25044	25148	25253	25359	25465
2	199.40	199.42	199.43	199.45	199.46	199.47	199.47	199.48	199.49	199.51
3	43.686	43.387	43.085	42.778	42.622	42.466	42.308	42.149	41.989	41.829
4	20.967	20.705	20.438	20.167	20.030	19.892	19.752	19.611	19.468	19.325
5	13.618	13.384	13.146	12.903	12.780	12.656	12.530	12.402	12.274	12.144
6	10.250	10.034	9.8140	9.5888	9.4741	9.3583	9.2408	9.1219	9.0015	8.8793
7	8.3803	8.1764	7.9678	7.7540	7.6450	7.5345	7.4225	7.3088	7.1933	7.0760
8	7.2107	7.0149	6.8143	6.6082	6.5029	6.3961	6.2875	6.1772	6.0649	5.9505
9	6.4171	6.2274	6.0325	5.8318	5.7292	5.6248	5.5186	5.4104	5.3001	5.1875
10	5.8467	5.6613	5.4707	5.2740	5.1732	5.0705	4.9659	4.8592	4.7501	4.6385
11	5.4182	5.2363	5.0489	4.8552	4.7557	4.6543	4.5508	4.4450	4.3367	4.2256
12	5.0855	4.9063	4.7214	4.5299	4.4315	4.3309	4.2282	4.1229	4.0149	3.9039
13	4.8199	4.6429	4.4600	4.2703	4.1726	4.0727	3.9704	3.8655	3.7577	3.6465
14	4.6034	4.4281	4.2468	4.0585	3.9614	3.8619	3.7600	3.6553	3.5473	3.4359
15	4.4236	4.2498	4.0698	3.8826	3.7859	3.6867	3.5850	3.4803	3.3722	3.2602
16	4.2719	4.0994	3.9205	3.7342	3.6378	3.5388	3.4372	3.3324	3.2240	3.1115
17	4.1423	3.9709	3.7929	3.6073	3.5112	3.4124	3.3107	3.2058	3.0971	2.9839
18	4.0305	3.8599	3.6827	3.4977	3.4017	3.3030	3.2014	3.0962	2.9871	2.8732
19	3.9329	3.7631	3.5866	3.4020	3.3062	3.2075	3.1058	3.0004	2.8908	2.7762
20	3.8470	3.6779	3.5020	3.3178	3.2220	3.1234	3.0215	2.9159	2.8058	2.6904
21	3.7709	3.6024	3.4270	3.2431	3.1474	3.0488	2.9467	2.8408	2.7302	2.6140
22	3.7030	3.5350	3.3600	3.1764	3.0807	2.9821	2.8799	2.7736	2.6625	2.5455
23	3.6420	3.4745	3.2999	3.1165	3.0208	2.9221	2.8198	2.7132	2.6016	2.4837
24	3.5870	3.4199	3.2456	3.0624	2.9667	2.8679	2.7654	2.6585	2.5463	2.4276
25	3.5370	3.3704	3.1963	3.0133	2.9176	2.8187	2.7160	2.6088	2.4960	2.3765
26	3.4916	3.3252	3.1515	2.9685	2.8728	2.7738	2.6709	2.5633	2.4501	2.3297
27	3.4499	3.2839	3.1104	2.9275	2.8318	2.7327	2.6296	2.5217	2.4078	2.2867
28	3.4117	3.2460	3.0727	2.8899	2.7941	2.6949	2.5916	2.4834	2.3689	2.2469
29	3.3765	3.2111	3.0379	2.8551	2.7594	2.6601	2.5565	2.4479	2.3330	2.2102
30	3.3440	3.1787	3.0057	2.8230	2.7272	2.6278	2.5241	2.4151	2.2997	2.1760
40	3.1167	2.9531	2.7811	2.5984	2.5020	2.4015	2.2958	2.1838	2.0635	1.9318
60	2.9042	2.7419	2.5705	2.3872	2.2898	2.1874	2.0789	1.9622	1.8341	1.6885
120	2.7052	2.5439	2.3727	2.1881	2.0890	1.9839	1.8709	1.7469	1.6055	1.4311
∞	2.5188	2.3583	2.1868	1.9998	1.8983	1.7891	1.6691	1.5325	1.3637	1.0000

Appendix A.5 Upper Quantiles of the Distribution of Wilcoxon-Mann-Whitney Statistic

n_1	α	$n_2 = 2$	3	4	5	6	7	8	9	10	11	12	13	14	15	16	17	18	19	20
2	0.001	0	0	0	0	0	0	0	0	0	0	0	0	0	0	0	0	0	0	0
	0.005	0	0	0	0	0	0	0	0	0	0	0	0	0	0	0	0	0	1	1
	0.01	0	0	0	0	0	0	0	0	0	0	0	1	1	1	1	1	1	2	2
	0.025	0	0	0	0	0	0	1	1	1	1	2	2	2	2	2	3	3	3	3
	0.05	0	0	0	1	1	1	2	2	2	2	3	3	4	4	4	4	5	5	5
	0.10	0	1	1	2	2	2	3	3	4	4	5	5	5	6	6	7	7	8	8
3	0.001	0	0	0	0	0	0	0	0	0	0	0	0	0	0	0	1	1	1	1
	0.005	0	0	0	0	0	0	0	1	1	1	2	2	2	3	3	3	3	4	4
	0.01	0	0	0	0	0	1	1	2	2	2	3	3	3	4	4	5	5	5	6
	0.025	0	0	0	1	2	2	3	3	4	4	5	5	6	6	7	7	8	8	9
	0.05	0	1	1	2	3	3	4	5	5	6	6	7	8	8	9	10	10	11	12
	0.10	1	2	2	3	4	5	6	6	7	8	9	10	11	11	12	13	14	15	16
4	0.001	0	0	0	0	0	0	0	0	1	1	1	2	2	2	3	3	4	4	4
	0.005	0	0	0	0	1	1	2	2	3	3	4	4	5	6	6	7	7	8	9
	0.01	0	0	0	1	2	2	3	4	4	5	6	6	7	8	8	9	10	10	11
	0.025	0	0	1	2	3	4	5	5	6	7	8	9	10	11	12	12	13	14	15
	0.05	0	1	2	3	4	5	6	7	8	9	10	11	12	13	15	16	17	18	19
	0.10	1	2	4	5	6	7	8	10	11	12	13	14	16	17	18	19	21	22	23

Source: Table 1 of Extended Tables of Critical Values for Wilcoxon's Test Statistic by L. R. Verdooren, *Biometrika* (1963), Vol. 50, pp. 177–186.

Appendix A.5 Continued

n_1	α	$n_2 = 2$	3	4	5	6	7	8	9	10	11	12	13	14	15	16	17	18	19	20
5	0.001	0	0	0	0	0	0	1	2	2	3	3	4	4	5	6	6	7	8	8
	0.005	0	0	0	1	2	2	3	4	5	6	7	8	8	9	10	11	12	13	14
	0.01	0	0	1	2	3	4	5	6	7	8	9	10	11	12	13	14	15	16	17
	0.025	0	1	2	3	4	6	7	8	9	10	12	13	14	15	16	18	19	20	21
	0.05	1	2	3	5	6	7	9	10	12	13	14	16	17	19	20	21	23	24	26
	0.10	2	3	5	6	8	9	11	13	14	16	18	19	21	23	24	26	28	29	31
6	0.001	0	0	0	0	0	0	2	3	4	5	5	6	7	8	9	10	11	12	13
	0.005	0	0	1	2	3	4	5	6	7	8	10	11	12	13	14	16	17	18	19
	0.01	0	0	2	3	4	5	7	8	9	10	12	13	14	16	17	19	20	21	23
	0.025	0	2	3	4	6	7	9	11	12	14	15	17	18	20	22	23	25	26	28
	0.05	1	3	4	6	8	9	11	13	15	17	18	20	22	24	26	27	29	31	33
	0.10	2	4	6	8	10	12	14	16	18	20	22	24	26	28	30	32	35	37	39
7	0.001	0	0	0	0	1	2	3	4	6	7	8	9	10	11	12	14	15	16	17
	0.005	0	0	1	2	4	5	7	8	10	11	13	14	16	17	19	20	22	23	25
	0.01	0	1	2	4	5	7	8	10	12	13	15	17	18	20	22	24	25	27	29
	0.025	0	2	4	6	7	9	11	13	15	17	19	21	23	25	27	29	31	33	35
	0.05	1	3	5	7	9	12	14	16	18	20	22	25	27	29	31	34	36	38	40
	0.10	2	5	7	9	12	14	17	19	22	24	27	29	32	34	37	39	42	44	47
8	0.001	0	0	0	1	2	3	5	6	7	9	10	12	13	15	16	18	19	21	22
	0.005	0	0	2	3	5	7	8	10	12	14	16	18	19	21	23	25	27	29	31
	0.01	0	1	3	5	7	8	10	12	14	16	18	21	23	25	27	29	31	33	35
	0.025	1	3	5	7	9	11	14	16	18	20	23	25	27	30	32	35	37	39	42
	0.05	2	4	6	9	11	14	16	19	21	24	27	29	32	34	37	40	42	45	48
	0.10	3	6	8	11	14	17	20	23	25	28	31	34	37	40	43	46	49	52	55

	0.001	0	0	0	2	3	4	6	8	9	11	13	15	16	18	20	22	24	26	27
	0.005	0	1	2	4	6	8	10	12	14	17	19	21	23	25	28	30	32	34	37
9	0.01	0	2	4	6	8	10	12	15	17	19	22	24	27	29	32	34	37	39	41
	0.025	1	3	5	8	11	13	16	18	21	24	27	29	32	35	38	40	43	46	49
	0.05	2	5	7	10	13	16	19	22	25	28	31	34	37	40	43	46	49	52	55
	0.10	3	6	10	13	16	19	23	26	29	32	36	39	42	46	49	53	56	59	63
	0.001	0	0	1	2	4	6	7	9	11	13	15	18	20	22	24	26	28	30	33
	0.005	0	1	3	5	7	10	12	14	17	19	22	25	27	30	32	35	38	40	43
10	0.01	0	2	4	7	9	12	14	17	20	23	25	28	31	34	37	39	42	45	48
	0.025	1	4	6	9	12	15	18	21	24	27	30	34	37	40	43	46	49	53	56
	0.05	2	5	8	12	15	18	21	25	28	32	35	38	42	45	49	52	56	59	63
	0.10	4	7	11	14	18	22	25	29	33	37	40	44	48	52	55	59	63	67	71
	0.001	0	0	0	3	5	7	9	11	13	16	18	21	23	25	28	30	33	35	38
	0.005	0	1	3	6	8	11	14	17	19	22	25	28	31	34	37	40	43	46	49
11	0.01	0	2	5	8	10	13	16	19	23	26	29	32	35	38	42	45	48	51	54
	0.025	1	4	7	10	14	17	20	24	27	31	34	38	41	45	48	52	56	59	63
	0.05	2	6	9	13	17	20	24	28	32	35	39	43	47	51	55	58	62	66	70
	0.10	4	8	12	16	20	24	28	32	37	41	45	49	53	58	62	66	70	74	79
	0.001	0	0	1	3	5	8	10	13	15	18	21	24	26	29	32	35	38	41	43
	0.005	0	2	4	7	10	13	16	19	22	25	28	32	35	38	42	45	48	52	55
12	0.01	0	3	6	9	12	15	18	22	25	29	32	36	39	43	47	50	54	57	61
	0.025	2	5	8	12	15	19	23	27	30	34	38	42	46	50	54	58	62	66	70
	0.05	3	6	10	14	18	22	27	31	35	39	43	48	52	56	61	65	69	73	78
	0.10	5	9	13	18	22	27	31	36	40	45	50	54	59	64	68	73	78	82	87

Appendix A.5 Continued

n_1	α	$n_2 = 2$	3	4	5	6	7	8	9	10	11	12	13	14	15	16	17	18	19	20
13	0.001	0	0	2	4	6	9	12	15	18	21	24	27	30	33	36	39	43	46	49
	0.005	0	2	4	8	11	14	18	21	25	28	32	35	39	43	46	50	54	58	61
	0.01	1	3	6	10	13	17	21	24	28	32	36	40	44	48	52	56	60	64	68
	0.025	2	5	9	13	17	21	25	29	34	38	42	46	51	55	60	64	68	73	77
	0.05	3	7	11	16	20	25	29	34	38	43	48	52	57	62	66	71	76	81	85
	0.010	5	10	14	19	24	29	34	39	44	49	54	59	64	69	75	80	85	90	95
14	0.001	0	0	2	4	7	10	13	16	20	23	26	30	33	37	40	44	47	51	55
	0.005	0	2	5	8	12	16	19	23	27	31	35	39	43	47	51	55	59	64	68
	0.01	1	3	7	11	14	18	23	27	31	35	39	44	48	52	57	61	66	70	74
	0.025	2	6	10	14	18	23	27	32	37	41	46	51	56	60	65	71	75	79	84
	0.05	4	8	12	17	22	27	32	37	42	47	52	57	62	67	72	78	83	88	93
	0.10	5	11	16	21	26	32	37	42	48	53	59	64	70	75	81	86	92	98	103
15	0.001	0	0	2	5	8	11	15	18	22	25	29	33	37	41	44	48	52	56	60
	0.005	0	3	6	9	13	17	21	25	30	34	38	43	47	52	56	61	65	70	74
	0.01	1	4	8	12	16	20	25	29	34	38	43	48	52	57	62	67	71	76	81
	0.025	2	6	11	15	20	25	30	35	40	45	50	55	60	65	71	76	81	86	91
	0.05	4	8	13	19	24	29	34	40	45	51	56	62	67	73	78	84	89	95	101
	0.10	6	11	17	23	28	34	40	46	52	58	64	69	75	81	87	93	99	105	111
16	0.001	0	0	3	6	9	12	16	20	24	28	32	36	40	44	49	53	57	61	66
	0.005	0	3	6	10	14	19	23	28	32	37	42	46	51	56	61	66	71	75	80
	0.01	1	4	8	13	17	22	27	32	37	42	47	52	57	62	67	72	77	83	88
	0.025	2	7	12	16	22	27	32	38	43	48	54	60	65	71	76	82	87	93	99
	0.05	4	9	15	20	26	31	37	43	49	55	61	66	72	78	84	90	96	102	108
	0.10	6	12	18	24	30	37	43	49	55	62	68	75	81	87	94	100	107	113	120

	0.001	0	1	3	6	10	14	18	22	26	30	35	39	44	48	53	58	62	67	71
	0.005	0	3	7	11	16	20	25	30	35	40	45	50	55	61	66	71	76	82	87
17	0.01	1	5	9	14	19	24	29	34	39	45	50	56	61	67	72	78	83	89	94
	0.025	3	7	12	18	23	29	35	40	46	52	58	64	70	76	82	88	94	100	106
	0.05	4	10	16	21	27	34	40	46	52	58	65	71	78	84	90	97	103	110	116
	0.10	7	13	19	26	32	39	46	53	59	66	73	80	86	93	100	107	114	121	128
	0.001	0	1	4	7	11	15	19	24	28	33	38	43	47	52	57	62	67	72	77
	0.005	0	3	7	12	17	22	27	32	38	43	48	54	59	65	71	76	82	88	93
18	0.01	1	5	10	15	20	25	31	37	42	48	54	60	66	71	77	83	89	95	101
	0.025	3	8	13	19	25	31	37	43	49	56	62	68	75	81	87	94	100	107	113
	0.05	5	10	17	23	29	36	42	49	56	62	69	76	83	89	96	103	110	117	124
	0.10	7	14	21	28	35	42	49	56	63	70	78	85	92	99	107	114	121	129	136
	0.001	0	1	4	8	12	16	21	26	30	35	41	46	51	56	61	67	72	78	83
	0.005	1	4	8	13	18	23	29	34	40	46	52	58	64	70	75	82	88	94	100
19	0.01	2	5	10	16	21	27	33	39	45	51	57	64	70	76	83	89	95	102	108
	0.025	3	8	14	20	26	33	39	46	53	59	66	73	79	86	93	100	107	114	120
	0.05	5	11	18	24	31	38	45	52	59	66	73	81	88	95	102	110	117	124	131
	0.10	8	15	22	29	37	44	52	59	67	74	82	90	98	105	113	121	129	136	144
	0.001	0	1	4	8	13	17	22	27	33	38	43	49	55	60	66	71	77	83	89
	0.005	1	4	9	14	19	25	31	37	43	49	55	61	68	74	80	87	93	100	106
20	0.01	2	6	11	17	23	29	35	41	48	54	61	68	74	81	88	94	101	108	115
	0.025	3	9	15	21	28	35	42	49	56	63	70	77	84	91	99	106	113	120	128
	0.05	5	12	19	26	33	40	48	55	63	70	78	85	93	101	108	116	124	131	139
	0.10	8	16	23	31	39	47	55	63	71	79	87	95	103	111	120	128	136	144	152

Appendix B: SAS Programs

Appendix B.1 Procedures for Average Bioavailability

```
*******************************************************************;
*   Program:     BOOKEXAMPLE.SAS                                 *;
*   Author:      J.p. Liu                                        *;
*   Date:        June 24, 1991                                   *;
*   Description: This Program computes the test statistics and   *;
*                confidence intervals based on the methods       *;
*                for a 2 by 2 crossover design described in Chapter 4 *;
*                of the book entitled "Design and Analysis of    *;
*                Bioavailability and Bioequivalence Studies" by   *;
*                S.C. Chow and J.p. Liu and also examines        *;
*                the assumptions by the methods described in Chapter *;
*                8.                                              *;
* Methods:       Schuirmann's two one-sided tests procedure.     *;
*                Rodda and Davis's Bayesian procedure.           *;
*                Anderson-Huack's procedure.                     *;
*                Mandallaz-Mau's procedure.                      *;
*                Nonparametric two one-sided procedure.          *;
*                The classical shortest confidence interval.     *;
*                Westlake's symmetric confidence interval        *;
*                Locke's exact confidence interval for the ratio by *;
*                Fieller theorem.                                *;
*                Fixed Fieller's confidence interval for the ratio. *;
*                Nonparametric confidence interval.              *;
*                Normality and independence tests of             *;
*                Intrasubject/intersubject variabilities.        *;
* Dataset:       AUC data in Table 3.6.1  of Chapter 3.         *;
*******************************************************************;

libname out 'XXXXX:[XXXXX.XXXXX]';

*******************************************************************;
* Input data and print the raw data.                            *;
* Sequence = group(=1 and 2)                                    *;
* Subject = subject                                             *;
* Period = period(1=I and 2=II)                                 *;
* Tmt = formulation(1=reference and 2=test)                     *;
* y=raw AUC                                                     *;
*******************************************************************;

data one;
set out.ex361;
proc sort data=one;  by group subject tmt;
proc print data=one;

proc means n noprint data=one;  by group;
var y;
output out=no n=n;

*******************************************************************;
* Compute the subject totals.                                   *;
*******************************************************************;

proc means n noprint data=one;  by group subject;
var y;
output out=sum sum=ysum;

*******************************************************************;
* Dataset of the number of subjects in each sequence.           *;
*******************************************************************;

data no;  set no;
n=n/2;

proc sort data=no;  by group;
```

393

Appendix B.1 Continued

```
data one;  merge one no;  by group;

proc sort data=one;  by group subject tmt;

**********************************************************************;
* Perform analysis of variance using the full model with group as    *;
* carryover effect which is confounded with sequence effect.         *;
**********************************************************************;

proc glm data=one;
class tmt period group subject;
model y= group subject(group) period tmt/ss1 ss2 ss3 ss4 solution p ;
test h=group e=subject(group)/htype=3 etype=3;
means tmt period group;
lsmeans tmt period /stderr pdiff;
lsmeans group/stderr pdiff e=subject(group);

**********************************************************************;
* Perform analysis of variance using the reduced model without       *;
* carryover effect.                                                   *;
* Output dataset:                                                     *;
*               tanova - sums of squares and associate degrees of     *;
*                        freedom.                                      *;
*               lsmean1 - lsmeans of formulation and period means.    *;
*               pred - intrasubject residuals.                        *;
**********************************************************************;

proc glm data=one outstat=tanova;
class tmt period subject;
model y= subject period tmt/ss1 ss2 ss3 ss4 solution p ;
output out=pred p=yhat r=resid student=stresid;
means tmt period ;
lsmeans tmt period /stderr pdiff out=lsmean1;

proc print data=tanova;
proc print data=lsmean1;

**********************************************************************;
* Perform analysis of variance on subject totals to obtain the       *;
* intersubject residuals.                                            *;
**********************************************************************;

proc glm data=sum;
model ysum=group;
output out=predsum p=ysumhat r=residsum student=residsub;

data pred;  set pred;  if period=1;

**********************************************************************;
* Obtain the normal scores for intrasubject residuals, plot them vs. *;
* predicted values and normal scores, and perform Shapiro-Wilk's test *;
* for normality                                                      *;
**********************************************************************;

proc rank out=rpred normal=blom data=pred;
var stresid;
ranks rankr;
proc sort data=rpred;  by rankr;

proc plot data=rpred;
plot stresid*yhat;
plot stresid*rankr;

proc univariate normal plot data=pred;
```

394

Appendix B.1 Continued

```
var stresid;



************************************************************************;
* Obtain the normal scores for intersubject residuals, plot them    *;
* vs.  normal scores, and perform Shapiro-Wilk's test               *;
* for normality                                                     *;
************************************************************************;

proc rank out=rpredsum normal=blom data=predsum;
var residsub;
ranks ranksum;
proc sort data=rpredsum;   by ranksum;

proc plot data=rpredsum;
plot residsub*ranksum;

proc univariate normal plot data=predsum;
var residsub;

************************************************************************;
*   Test the assumption of independence between intrasubject and     *;
*   intersubject variabilities.                                      *;
************************************************************************;

data rpred1;   set rpred;
keep subject stresid;
proc sort data=rpred1;   by subject;

data rpredsum;   set rpredsum;
keep subject residsub;
proc sort data=rpredsum;   by subject;

data indep;    merge rpred1 rpredsum;   by subject;
proc corr pearson spearman data=indep;
var stresid residsub;

************************************************************************;
* Create datasets for df, mse, number of subjects, and least squares*;
* means of formulation.                                             *;
************************************************************************;

data error;   set tanova;   if _SOURCE_='ERROR';
mse=ss/df;
sse=ss;
id=1;
keep id mse sse df;
proc sort data=error;  by id;

data no1;   set no; if group=1;
n1=n;   id=1;
keep n1 id;
proc sort data=no1;   by id;

data no2;   set no;   if group=2;
n2=n;   id=1;
keep n2 id;
proc sort data=no2;   by id;

data lsmean11;   set lsmean1;   if tmt=1;
my1=lsmean;
```

395

Appendix B.1 Continued

```
id=1;
keep id my1;
proc sort data=lsmean11;   by id;

data lsmean12;   set lsmean1;   if tmt=2;
my2=lsmean;
id=1;
keep id my2;
proc sort data=lsmean12;   by id;

*********************************************************************;
* Perform Schuirmann's two one-side tests procedure, compute the   *;
* 90% classic shortest confidence interval, and conduct Rodda and  *;
* Bayesian procedure based on the summary statistics extracted from *;
* PROC GLM.                                                         *;
* t: T-statistic for equality(with p-value = p).                   *;
* t1: T-statistics for testing that test formulation is not too low *;
*     (with p-value = p1).                                         *;
* t2: T-statistics for testing that test formulation is not too high *;
*     (with p-value = p2).                                         *;
* Lower(upper): Lower(upper) 90% confidence limits.                *;
* Plower(Pupper): Lower(upper) 90% confidence limits expressed as  *;
*                 percentage of the estimated reference mean.      *;
* prd: estimated posterior probability between equivalence limits. *;
*********************************************************************;

data twoside;   merge lsmean11 lsmean12 no1 no2 error;   by id;
spool=mse*(0.5)*((1/n1)+(1/n2));
se=sqrt(spool);
diff=my2-my1;
t=(my2-my1)/se;
t1=(my2-my1+(0.2*my1))/se;
t2=-(((0.2*my1)-(my2-my1))/se;
ct=tinv(0.05,df,0);
absct=abs(ct);
lower=(my2-my1)-(absct*se);
upper=(my2-my1)+(absct*se);
plower=((lower/my1)+1)*100;
pupper=((upper/my1)+1)*100;
p=2*(1-probt(abs(t),df));
p1=1-probt(t1,df);
p2=probt(t2,df);
prd=probt(-t2,df)-probt(-t1,df);
proc print;
var my1 my2 n1 n2 df mse diff t t1 t2 lower upper plower pupper p p1 p2 prd;

*********************************************************************;
* Perform Anderson-Hauck's  procedure based on the summary statistics*;
* extracted from PROC GLM.                                         *;
*********************************************************************;

data andhau;   merge lsmean11 lsmean12 no1 no2 error;   by id;
spool=mse*(0.5)*((1/n1)+(1/n2));
se=sqrt(spool);
diff=my2-my1;
bll=(-0.2)*my1;
bul=-bll;
t=(my2-my1-(0.5)*(bll+bul))/se;
del=(0.5)*(bul-bll)/se;
u=abs(t)-del;
l=-abs(t)-del;
pu=probt(u,df,0);
pl=probt(l,df,0);
p=pu-pl;
```

Appendix B.1 Continued

```
proc print;
var my1 my2 n1 n2 df mse diff p ;

******************************************************************;
* Perform Mandallaz-Mau's procedure based on the summary statistics  *;
* extracted from PROC GLM.                                           *;
******************************************************************;

data ManMau;   merge lsmean11 lsmean12 no1 no2 error;  by id;
spool=mse*(0.5)*((1/n1)+(1/n2));
se=sqrt(spool);
diff=my2-my1;
meansq1=sqrt(mse*(1+0.8*0.8)/(n1+n2));
meansq2=sqrt(mse*(1+1.2*1.2)/(n1+n2));
t1=(my2-(0.8*my1))/meansq1;
t2=(my2-(1.2*my1))/meansq2;
pu=probt(t1,df,0);
pl=probt(t2,df,0);
p=pu-pl;
proc print;
var my1 my2 n1 n2 df mse t1 t2 pl pu p;

******************************************************************;
* Compute the Westlake's symmetric 90% confidence interval based on  *;
* the summary statistics extracted from PROC GLM.                    *;
* Lower(upper): Lower(upper) 90% confidence limits.                  *;
* Plower(Pupper): Lower(upper) 90% confidence limits expressed as    *;
*                 percentage of the estimated reference mean.        *;
******************************************************************;

data westlake;   merge lsmean11 lsmean12 no1 no2 error;  by id;
spool=mse*(0.5)*((1/n1)+(1/n2));
se=sqrt(spool);
diff1=my1-my2;
diff2=diff1;
if diff2=0 then diff2=0.001*my1;
sumk=2*diff2/se;
inc=0.1;
k1=2*sumk;
if sumk > 0 then k1=-2*sumk;
again:    k2=sumk-k1;
pr1=probt(k1,df,0);
pr2=1-probt(k2,df,0);
prob=pr1+pr2;
if pr1 > 0.1 then k1=2*k1;
if pr1 > 0.1 then goto again;
if pr2 >0.1 then k1=k1+sumk;
if pr2 >0.1 then goto again;
start:
k1=k1+inc;
k2=k2-inc;
pr1=probt(k1,df,0);
pr2=1-probt(k2,df,0);
prob=pr1+pr2;
if prob < 0.1 then goto start;
k1=k1-inc;
k2=k2+inc;
inc=inc/5;
if inc < 0.0001 then goto end;
goto start;
end: diff=-diff1;
upper=diff+k1*se;
lower=diff+k2*se;
pupper=100*upper/my1;
```

Appendix B.1 Continued

```
plower=100*lower/my1;
proc print;
var my1 my2 n1 n2 df mse diff k1 k2 lower upper plower pupper;

proc sort data=one;    by tmt period;

**************************************************************;
* Generate the dataset of period differences and subject totals *;
**************************************************************;

data one1;    set one;
if period=1;
y1=y;
drop period y;
data one2;    set one;
if period=2;
y2=y;
drop period y;
proc sort data=one1;    by group subject;
proc sort data=one2;    by group subject;
data work;    merge one1 one2;    by group subject;
id=1;
proc sort data=work;    by id group subject;
data work;    merge work lsmean11;    by id;
w1=.;
w2=.;
if group=2 then w1=y1;
if group=1 then w1=y2;
if group=2 then w2=y2;
if group=1 then w2=y1;
diffw=w1-w2;
sumw=w1+w2;
d=(y2-y1);
d1=.;
if group=1 then d1=d+(2*(0.2*my1));
if group=2 then d1=d;
d12=d1/2;
d2=.;
if group=1 then d2=d-(2*(0.2*my1));
if group=2 then d2=d;
d22=d2/2;
t=y1+y2;
p=.;
if group=1 then p=d;
if group=2 then p=-d;
proc print;
proc sort data=work;    by group;

**************************************************************;
* Obtain STT, STR, and SRR using PROC CORR.              *;
**************************************************************;

proc corr csscp data=work outp=sscp;    by group;
var w1 w2;
proc print data=sscp;


data sscp21;    set sscp;    if group=1;
if _type_='CSSCP' and _name_='W1';
ss22=w1;    ss12=w2;
keep group ss22 ss12;
proc sort data=sscp21;    by group;
data sscp22;    set sscp;    if group=1;
if _type_='CSSCP' and _name_='W2';
```

Appendix B.1 Continued

```
ss11=w2;
keep group ss11;
proc sort data=sscp22;   by group;
data sscp2;   merge sscp21 sscp22;  by group;

data sscp11;   set sscp;   if group=2;
if _type_='CSSCP' and _name_='W1';
ss22=w1;   ss12=w2;
keep group ss22 ss12;
proc sort data=sscp11;   by group;
data sscp12;   set sscp;   if group=2;
if _type_='CSSCP' and _name_='W2';
ss11=w2;
keep group ss11;
proc sort data=sscp12;   by group;
data sscp1;   merge sscp11 sscp12;  by group;

data sscpf;   set sscp1 sscp2;
proc means noprint sum data=sscpf;
var ss11 ss22 ss12;
output out=sscps sum=css11 css22 css12;
data sscps;   set sscps;   id=1;
proc sort data=sscps;   by id;

************************************************************************;
* Compute the Locke 90% confidence interval for the ratio based on   *;
* the summary statistics extracted from PROC GLM and PROC CORR.      *;
* flowerp(fupperp): Lower(upper) 90% confidence limits for the ratio *;
*                    by Fieller theorem.                             *;
* fflowerp(ffupperp): Lower(upper) 90% confidence limits for the     *;
*                    ratio  by fixed Fieller method.                 *;
************************************************************************;

data locke;   merge lsmean11 lsmean12 no1 no2 error sscps;  by id;
spool=mse*(0.5)*((1/n1)+(1/n2));
se=sqrt(spool);
ms11=css11/df;
ms12=css12/df;
ms22=css22/df;
r=my2/my1;
ct=tinv(0.05,df,0);
absct=abs(ct);
w=((1/n1)+(1/n2))/4;
g2=(absct/(my1/sqrt(w*mse)))**2;
g1=(absct**2)/((my1**2)/(w*ms11));
sr1=ms22/ms11;
sr2=ms12/ms11;
k=(r**2)+sr1*(1-g1)+sr2*((g1*sr2)-(2*r));
flower=.;
fupper=.;
fflower=.;
ffupper=.;
if g1 <1 and k ge 0 then
flower=((r-(g1*sr2))-(absct*sqrt(ms11*w)/my1)*sqrt(k))/(1-g1);
if g1 <1 and k ge 0 then
fupper=((r-(g1*sr2))+(absct*sqrt(ms11*w)/my1)*sqrt(k))/(1-g1);
k1=r**2+(1-g2);
if g2 < 1 and k1 ge 0 then
fflower=(r-(absct*sqrt(mse*w)/my1)*sqrt(k1))/(1-g2);
if g2 < 1 and k1 ge 0 then
ffupper=(r+(absct*sqrt(mse*w)/my1)*sqrt(k1))/(1-g2);
flowerp=100*flower;
fupperp=100*fupper;
fflowerp=100*fflower;
```

Appendix B.1 Continued

```
ffupperp=100*ffupper;
proc print;
var my1 my2 n1 n2 df ms11 ms12 ms22 r flowerp fupperp fflowerp ffupperp;

****************************************************************;
* Perform Schuirmann's  two one-sided tests procedure based on the    *;
* period differences.                                                 *;
* d: equality of formulation effects.                                 *;
* d12: for hypothesis that the test formulation is not too low.       *;
* d22: for hypothesis that the test formulation is not too high       *;
* p: equality of period effects.                                      *;
* t: equality of carryover effect.                                    *;
* The results of d, p, and t should be the same as those from GLM.    *;
* The results of d12 and d22(i. e., p-values) should be the same as   *;
* those in the dataset "twoside".                                     *;
****************************************************************;
proc ttest data=work;
class group;
var d d12 d22 p t;

****************************************************************;
* Perform nonparametric two one-sided tests procedure based on the    *;
* period differences.                                                 *;
* d: equality of formulation effects.                                 *;
* d12: for hypothesis that the test formulation is not too low.       *;
* d22: for hypothesis that the test formulation is not too high       *;
* p: equality of period effects.                                      *;
* t: equality of carryover effect.                                    *;
****************************************************************;

proc nparlway wilcoxon data=work;
class group;
var d d12 d22 p t;

****************************************************************;
* Compute nonparametric 90% confidence interval based on all possible*;
* differences of the period differences between two sequences.       *;
****************************************************************;

data out1;    set work;
if group=1;
d1=d/2;
keep subject d1;
data out2;    set work;
if group=2;
d=d/2;
id=subject;
keep id d;
proc sort data=out2;  by d;
data out2;    set out2;
do subject=1 to 12;
  d2=d;
  id=id ;
 output;
end;
keep id subject d2;
proc sort data=out1;  by subject;
proc sort data=out2;  by subject id;
data diff;    merge out1 out2;  by subject;
diff=d1-d2;
index=1;
proc univariate normal plot data=diff;
var diff;
output out=median n=ncom median=median;
```

400

Appendix B.1 Continued

```
data median;  set median;  index=1;
proc sort data=median; by index;
proc sort data=diff;  by index;
data diff;  merge diff median;  by index;
proc rank out=rdiff data=diff;
var diff;
ranks rdiff;
data rdiff;  set rdiff;
w05=43;
w95=144-43+1;
if rdiff=w05 or rdiff=w95;
proc sort data=rdiff;   by rdiff;
proc print;
```

Appendix B.2   Bayesian Method by Grieve

```
************************************************************************;
*   Program:      BOOKEXGR.SAS                                        *;
*   Author:       J.p. Liu                                            *;
*   Date:         June 24, 1991                                       *;
*   Description: This program computes the test statistics and estimated *;
*                 posterior probability between equivalence limits by   *;
*                 Grieve's Bayesian method for a 2 by 2 crossover design *;
*                 described in Chapter 4 of the book entitled "Design and *;
*                 Analysis of Bioavailability and Bioequivalence Studies" *;
*                 by S.C. Chow and J.p. Liu.                          *;
*   Methods:      Bayesian method by Grieve.                          *;
*   Dataset:      AUC data in Table 3.6.1 of Chapter 3.               *;
************************************************************************;

************************************************************************;
* Compute the test statistics and posterior probability in presence of *;
* carryover effects.                                                  *;
* Input: Intersubject/intrasubject sums of squares from the full model *;
*         with carryover effect, estimates of formulation and          *;
*         carryover effects, estimated reference mean.                *;
************************************************************************;

data one;
sse=3679.43;
ssp=16211.49;
rmean=82.5594;
eqlimit=0.1*rmean;
fhat=-2.2875;
rhat=-4.7958;
mean=(fhat/2)+(rhat/2);
n1=12; n2=12;
n=n1+n2;
f=(((sse+ssp)**2)*(n-6))/(sse**2+ssp**2);
f=f+4;
h=((f-2)*(sse+ssp))/(n-4);
df=f;
m=((1/n1)+(1/n2));
sq=(f*(m*h)/(8*f))/(f-2);
s=sqrt(sq);
df1=(df+1)/2;
df2=df/2;
df3=1/2;
t=abs(tinv(0.05,df,0));
l=mean-(t*s);
u=mean+(t*s);
l2=2*l;
u2=2*u;
tl=(-eqlimit-mean)/s;
tu=(eqlimit-mean)/s;
pu=probt(tu,df,0);
pl=probt(tl,df,0);
p=pu-pl;
proc print;

************************************************************************;
* Compute the test statistics and posterior probability in absence of *;
* carryover effects.                                                  *;
* Input: Intrasubject sum of squares from the reduced model without    *;
*         carryover effect, estimates of formulation effect, estimated *;
*         reference mean.                                             *;
************************************************************************;

data two;
sse=3679.43;
```

402

Appendix B.2 Continued

```
rmean=82.5594;
eqlimit=0.1*rmean;
fhat=-2.2875;
mean=(fhat/2);
n1=12; n2=12;
n=n1+n2;
df=n-2;
m=((1/n1)+(1/n2));
sq=df*m*sse/((df-2)*8*(n-2));
s=sqrt(sq);
df1=(df+1)/2;
df2=df/2;
df3=1/2;
t=abs(tinv(0.05,df,0));
l=mean-(t*s);
u=mean+(t*s);
l2=2*l;
u2=2*u;
tl=(-eqlimit-mean)/s;
tu=(eqlimit-mean)/s;
pu=probt(tu,df,0);
pl=probt(tl,df,0);
p=pu-pl;
proc print;
```

Appendix B.3 MVUE, MLE, RI, MIR for a 2 × 2 Crossover Design

```
********************************************************************;
*    Program:            BOOKLOG.SAS                              *;
*    Author:             J.p. Liu                                 *;
*    Date:               June 25, 1991                            *;
*    Description:        This program computes the minimum variance *;
*                        unbiased estimator, maximum likelihood    *;
*                        estimator of direct formulation effect    *;
*                        on the original scale as well as mean of  *;
*                        individual ratios, ratio of least squares *;
*                        means. In addition, this program also     *;
*                        calculates the estimated variance of MVUE, *;
*                        estimated bias, mse, variance of MLE and RI *;
*                        by the methods described in Chapter 6 of  *;
*                        the book entitled "Design and Analysis of *;
*                        Bioavailability and Bioequivalence Studies" by *;
*                        S.C. Chow and J.p. Liu.                   *;
*    Methods:            MVUE, MLE, RI, MIR for 2 by 2 crossover   *;
*                        design with equal sample size for both    *;
*                        sequences. This program can be easily     *;
*                        modified for unequal sample sizes.        *;
*    Dataset:            Clayton-Leslie AUC data in Table 6.9.1 of *;
*                        Chapter 6.                                *;
********************************************************************;
libname out 'XXXXX:[XXXXX.XXXXX]';

********************************************************************;
*   Input data and rearrange data so that y11 and y12 represent the *;
*   responses of two periods for a subject. Do the logarithmic    *;
*   transformation. Compute period differences of log-responses.  *;
********************************************************************;

data book;    set out.clayton;
resp=y;
data one1;    set book;
if seq=1;
data one11;   set one1;
if tmt=2;
y11=resp;
keep subject seq y11;
proc sort data=one11;  by subject;

data one12;   set one1;
if tmt=1;
y12=resp;
keep subject seq y12;
proc sort data=one12; by subject;
data one;    merge one11 one12;

data one2;    set book;
if seq=2;
data one21;   set one2;
if tmt=1;
y11=resp;
keep subject seq y11;
proc sort data=one21;  by subject;

data one22;   set one2;
if tmt=2;
y12=resp;
keep subject seq y12;
proc sort data=one22; by subject;
data two;    merge one21 one22;

data work;    set one two;
```

404

Appendix B.3 Continued

```
if seq =1 then y=(y11/y12);
if seq =2 then y=(y12/y11);
x11=log(y11);
x12=log(y12);
x=(x11-x12)/2;
id=1;

*proc print;
proc sort data=work;   by id seq;
data work1;   set work; by id seq;
x1=x;
if seq=2 then x1=-x;
y1=y11;
y2=y12;

**********************************************************************;
*   Compute the formulation effect on the logarithmic scale, mean of   *;
*   individual subject ratios, and ratio of the least squares means.   *;
**********************************************************************;

proc univariate noprint normal plot data=work1;   by id seq;
var x y y1 y2  ;
output out=outp n=nx ny ny1 ny2 mean=meanx meany meanry1 meanry2 css=cssx cssy
cssy1 cssy2;
data outp;   set outp;
n1=ny-1;
meanx1=meanx;
meany1=meany/2;
if seq=2 then meanx1=-meanx1;
cssx1=cssx;
meanrry1=.;
meanrry2=.;
if seq=1 then meanrry1=meanry1;
if seq=2 then meanrry1=meanry2;
if seq=1 then meanrry2=meanry2;
if seq=2 then meanrry2=meanry1;
drop ny1 ny2 cssy1 cssy2;
proc sort data=outp;   by id;
proc means sum noprint data=outp;   by id;
var meanx1 meany1 meanrry1 meanrry2 n1 ny cssx1 ;
output out=out sum=meanx meany meanry1 meanry2 df n cssx;
data out;   set out;
meanry1=meanry1/2;
meanry2=meanry2/2;
meanrry=(meanry1)/(meanry2);
n=n/2;
f1=df/2;
a=2/(n);
w=2/n;
a=-a/4;
a1=4*a;
t=abs(tinv(0.95,df,0));
sq=cssx/df;
sem=sqrt(sq*((1/n)+(1/n)));
tsem=t*sem;
l=meanx-tsem;
u=meanx+tsem;
proc sort data=out;   by id;
```

Appendix B.3 Continued

```
*****************************************************************;
*   Compute th Phi function in (6.5.2) up to the 7th term and   *;
*   MVUE. Also calculate estimated variance of MVUE, estimated  *;
*   bias, variance, and mse of MLE and MIR.                     *;
*****************************************************************;

data out;   set out;   by id;
retain m;
m=0;
m1=0;
do j=0 to 7;
f2=f1+j;
j1=j;
if j le 1 then j1=1;
g1=gamma(f1);
g2=gamma(f2);
g3=gamma(j1);
m+(g1/(g2*g3))*((a*cssx)**j);
m1+(g1/(g2*g3))*((a1*cssx)**j);
end;
data out1;   set out;
mu=exp(0);
ri=meany;
rm=meanrry;
expx=exp(meanx);
mle=expx;
meanx1=expx*m;
mvue=meanx1;
lmle=exp(l);
umle=exp(u);
vmvue=exp(2*meanx)*((m**2)-m1);
bmle=expx*((exp((w/2)*sq))-1);
vmle=(exp(2*meanx))*exp(w*sq)*(((exp(w*sq))-1));
msemle=vmle+(bmle**2);
bperiod=(exp(0.14752)+exp(-0.14752));
bri=expx*((exp(2*sq))*(bperiod)-2)/2;
vr1=(exp((2*meanx)+(4*sq)))*((exp(4*sq))-1);
vr2=(1/n)*((exp(2*0.14752)) + (exp(-2*0.14752)));
vri=(0.25)*vr1*vr2;
mseri=(bri**2) + vri;
drop g1 g2 g3 f1 f2 j1;
proc print;
var mvue mle lmle umle rm ri vmvue bmle vmle msemle bri vri mseri;
```

```
****************************************************************;
*    Program:          BOOKEXVAR.SAS                          *;
*    Author:           J.p. Liu                               *;
*    Date:             June 27, 1991                          *;
*    Description:      This program computes parametric       *;
*                      and nonparametric versions of          *;
*                      Pitman-Morgan tests based on residuals *;
*                      from sequence-by-period means and test *;
*                      statistics for equivalence in intrasubject *;
*                      variabilities based on formulas (7.5.7) *;
*                      and (7.5.8) in the book entitled "Design *;
*                      and Analysis of Bioavailability and    *;
*                      Bioequivalence Studies" by S.C. Chow and *;
*                      J.p. Liu.                              *;
*    Methods:          Parametric and nonparametric versions of *;
*                      Pitman-Morgan test.                    *;
*                      Parametric and nonparametric version of *;
*                      the tests for equivalence in intrasubject *;
*                      variablities.                          *;
*    Dataset:          AUC data in Table 3.6.1 of Chapter 3.  *;
****************************************************************;


****************************************************************;
* Input data set and print the raw data.                      *;
* Compute the residuals from sequence-by-period means.        *;
* Rearrange data so that y11 and y12 represent the two        *;
* residuals of a subject.                                     *;
* Compute Vik, Uik, Ulik, and Uuik.                           *;
****************************************************************;

libname out 'XXXXX:[XXXXX.XXXXX]';
data one;
set out.ex361;
proc sort data=one;  by group subject tmt;
proc print data=one;
proc sort data=one;  by group period subject;
proc means noprint mean data=one;  by group period;
var y;
output out=mean mean=meany;
proc sort data=mean;  by group  period;
data one;   merge one mean;  by group period;
resp=y;
y=resp-meany;
data one1;  set one;
if period=1;
y1=y;
meany1=meany;
drop period y meany;
data one2;   set one;
if period=2;
y2=y;
meany2=meany;
drop period y meany;
proc sort data=one1;   by group subject;
proc sort data=one2;   by group subject;
data work;  merge one1 one2;  by group subject;
id=1;
proc sort data=work;   by id group subject;
data work;  set work;
w1=.;   w2=.;
if group=2 then w2=y1;
if group=1 then w2=y2;
if group=2 then w1=y2;
if group=1 then w1=y1;
```

407

```
meanw1=.;
meanw2=.;
if group=2 then meanw2=meany1;
if group=1 then meanw2=meany2;
if group=2 then meanw1=meany2;
if group=1 then meanw1=meany1;
diffw=w2-w1;
sumw=w1+w2;
ll=0.8;
uu=1.2;
sumwl=(ll*w1)+w2;
sumwu=(uu*w1)+w2;
proc print;

**************************************************************;
* Obtain Pearson and Spearman correlation coefficients   *;
* by PROC CORR and output sum of squares and            *;
* crossproducts of corresponding Vik, Uik, Ulik, and    *;
* Uuik.                                                 *;
**************************************************************;

proc corr pearson cov data=work;
var w1 w2;
proc corr pearson noprint csscp out=css1 data=work;
var w1 w2;
proc corr pearson cov data=work;
var diffw sumw;
proc corr pearson  csscp out=css2 data=work;
var diffw sumw;
proc corr spearman  data=work;
var diffw sumw;
proc corr pearson csscp out=css3 data=work;
var diffw sumwl;
proc corr spearman  data=work;
var diffw sumwl;
proc corr pearson csscp out=css4 data=work;
var diffw sumwu;
proc corr spearman  data=work;
var diffw sumwu;

**************************************************************;
* Perform the tests(parametric) mentioned at the       *;
* introduction of this program by using STT, STR, and  *;
* SRR.                                                 *;
**************************************************************;

data css11;  set css1;
if _type_="CSSCP" and _name_="W1";
sswll=w1;   ssw12=w2;
id=1;
keep id sswll ssw12;
proc sort data=css11;   by id;
data css12;  set css1;
if _type_="CSSCP" and _name_="W2";
ssw22=w2;
id=1;
keep id ssw22;
proc sort data=css12;   by id;
data cssraw;  merge css11 css12;
ll=0.8;
uu=1.2;
r=ssw12/sqrt(sswll*ssw22);
f=ssw22/sswll;
mswll=sswll/22;
```

Appendix B.4 Continued

```
msw12=ssw12/22;
msw22=ssw22/22;
dem=(msw11*msw22)-(msw12**2);
numl=msw22-(ll*msw11)+((ll-1)*msw12);
numu=msw22-(uu*msw11)+((uu-1)*msw12);
fl=(21*(numl**2))/(((ll+1)**2)*dem);
fu=(21*(numu**2))/(((uu+1)**2)*dem);
fpm=(((f-1)**2)*21)/(4*f*(1-(r**2)));
fp=1-probf(fpm,1,21,0);
t=sqrt(fpm);
tp=2*(1-probt(abs(t),21,0));
tl=sqrt(fl);
tu=sqrt(fu);
if numu lt 0 then tu=-tu;
tpl=1-probt(abs(tl),21,0);
tpu=1-probt(abs(tu),21,0);
proc print data=cssraw;
var fpm t fp tp fl tl tpl fu tu tpu;

*************************************************************;
* Perform the Pitman-Morgan test(parametric) based on   *;
* the correlation computed from the residuals.          *;
* The results should be the same as those in dataset    *;
* cssraw.                                                *;
*************************************************************;

data css21;  set css2;
if _type_="CSSCP" and _name_="DIFFW";
ssw11=diffw;    ssw12=sumw;
id=1;
keep id ssw11 ssw12;
proc sort data=css21;  by id;
data css22;  set css2;
if _type_="CSSCP" and _name_="SUMW";
ssw22=sumw;
id=1;
keep id ssw22;
proc sort data=css22;   by id;
data csssum;   merge css21 css22;
r=ssw12/sqrt(ssw11*ssw22);
msw11=ssw11/22;
msw12=ssw12/22;
msw22=ssw22/22;
f=(21*r**2)/(1-(r**2));
fp=1-probf(f,1,21,0);
t=sqrt(f);
tp=2*(1-probt(abs(t),21,0));
proc print data=csssum;
var r f fp t tp;

*************************************************************;
* Perform the test(parametric) for the subhypothesis    *;
* that the intrasubject variability of the test         *;
* formulation is not too low,   based on                *;
* the correlation computed from the residuals.          *;
* The results should be the same as those in dataset    *;
* cssraw.                                                *;
*************************************************************;

data css31;  set css3;
if _type_="CSSCP" and _name_="DIFFW";
ssw11=diffw;    ssw12=sumwl;
id=1;
keep id ssw11 ssw12;
```

409

Appendix B.4   Continued

```
proc sort data=css31;   by id;
data css32;   set css3;
if _type_="CSSCP" and _name_="SUMWL";
ssw22=sumwl;
id=1;
keep id ssw22;
proc sort data=css32;   by id;
data csssum1;   merge css31 css32;  by id;
r=ssw12/sqrt(ssw11*ssw22);
rl=r;
msw11=ssw11/22;
msw12=ssw12/22;
msw22=ssw22/22;
fl=(21*r**2)/(1-(r**2));
tl=r/sqrt((1-(r**2))/21);
tpl=1-probt(tl,21,0);
proc print data=csssum1;
var rl fl tl tpl;

*************************************************************;
* Perform the test(parametric) for the subhypothesis   *;
* that the intrasubject variability of the test        *;
* formulation is not too high,  based on               *;
* the correlation computed from the residuals.         *;
* The results should be the same as those in dataset   *;
* cssraw.                                              *;
*************************************************************;

data css41;   set css4;
if _type_="CSSCP" and _name_="DIFFW";
ssw11=diffw;   ssw12=sumwu;
id=1;
keep id ssw11 ssw12;
proc sort data=css41;   by id;
data css42;   set css4;
if _type_="CSSCP" and _name_="SUMWU";
ssw22=sumwu;
id=1;
keep id ssw22;
proc sort data=css42;   by id;
data csssum3;   merge css41 css42;  by id;
r=ssw12/sqrt(ssw11*ssw22);
ru=r;
msw11=ssw11/22;
msw12=ssw12/22;
msw22=ssw22/22;
fu=(21*r**2)/(1-(r**2));
tu=r/sqrt((1-(r**2))/21);;
tpu=probt(tu,21,0);
proc print data=csssum3;
var ru fu tu tpu;
```

Index